The Range Men

Pioneer Ranchers of Alberta

Leroy Victor Kelly

Foreword by Sid Marty

VICTORIA • VANCOUVER • CALGARY

Heritage House Publishing Company Ltd
#108 – 17665 66A Avenue
Surrey, BC V3S 2A7
www.heritagehouse.ca

Library and Archives Canada Cataloguing in Publication

Kelly, L. V. (Leroy Victor), 1880–1956
     The range men: pioneer ranchers of Alberta / Leroy Victor Kelly. —1st Heritage House ed.

Originally published: Toronto: W. Briggs, 1913.
ISBN 978-1-894974-94-3

     1. Ranchers—Alberta—History. 2. Ranches—Alberta—History. 3. Ranch life—Alberta—History. 4. Indians of North America—Alberta—History. 5. Alberta—History. I. Title.

FC3670.R3K45 2009   971.23'02   C2009-904509-5

Cover and interior design: Pete Kohut
Front-cover photo: Cowboys in Claresholm, Alberta, *circa* 1908–10,
     Glenbow Archives NA-2852-1
Back-cover photo: Owners of St. Ann Ranch Trading Company, Trochu Valley, Alberta,
     circa 1908, Glenbow Archives NA-397-4

Printed in Canada

Mixed Sources
Cert no. SW-COC-001271
© 1996 FSC
FSC
The interior of this book was printed on 100% post-consumer recycled paper, processed chlorine free and printed with vegetable-based inks.

Heritage House acknowledges the financial support for its publishing program from the Government of Canada through the Book Publishing Industry Development Program (BPIDP), Canada Council for the Arts and the province of British Columbia through the British Columbia Arts Council and the Book Publishing Tax Credit.

BRITISH COLUMBIA
ARTS COUNCIL
Supported by the Province of British Columbia

Canada Council   Conseil des Arts
for the Arts   du Canada

CONTENTS

FOREWORD by Sid Marty ix

PREFACE • xiii

Foreword

SINCE its first appearance in 1913, L.V. Kelly's *The Range Men* (originally subtitled *The Story of the Ranchers and Indians of Alberta*) has been an essential reference for any reader interested in the early history of Alberta, particularly the region south of Red Deer to the "Medicine Line" and the bordering state of Montana.

The book is an ambitious attempt by an amateur historian and newspaper writer to chronicle a stretch of time from the origins of the earliest First Nations inhabitants and the arrival of the first known European in the region (the Hudson's Bay Company's Henry Kelsey, in 1691) to the coming of Chevalier de Niverville in 1751 through to the age of the automobile and the first Calgary Stampede in 1912—which marks, by way of celebrating what was even then an already bygone day, the culmination of the pioneering ranch culture that once dominated southern Alberta's economy.

From that last summation, the reader might assume that this book is wide-ranging in scope; it is indeed that, and it is also breezy in tone. "The cattle business, once started, went with a swing," writes Kelly—and once started, so does his book. No doubt it would have benefited from more skilful editing than its first publisher afforded it; nevertheless, the vibrancy of the frontier flows through, refusing, like the prairie wind, to be contained. Its great advantage is one of perspective, since Kelly, a reporter with the *Calgary Herald* from 1902 to 1905, was writing at a time when many noted pioneering ranchers, Mounties, homesteaders and some of the more famous warriors of the Plains were still alive and able to tell their stories in the flesh. This gives the book an immediacy and an authenticity that can only be approached today by the writer of historical novels and creative non-fiction.

This is not to say that Kelly's text is light on fact and detail. Concrete information on the depravation of the Plains tribes by the whisky traders, improvements in their lot after the coming of the North West Mounted Police to Rupert's Land in 1874, and the year-by-year account of the developing ranching industry—complete with a reporting on its economic underpinnings and advances after the arrival of the railroad—are all here, and provide valuable insights for the history buff. If you want to know when the biggest roundup in Alberta occurred, or what the wages of a cowhand were in the 1880s, or how many cows, bulls, sheep and horses were on the range, you will learn that here.

Kelly, a self-effacing author who seems, as a person, to have dropped out of our collective memory long ago (after moving to Vancouver to write for the *Daily Province* from 1915 to 1949), was a man who knew a good story when he heard or read one, and he tells scores of good ones here. What emerges, after an interesting preamble on Aboriginal origins and early explorers, is a cultural portrait of the time from around 1875 to 1913. Cowboys, outlaws (both White and Native), celebrated chieftains, mere murderers, disappearing buffalo, whisky peddlers, the first Black and Chinese Albertans, missionaries, cattle barons and brave policemen move in a continuous cavalcade through the magic-lantern show of our imagination in these pages.

In reading accounts such as this, we should beware the tendency to judge the past through the lens of current social fashions, or we will tend to underestimate the nature of historical figures and risk skewing our critical evaluation of the import of past events. It is true that, from our present-day perspective, Kelly reflects his era with the hide on, and also with the presumptive condescension of his time toward anyone not of the White race, and that women, although present, are rather shadowy figures here.

We may take some wincing comfort, I hope, in his inclination to celebrate good character in whatever skin tone or gender he finds it, as well as his general sympathy for the lot of all his subjects, whether ravaged by violent storm and flood or brutalized by the enmity and violence of men. Nor does he fail to sympathize with dumb beasts caught, for example, in the terrible blizzards of 1886–87, steers that could not reach through the snow to find grass, who chewed the hair off their dead herd mates or died with their stomachs punctured by splinters of wood they had gnawed in freezing desperation.

The book then, rich in explicatory detail, is also a feast of anecdotes that leaves us with more questions than answers, such as a duel on the Walrond Ranch, in 1895, between John Lamar and Gilbert McKay, "... both cowmen, and the former the quickest and surest man with a six-shooter in the country," or the origin and final disposition of some "half-breed" outlaws who roamed the country after the Riel rebellion of 1885. Despite their brevity, the stories constantly open other windows on that time. For example, it seems the pioneers in 1888, a period of prohibition, were all suddenly gripped by a longing for religion, for they began buying Bibles at a wondrous rate. But the Bibles, carted in from Montana in book-shaped metal casks stamped "Holy Bible," contained not the scripture, but demon whisky. This attempt at prohibition, as with later attempts in our history, was doomed to failure.

Another story touches on the diverse roots of Alberta's settlers. Books have been written about the noble, humble or honest origins of our pioneers, but Kelly tells of the members of the Kingfisher band of cattle thieves, who fled the law in Texas and headed north to become "splendid stock-hands and . . . respectable and honored citizens" in Alberta.

Like Paul Sharp's landmark work *Whooping it Up in Chinook Country*, *The Range Men* constantly reminds us how fluid the international boundary, or even citizenship, was at that time. Like the last herds of Plains buffalo drifting with the seasons, most people simply ignored its presence, while to the Plains Indians, it was another meaningless White man's diversion. Before 1883 and the coming of the railway, a letter posted in Fort Calgary bound for Ottawa needed American postage, since it travelled first to Fort Benton, Montana, and down the Missouri by steamboat to the nearest railhead before wending its way east. Life in southern Alberta and northern Montana, especially in the ranching culture, evolved from the same Spanish and Texan roots, and the cowboys and cattle barons moved from one national stem to the other without much concern about politics or tariffs for much of the period, until federal governments finally intervened to regulate the customs process. People of goodwill on both sides of the border conducted business with a handshake, and a man's word truly was his bond.

United by the same hardships of pioneer life, Montanans and Albertans in that era dealt directly with those who saw the new land as a place free of the law, where they could indulge in vicious acts without restraint, though law enforcement under the Americans was less prone to ceremony than it was under the jurisdiction of the redcoats.

One frontier thug, a man named Long, bludgeoned and shot a Mountie named Richardson in Middle Buttes, Montana, where the officer had gone to look for stray horses. While Long celebrated his victory with his pals, a kindly American rancher immediately harnessed his team, loaded the wounded officer in a wagon and drove him back to Fort Macleod, Alberta. When they arrived, they learned by telegraph that Justice Brown in Montana had already dealt with Long, who had drunkenly made his way to Brown's home and ordered him to "come out at once and be shot."

"Quiet men often deceive bullies like Long," wrote Kelly, "and this one proved to be of this sort, for he appeared very promptly with a shotgun at his shoulder and blew the belligerent Long into something resembling a colander."

The Range Men, valuable for the insights it gives into the history of Alberta and the roots of the western culture that still influence the life we live today, has been a resource for Alberta writers including myself over the years, and it will be interesting to see, as this reprint takes its place on the shelves and perhaps on the Internet, how many of the stories given all too briefly within these pages will inspire new insights and other narratives in the years to come.

Sid Marty
Willow Valley Alberta
July 2009

Preface

THE following work has been compiled with great care and all facts have been gathered from authentic sources. The story of the ranches of Alberta is one that requires no romancing to make it of interest, and the author of this work has taken his data from the reports of the North West Mounted Police, government reports of William Pearce, files of the *Calgary Herald,* the *Macleod Gazette* and other publications. Also, much valuable information was supplied by old-time stockmen of high standing, and, all told, the work—though containing many remarkable anecdotes and experiences—is based upon facts alone.

Thanks are due to E.H. Maunsell and Robert Patterson, MPP, of Macleod; George Lane, A.E. Cross, P. Burns, H.C. McMullen, of Calgary; Hon. A.J. McLean, MPP for Taber; R.L. Shaw, MPP for Stettler; George Emerson, of Brooks; Walter Huckvale, of Medicine Hat; G.E. Goddard, of Cochrane; Frank Ricks, of Ghost Pine; H.A. Kanouse, of Chehalis, Washington; J.D. McGregor, of Brandon; Duncan S. McIntosh, of Calgary; John Herron, ex-MP, of Pincher Creek; W.F. Parker, of Macleod; W.F. Cochrane, of Nelson, BC; J.S. Dennis, of the Canadian Pacific Railway; Dr. J.G. Rutherford, of Calgary; and Rufus Kimpton, of Windermere, BC, for assistance rendered. The author is also obligated to Tom Wilson, of Banff, for help he has given in the gathering of data.

In some cases this work might appear incomplete, but the collection of the full story of every ranch would be an impossible task. Ranches have started, flourished and died. Their very names have been forgotten, the old ranchers are scattered and passing away, and thus much information has been lost. But this work is the result of wide trips through the ranch country, careful studying of conditions, and particular pains in preparing only bona fide information.

—*Leroy Victor Kelly*

CHAPTER I

Generally Descriptive

MILLIONS of acres of luxuriant grasses, hundreds of sparkling rivers, thousands of pure springs, rolling hills and bluffs of brush, deep valleys and snug, grassy, well-protected river bottoms—that was Alberta when the Indians owned it and the buffalo chose it as their breeding grounds. That was Alberta when the ranchers picked it as the last and best range country. But it is not so today. The rolling hills remain, the snug river bottoms, the springs and streams and rivers, but the range is gone forever, cut up by the fences of the farmer, and the railroad. A grand country, the rich mixed-farming province of the Dominion, the wheat empire, the land of the future. The ranches have gone, the open range is enclosed in fences, the wild cattle no longer roam at will across the broad sweeps of the prairies.

The story of the range stockmen of Alberta forms an engrossing chapter in the history of the province. These are the men who first braved the frontier in an effort to establish legitimate business, the men who really carved the way and proved that the country was one of vast realization. They lived on the outskirts of the farthest police patrol, they herded their stock and guarded it against the untamed Indians and the wild beasts of the mountains and hills, they lived in mud-roofed shacks, eating rough foods, without any of the luxuries that are today considered necessities. A wild land, with wild natives inhabiting it, wild cattle to herd, wild animals to contend against, it is no wonder that the cattlemen were hard, prompt, ready men whose adventurous lives were of deep interest to the following thousands who settled in the country where the stockmen pioneered. Traders were in the range country before the cattlemen, but the traders were not established, and though they had livelier times with the red men than the stockmen ever had, yet they did not leave any lasting developments or improvements, to say the least. Life among the southern Alberta Indians was adventurous at the most promising times in the very early days, and the free traders who supplied them with the merchandise of the time did not conduct their business along a line that tended to make it less adventurous.

It was the cattlemen, their undertaking somewhat assured by the Mounted Police, who started the development. They came with their herds, establishing in Alberta that old, old industry of the breeding of livestock. 'Tis the oldest honorable business in the world, the occupation of the great men of most ancient history; 'tis a healthful, vigorous, open contest against weather, men, conditions and disease. It required mentality, physical ability and perseverance, and the rewards were appropriate to the energy and ability displayed.

No business in the world can recuperate from losses that the cattle industry receives and recovers from. No known legitimate undertaking could meet the blows that the best ranches of this province came smilingly and hopefully through. No business, without insurance, could within three or four years be wealthier than it was before half its capital stock was utterly lost. What business, other than ranching, could have survived such damages as were experienced by the stockmen in 1886–87, in 1893, in 1906–07?

The story of the stockmen and the ranches is one that presents most interesting characters and characteristics, men of ability, of judgment, of superiority in many ways: some were rough, uncultured, crude, but with a mentality that often placed them superior to the most highly polished, for ranching was a big business and it required big men to handle it most successfully. The rancher might consider it was legitimate to brand a maverick, to round up a stray calf and put the sign of ownership on it, but his word was usually unquestioned and his statements were pledges. John Cowdry, who conducted a private banking business in the town of Macleod from 1886 until 1905, through the heyday of the best years of the ranching business, never had a word of trouble with any of his customers. Often he advanced money on a man's mere word, and never was his confidence abused. A man's word was his bond in those primitive days. The cowboys and ranchers were honest, clean and straight in all their money dealings, even though they might pick up a calf. Cowdry never had trouble except in one instance, and that was not with a cattleman, the objector being a member of the legal fraternity who found a quibble and dwelt on it, as is the prerogative of those educated in legal lore. When Cowdry sold his private banking business (Cowdry Brothers) to the Canadian Bank of Commerce in 1905, the bank took over from the Cowdrys some four hundred and eighty-seven thousand dollars in loans, of which amount the Cowdrys had to guarantee fifty-two thousand dollars for eighteen months. After the expiration of this time, Cowdry was released from this guarantee, and out of that fifty-two thousand dollars he had to pay something under two hundred dollars, an astonishingly pleasing average to any banker.

The cowboys were of a class that was equally satisfactory. They might go on hilarious "busts" when in town, they might "shoot-up" a barroom and smash every light in the place, they might ride into stores on the backs of frantic horses, but they were good men, the kind who worked for their employers. If a flooded river must be crossed to save some of the cattle carrying the brand of the outfit they worked for, they plunged in and braved the torrents with their driftwood and their deadly "drag." If a fifty-mile ride was necessary in order to save a horse, they took it. If an all-night vigil beside a herd of freezing stock was necessary to save those animals, the vigil was cheerfully undertaken. Many and many a night the cowboys sat on their horses, bundled to the ears, while the bitter winds of ten and twenty below zero swept across the prairies; and there are known instances where cowboys whose feet and legs were frozen remained by the herd and pulled it through until relief came. The employer's stock was their own; a theft was a personal loss and a mark against the character of the cowboys working for that outfit. They were "good" men—rough, ready, and true. The cowboy of the present day is not of the same class, due partly to present economic conditions and half-education along socialistic lines, due partly also to the innumerable published romances of cowboys, bucking horses, gallant "gunmen" and a picturesque West. Today a cowboy thinks more of getting to town and riding a prancing horse through the streets, with big spurs clanking, chaps flapping, and wide hat jauntily on head. Cowboys of the old school still exist, but they are prominent because of their scarcity, and they are treasured by the rancher who is fortunate enough to have them on his payroll. Just as an instance of the change in the characteristics of the stock-hands, an incident which occurred in 1912 will be illustrative.

A ranch owner in the southern portion of the province had sold seven hundred head of "rough stuff"—any kind of matured stock—to P. Burns, and he went to his range to oversee the shipping of them. Several thousand head of cattle were on the range, watched over by sixteen cowboys of the new school, these sixteen doing a work that six of the old-time cowboys would have done and have had plenty of time left over to loaf. Near the shipping point was a goodly herd of "tops," prime beef, suitable for export to the most select market, but the rancher wanted none of these, as the price for the "roughs" was not as high as the "tops" would draw. Early in the morning of the day that the cars drew into the siding to wait for the seven hundred head, the sixteen cowboys were sent forth to comb the hills and coulees for enough rough cattle to make up the shipment. All day long the rancher watched the draws to see the animals commence to straggle in. All day long he gnawed his moustache in impatience, muttering strange words of strong import, and all day long no cattle came. Toward evening he was forced to fill his shipment with the "tops" that were handy. Then the sixteen cowpunchers came gaily back, happy in the knowledge of a day's work well done. Their saddles sagged with the dead bodies of rattlesnakes, their faces beamed with satisfaction, for when searching for their employer's rough stock they had stumbled upon a snakepit, and spent a lucrative day in slaughtering the reptiles, every snake killed being one dollar in the pocket of a cowboy, that being the price which a curio dealer in Calgary was paying for the skins. "Those snakes have cost me ten thousand dollars," said the rancher, as he proceeded to tell his sixteen hired hands just what the duties of a real cowboy consisted of.

Though the old-time rancher was generally an honest man, there came, in time, to be a goodly number of "rustlers," cattle thieves, men who would "lift" any animal they thought they could get away with. Their methods were often crude, but more often markedly clever, and they worked at times with a boldness and skill that brought a reward in fast-growing herds to which little suspicion was pointed. There are men alive today who thank the "running-iron" and the skill of their lone rope-arm for a comfortable competence. Acid, fire, and knife assisted these lawless men, and no one can tell the damage they did to the great herds that dotted the green prairies. Many were caught in the toils of the ever-vigilant Mounted Police, but many escaped to enjoy their profits. Brands were the mark of ownership, a method that dated from the earliest years of ranching in Texas. Animals were branded in the side with a red-hot iron—usually on the left side, owing to a peculiarity of cattle-nature. A right-side brand was never considered as good as one on the left side, because when stock was driven into a corral and a man sat on his horse in the middle to see the brands on the animals as they walked around the outer edges of the enclosed space, the animals always moved from right to left, thus exposing their left sides to the person in the centre. Out of sixty-seven brands in the Macleod district in 1888, there were only nine on the right sides. Sometimes expert brand-alterers would do a very satisfactory job with milkweed, pricking the skin of the captured beasts and rubbing the milkweed into the wounds, effectively changing the legitimate mark. But the running-iron, or round iron, was most favored, as it was easily and secretly made by cutting a wagon-iron in two and using the rounded end. This implement was exceptionally simple of concealment, also, as it could be slipped inside a long bootleg and kept out of sight until necessity or opportunity for its use arose. The sure way of detecting the work of a running-iron was to kill the suspected beast and examine the inside of the hide, soaking being sometimes necessary. Though a perfectly made "O," or any other brand, might appear on the outside, it would only appear on the inside as a quarter circle, or some segment of the outside burn, for the

This 1884 illustration of some of the cattle brands used by southwestern Alberta ranchers was published in the *Macleod Gazette*.

brand "run on" with the round iron never shows through, while the one pressed on with the flat-faced brand is ever evident on the inside. The "hair brand" was also a strong favorite, this method being simply the taking off of the hair with knife or acid, thus temporarily changing the appearance of the original brand.

Mounted Police officers say there was little trouble from cattle rustlers in the country south of Lethbridge until the Mormons settled. In the range country of the States, and in most districts of Alberta, a maverick on the range is anyone's animal, be it horse or "critter." Though this was generally understood, there were many ranchers who were not particular in branding their horses every year, the habit being to brand only occasionally, when they felt like it. This rather slipshod method resulted in many unbranded horses being at large, the ranchmen usually considering as their property the stock that was running with the animals that carried their brand. Over in the United States, where ranching was older than in Canada, the maverick question had been long established by custom, any unbranded animal over a year old being the property of the man who clapped the first iron on it, and the Mormons, or some of them at least, brought this habit with them. They rode the range and picked up the mavericks, much to the consternation of many ranchers, and much to the annoyance of the Mounted Police, whose already excessive work was thus increased. Many head of good, bad, and indifferent horses, and even cattle, went to enrich Mormon herds, though such deeds decidedly did not lie entirely among these people, who, in many cases, believed they were doing quite the correct thing in taking up such stock, and did it cheerfully, for it meant a gain to them.

But long before the Mormons came, there were wide stealings in the Macleod country, half-breeds, Indians, and white men "lifting" horses, cattle, and young stock. It was often very difficult to punish these thieves, because of the insufficiency of evidence, and men might be morally certain they had captured thieves, but had to let them off. Lieutenant-Colonel Macleod, CMG, the first officer to command a detachment of Mounted Police in the south, the man who, assisted by Captain Winder and the first Mounted Police force, established law and order in the lawless land, was often hard-pressed in order to hand out punishments that were certainly owing to certain persons but were not administrable because of legal weaknesses in the prosecutions. He endeavored always to give justice and due meed to all, and one case will demonstrate how this must sometimes come about.

A.P. Patrick, DLS, who surveyed the southern Indian reservations after the treaty of 1877, was staying at Macleod about that time, and owned a grey saddle horse of which he was very proud. Other horses belonging to his pack trains roamed the neighboring prairies, and one day he decided it was time to get these horses in and have them ready for work. Hiring a half-breed, he mounted him on the treasured grey and sent him out. A week went by, ten days, no half-breed. Patrick was beginning to wonder, and then the breed came back—in charge of a Mounted Policeman well known as "Buffalo" Heeney. Heeney had been near the boundary line and had there seen the breed and a band of Indians heading south. Recognizing the grey and knowing that Patrick would not be likely to have sold it, he arrested the rider and brought animal and man back to Macleod on suspicion. The man was tried for theft of the horse, but of course was dismissed, as he had been sent out with the animal by Patrick himself. The dismissal was more or less reluctant, because everyone was morally certain that the breed had intended theft. Nevertheless, he was released and arose to leave the courtroom, when an interruption came from Tony Lachappelle, the keeper of a small store at Macleod. Tony volunteered the interesting information that on the day the breed had gone away on the grey horse, he had visited Lachappelle's store and secured a bottle of ginger, telling the proprietor that he was working for Patrick and the goods were to be thus charged. A new charge was laid against the breed; Patrick was called and denied giving any order to charge merchandise to him, and Colonel Macleod at once sentenced the prisoner to one year in jail for obtaining goods by false pretenses.

All sorts and conditions of men came into Alberta once the Mounted Police were established—red men, white men, breeds, yellow and black. The best speci- men of the Negro race who came was an expert cowman and bronco-twister named John Ware, a splendid roper and a grand roughrider. Ware was a white man in all but his skin, and his friendship was worth holding. "You ask any old-timer who he'd go to if he was caught late at night on the prairie with John Ware's house ten miles away and some white man's house five miles distant. They'd sure say they'd pick John Ware's, for he was a 'white' man," declared an old stockman. Ware came during the very early eighties, with the Bar U cattle, and before him there had been only two or three Negroes seen in the country. One was a whisky trader named Bond, another was a former slave named Green, who came from Idaho. After Ware, another Negro came—Jesse Williams, who murdered a man and was hanged. The first Negro in the Pincher Creek section was a blacksmith named Dyson, who came from near Chatham, Ontario, and carried on his trade for many years at Pincher Creek, where he still resides in very comfortable circumstances.

With Colonel Macleod's household when he brought his family out to the frontier with the first police force was an old Negro "mammy," who was known to everyone in the family and in the country as "Auntie." Auntie was the nurse for Colonel Macleod's children, and considered herself as a pretty important portion of the frontier forces, being, as she often said, the "first white lady in the West." Next to the Negro Bond, who was arrested in 1874, she was probably the first colored person in western Alberta.

Ranchers always contend that, from the beginning to the end, the ranching business received little government encouragement since the first few years, but they cannot well say this of the Canadian Pacific Railway. Ranchmen have grumbled about the railroad—poor cars, poor handling at divisional points, poor feeding arrangements, poor supply of cars, and much slaughter of stock. But this is natural, and, if one inquires deeply, he will learn that most ranchers consider that the railway has usually been reasonably fair. Much to encourage the improvement of stock, to assist the development of the industry, was done where it could be done; great tracts of railway lands were leased to ranchers, animals for exhibitions were carried free, and purebred stock was brought in at half-rates. When the terrible winter of 1906–7 smote the stockmen, the CPR would not accept rental for their leases that year. In the beginning, when the road was poor and the equipment was mediocre, there were poor cars and sometimes slow service, but as wealth came and the improved cars were purchased and built, there was less room for complaint. Generally speaking, the railroad always tried to give the ranchers an "even break," and it did considerable good toward the development of the industry. It was to their own advantage to do so, and they usually worked to their own advantage as well as that of the ranchman.

At the annual national livestock convention that was held in the Canadian Building at Ottawa on February 5, 6 and 7, 1908, the ranchers were represented by John A. Turner, president of the Alberta Horse Breeders' Association; Walter Huckvale, president of the Western Stock Growers' Association, and R.G. Mathews, secretary, and A.B. Macdonald; George Lane, and W.B. Thorne, representing the Percheron breeders, and A.E. Cross. Other Alberta men interested in ranching who were present were P.C.H. Primrose, superintendent of the RNWMP–NWMP at Macleod; J.J. McHugh of Calgary; M.T. Morgan of High River; and John Herron, MP, of Pincher Creek.

The National Live Stock Association honored two Alberta representatives in the election of officers, appointing Walter Huckvale on the executive committee and John A. Turner on the board of directors.

An interesting address by H.C. McMullen, CPR livestock agent, and a general discussion that followed, give a very clear idea of just what grounds the livestock shippers of the West considered they had for complaining against the

Members of the Western Stock Growers' Association gather outside the Assiniboia Hotel in Medicine Hat in 1903. Among them are George Emerson, A.E. Cross, Patrick Burns, L.B. Cochrane, Charles Kettles, Charles Mitchell and W. Stewart.

railroad. Railway facilities, conditions, and the mange question arose and were exhaustively discussed, the discussions explaining conditions so clearly that all that is necessary is to reproduce them verbatim.

"Standing before you today," said Mr. McMullen, "and recognizing, as I do, men whose names stand for all that is best in their line, men whose fame as experts in their life work has reached the utmost confine of our great country, I am somewhat fearful that, in accepting your kind invitation, I have greatly exceeded my limitations.

"In discussing the very important questions of transportation and the relation of the railway companies to the commercial side of the livestock industry, I may say in explanation that I have been intimately connected with both for practically all of the active portion of my life, and have gained through experience some knowledge of the requirements of each.

"The problem presented in the safe, economical and expeditious handling of a nation's traffic is one that has engaged the master minds of those in the service of transportation companies ever since the industrial world found it necessary to exchange the products and manufactures of one section, community or nation, for those of another.

"Railways, steamship lines, electric roads, stage, freight and pack routes, are each and all but interchangeable parts of one great system by which the requirements of hundreds of millions of people living in widely scattered communities, and under widely varying conditions, are served, and the necessities and complexities of existing conditions stimulated by the keen competition attendant on the unceasing warfare that is a recognized element of our industrial life demand fulfilment of the obligations expressed and implied between carrier and patron.

"Now, what are these obligations? What expressed, and what implied?

"First, that property and persons entrusted to the care of the carriers be safely forwarded, that undue risks be not incurred either to the person or the property accepted for furtherance to the point of destination designated.

"Secondly, that such person or property be transported as expeditiously as circumstances would warrant, always with due regard to the prime consideration—that of safety.

"Thirdly, that such service be performed as economically as is consistent with proper regard for the obligations of both parties to the contract, always bearing in mind physical conditions, cost of operation, risks involved, class of service required, and all the vital and material points that have a bearing on the ultimate result.

"I think it may safely be assumed that all corporations or individuals engaged in the business of furnishing transportation desire to carry on their trade with the minimum of friction, to avoid unnecessary controversy with the shipping public—in other words, to follow the lines of least resistance, which after all is but a natural law applied to modern business administration, and in the last analysis it would be but one form of business suicide for transportation companies or other great industrial concerns to wilfully antagonize the people whom they may elect to serve.

"Taking these things for granted, it is only fair to presume that each and every officer in charge of a department of a great system, and who, by reason of his demonstrated ability is placed in such position, and who is held responsible for its success, or lack of it, is honestly endeavoring to the best of his ability to acquit himself of that responsibility in such manner as shall give the largest measure of satisfaction to the public, without in any way sacrificing or endangering what he considers the best interests of his employers.

"There can be no question that the best interests of a transportation company are most surely served by any action that tends to increase the commercial prosperity of a community; that any harsh, unjust or oppressive methods would instantly react on the concern that, by its general policy, or through the acts of its servants, permitted or encouraged such unjust method in the form of decreased revenues and deserved unpopularity.

"Consequently, if these general propositions are admitted, it is only necessary to go a step further and inquire what class of service the public is now getting, and what, by comparison, [was] that rendered ten, twenty or thirty years ago. Is the service now given such as might consistently be demanded in view of the requirements of the business world of today? I venture to say it is.

"The vast increase of the world's population has made necessary the ever-widening centres of industrial activity, with consequent enlarged demands on the ability of the transportation companies to meet the heavier tax on their available plant and permanent way.

"The unexampled prosperity witnessed during the past ten years has loosened the purse strings of the world's money kings, and mills, factories and other industrial concerns have sprung up as if by magic; mines have been developed, forests explored, and that great factor in the material well-being of any community, the farmer, has been infected with the microbe of restlessness and is now seeking new worlds to conquer. What was a few years ago virgin prairie is now dotted with farms, and the trails of the now historic buffalo are being turned into roads that resound with the busy hum of the farmer's wagon. These things all have a definite and material effect on the relations of the public and the transportation companies, and while they have given added responsibilities and implied greater obligations they have also materially increased the difficulties of the old problem.

"Naturally, this convention is more keenly interested in the concrete question of livestock transportation than the abstract proposition of how the general public is served, and I will, by confining myself to the discussion of some of the most important features in this connection, endeavor to throw some light on the side that is viewed with most concern by those interested in the great ranching industry of the West.

"Each year an estimate is made of the probable volume of business to move from a given section of territory largely given over to the cattle ranches and the mixed farmers of the Western Provinces; also to the approximate space of time over which such movement is likely to be spread.

"Ordinarily this information is gathered and estimates made by those best qualified by long experience for such work, and in the main this information is correct, but in the handling of a perishable product, such as the beef cattle of the Western ranger, many very unfavorable conditions may and do arise—conditions that are not present in other lines of business.

"In the older Provinces, the business of breeding and preparing beef cattle for the market is carried on practically without intermission during the entire year, thus involving no great or unexpected strain on the resources of the carrier in the matter of rolling stock and motive power. Breeding grounds and feeding yards are situated in close proximity to market, with consequent short hauls and quick return of empties required for further shipments.

"But conditions in the West are radically different, and present difficulties and obstacles that are at times almost insurmountable. For instance, during the season of 1906 there were 115,000 head of beef cattle shipped from the range in a territory very little more than 300 miles long by 200 wide. Of these, 75,000 were destined for export, involving a rail haul of 2,500 miles to seaboard, from which point all cars must be returned empty, and with such despatch that other and equally important traffic was seriously disturbed. But the mere volume of business referred to was not the most vital part of the question, but the fact that seventy-five percent of the entire year's shipments were forced on the line within a period of ninety days, export shipments alone from the territory mentioned reaching the large total of 25,000 head per month, thus making an almost unbroken procession of loaded trains eastbound and one of empty trains westbound. In addition to trains in transit, there were at all times a large number of trains of empty cars lying at feeding stations out of commission while cattle were being fed and rested.

"This means delay from fifteen to thirty hours, during which time this large number of empty cars were not available for other shipments, as these must be held for instant use on requisition of shippers, who frequently call for their train on two hours' notice, to enable them to reach ship's side and load for sailing date.

"Again, during the season of 1907, in the month of October, notwithstanding a great falling off in the output for the season as compared with the previous year, the shipments exceeded the total for the corresponding month one year ago, thus demonstrating the unreliability of supposedly accurate forecasts.

"It will naturally occur to one to ask, what is the inference to be drawn from all this? That, while this company has for six months of the year a large surplus of stock cars, the method so long in vogue on the Western plains has the effect during the balance of the year of placing on the operating and car service departments an extraordinarily heavy strain. That, at one season everything else must be sacrificed to meet the requirements of the cattle trade, at another the company's equipment is lying idle for lack of suitable loading. This most wasteful state of affairs is the result of two influences, either or both of which may bring about the very condition the best interests of the transportation companies and ranchers alike would counsel our avoiding.

"First is the condition of the stock. After a winter on the open range, exposed to all the vicissitudes and uncertainties of our variable northern climate, with sometimes unfavorable conditions prevailing during the following summer, it frequently happens that stock will not reach marketable condition until late in the season. Then, all being subject to much the same influences, all finish, as far as they ever finish on the range, in thirty days, and are ready for sale. Buyers then are forced to take the tide at its flood, and buy when the cattle are ready, and not when, in their judgment, the market is most favorable for profit-making.

"The market, too, subject as it is to sudden and violent fluctuations, is an important factor in bringing about the congestion we are all so anxious to avoid, as on long-continued declines nobody will ship, but at the first indication of more favorable quotations every rancher and shipper on the range puts in orders for cars, which may have to be hurried from distant points to meet this sudden demand which the most skilled operator could not by any chance have foreseen.

"It must be obvious to even those who are wholly unacquainted with the proper and adequate equipment of a great railway system, and particularly one operating over such magnificent distances as the Canadian Pacific, that congestion is the inevitable result, with consequent delay, under conditions such as described. Aside from the question of car supply, efficiency of service is unfavorably affected in other directions and from the same causes.

"Coincident with the rush of cattle to the seaboard commences the grain movement to the head of the Great Lakes; coal dealers in every town between the mountains and Winnipeg are getting in supplies; lumber and all classes of building material are being rushed to the East to stock up for fall trade, and in a word every one wants to get to market at the same time.

"But we are only travelling in a circle, and with each round arrive at the same place. What is the remedy? How are we to go about effecting a change in this wasteful method of marketing a perishable commodity; wasteful alike to the producer and to the carrier, for under existing conditions, the cost of operation makes this, from a railroad standpoint, a non-remunerative traffic. I am of the opinion, however, that the process of evolution now at work in the livestock industry will gradually eliminate the element of waste and put it on a sound and profitable basis. This process should be hastened, and I believe will be, by the application of intelligent care and the exercise of reasonable foresight in preparing cattle to go on the market at the time the market demands finished beef in moderate quantities.

"Success in breeding and preparing cattle for the market is subject to the inexorable law of supply and demand, and the survival of the fittest is one of the results that do not require a prophet, or the son of a prophet, to foresee. The man who has sufficient enterprise to enable him to break away from the traditions of years, and go in for feeding for early sale, will not only inaugurate a movement for the relief of this periodical congestion, but will ensure for himself a profitable return for his labor and investment.

"It has been argued that it is not profitable to feed stock in the West, but this in the light of my experience is a fallacy, and is not susceptible of proof. It sounds incredible that the farmers of Ontario and Quebec can profitably engage in the business of breeding and feeding stock on land that is worth from fifty to seventy-five dollars per acre, with no free grazing, and such results be impossible to the rancher of the West, with cheap land and more or less open range available on which to run cattle during the summer season—land, a certain portion of which will be available for many years to come.

"Even with the rapid settlement of the prairie ranges now going on, there will always be in the Provinces of Alberta and Saskatchewan considerable areas of land unsuitable for cultivation that can and will be utilized for summer pastures, and with the almost unlimited quantities of natural hay available in Northern Alberta, together with the coarse grains that experience has shown will grow so abundantly, I cannot conceive of any more profitable method of marketing this feed than wrapped in the hide of a fat steer.

"Another feature of the case that is worthy of most serious consideration is that market quotations are nearly always one to two cents higher in May and June than in the latter part of the season when grass beef is being rushed to the front. The few ranchers who, during the last two or three years, have tried the experiment of winter feeding have found themselves amply repaid for the labor expended, and are now strong converts to the policy and methods that have made many an Eastern farmer rich.

"The glamor of romance that surrounded the cattle business in the old days still exerts a strong influence in the West, and it is difficult work to convince the graduates of the old school that modern methods are more profitable in the end. But the laws of the Medes and the Persians were not more unalterable than that governing supply and demand, and if the rancher or mixed farmer of the West will not conform to this law sufficiently to produce the quantity and quality of supply to meet the demand, he will find the markets drifting further away each year until the man with the gold will seek new sources of supply, and thus again demonstrate that, after all, and in all branches of trade, we follow the lines of least resistance.

"The livestock trade, as carried on in the old days, presented many features strange and perplexing to the mind of the average man accustomed to modern methods. The pioneers in the ranching business in the great West were men possessed of all the requisites for making a complete success of their life work—indomitable will, unlimited courage, and an abiding faith in the correctness of their methods—but, I submit, with all due respect for those grand old men, they were mistaken. In the early seventies such men as Gibb Seewright, Matt Ryan, Billy Lang and Sam Devine were the kings of the southern trail, and the blaze of their campfires lighted the way clear from El Paso to Abilene, the plains were dotted with their immense herds, and they reigned pre-eminent as the monarchs of that now historic range.

"In the north, that prince of them all, John Quincy Shirley, with Frank Perley, Tony Nodyne and Billy Cotent, kept the trail lined with cattle, from the sink of the 'Humboldt' and Matt Taylor's bridge on Snake River to the Cheyenne Wells. These men held undisputed sway over a territory as large as an empire, and the dust of one herd had not cleared away before the lead steers of the next bunch were in sight on the long reach to the markets of the East. Men bought, sold and traded in herds of 20,000 head; transactions of enormous magnitude were everyday occurrences, and the man who had only a paltry 5,000 or 10,000 head of cattle was considered in genteel poverty. The famous J.Q.S., the Goosegg, and other well-known brands spread over hundreds of miles and were carried with the drift into half a dozen states and territories. Prosperity seemed unbounded, and in a measure it was, for in those days material wealth was based on the number of cattle carrying the brand. When a calf was branded, no matter how scrubby, he was turned loose with the remark: "Oh, well, he counts one in the herd," and "there was the rub." He would count one in the herd, but what would he count in the market? What would be left for the rancher after the sale was effected?

"They used to tell a story in the old days of a bunch of 2,000 of those steers that came up from the Southwest over the old Santa Fe trail and were held up in Kansas for sixty days' feed, with instructions to the commission house to pay feed charges and throw the cattle on the market, that when the owner received his account of sales he found he was debited with twenty-five cents a head.

"But a change came o'er the spirit of this dream of millions, a change brought about through settlement of the great ranges, and a demand for a better class of steers consequent on the keen competition inaugurated in the great central markets. The sleek, fat, well-bred, well-fed steer came into the fight, and, with a confidence founded on conscious merit, demanded recognition, and, sir, he got it. Consternation reigned in the big camp, prices slowly but surely receded, and ruin stared the old-timer in the face. Bankruptcy sat grinning like a hideous spectre on the gatepost of every corral in the Southwest. The sun of the Longhorn had set. There could be but one end to that sort of thing, and the men who failed to recognize the fact that modern conditions demanded new and better methods went to the wall, and, sir, they are still going to the wall. Without intentional disparagement of the men of the old guard, those long dead and gone heroes of the trail, men to whose memory I take off my hat and bow my head in honor, men who carried the banner of civilization through every mountain pass in the great West, I say that, pursuing the same policy and with the same sublime disregard of the essentials, these men would fall by the wayside in the race for supremacy today. In the fight for a place in the livestock world, competition is as fierce as in any other recognized industry, and success comes only to those equipped for the battle. But even those of the old school, and who once reckoned values by numbers rather than quality, and who, though a steer was built like an elk, and could outrun a quarter horse, still counted him 'one in the herd,' are getting their eyes opened, and taking a different view. The ultimate goal of all progressive stockmen should be quality, breeding and feeding, not so much the size of the herd as the size of the check an individual steer will fetch.

"Now, I have the most profound respect for our old friend, the 'steer.' I have a keen appreciation of his worth as a moneymaker, but he is not the only breadwinner in the family.

"There is another, a very industrious little chap, who has lifted more mortgages and built more homes than he is usually given credit for the despised, unloved, homely sheep. And, sir, we are not giving this branch of the industry the consideration and attention the facts would warrant.

"Now, sir, what are the facts. In the Provinces of Alberta and Saskatchewan there are, roughly speaking, 200,000 sheep. In the State to the south of us, and divided by an imaginary line, there are 6,000,000. Is this not a state of affairs worth enquiring into? There are a few other facts in this connection to which I desire to call your attention.

"We import from Oregon and Washington each year 70,000 mutton sheep for local consumption at our coast cities; on these we pay an *ad valorem* duty of twenty-five percent. We import each year, from Australia and New Zealand, thousands of carcasses of chilled mutton in cold storage ships, one well-known firm alone handling 25,000. We sell our wool for 14 to 16 cents per pound, while our neighbor in Montana gets 22 to 25 cents for the same grade.

"Should one of you decide to go into the sheep business and invest, say $50,000, placing half in Southern Alberta and the other half in Montana, you will find yourself in the very peculiar position of drawing from your Montana flock just double the returns of the one in Alberta, and looking around for an explanation, you find an imaginary line—nothing more.

"The feeding grounds of Milk River are just as good as twenty miles south in the 'Sweetgrass Hills,' the water is just as pure, the climate cannot be radically different in that short distance, then why is it not possible to each year increase our flocks instead of seeing the numbers grow steadily smaller?

"During the year 1906, we shipped from the sheep range 57,000 mutton sheep; in 1907, the figures showed 28,450, a falling off that should give us grounds for enquiring why.

"I am not a political economist. I do not claim to be an expert in solving these fine points. I have no intention of declaring this policy right, or that one wrong, but I do know that when I go over these sheep ranges of Montana each year, as I do, in an effort to secure the immigration of a few of the large growers of that State, and am asked what advantages we can offer, and am obliged to remain silent, I think it is time that a deliberative body, such as is here assembled, should devote some leisure to framing an answer to the Montana sheepman's question.

"Now, in conclusion, I want to say a few words with reference to conditions obtaining in the horse trade in the West. While there are many enterprising men in the prairie and mountain Provinces who have displayed a surprising amount of that pluck and optimism with which our country is so liberally endowed, there is yet vast room for improvement in the matter of placing our horse herds on the high plane the latent wealth and resources of this great Dominion would warrant.

"I am moved to make a special plea on behalf of these public-spirited men who have invested largely in their individual efforts to better conditions in this respect, and I would bespeak a more liberal encouragement and a more pronounced recognition of their worth in the form of moral and financial support from the older communities in the East, and the parent organizations formed to foster and promote the breeding of the very best to be found in the horse markets of the North American continent.

"An arrangement whereby the prize list at our Western fairs and horse shows would receive even more liberal aid from the East, which has been fairly generous in the past, would do much to fortify our horsemen in their resolution to not stop short of the very top.

"Mr. Chairman and Gentlemen, I thank you for the cordial welcome extended, and the patient hearing accorded me, and trust when this convention shall have concluded its labors much good will have been accomplished, and the interests of the great industry of which we all are so justly proud will be placed in a much sounder position than ever before."

"Mr. Chairman, I wish I had the faculty of making people as well satisfied with themselves as Mr. McMullen does," said Mr. Frank Whiteside, of Stettler. "Two of the most important things we came here to have dealt with have been overlooked. Mr. McMullen went into the relation of the transportation companies to the livestock industry. If a stranger had come in and had a copy of our programme to refer to, I think he would have been at a loss to know whether Mr. McMullen was speaking for the transportation companies or for the livestock men. It seems to me that, while he would have us believe that time will rectify all the existing difficulties in connection with the transportation of our stock, we, on the contrary, have found, by the public discussion, both in our meetings and in the public press, that there are grievances of the shippers of the West that have not been settled. Mr. McMullen told us that his company had at its disposal something like 3,000 stock cars, and that for six months of the year these were lying on the sidings rusting. He also told us that it took them ninety days to carry all the stock from the West—something like 75,000 head of beef cattle. Surely, had these cars been properly distributed and loaded, with seventeen or eighteen head of cattle in the car, the movement of the cattle could have been effected in a much shorter time. These cars, if loaded with eighteen head of cattle each, could move 54,000 head in one trip. On this basis, it ought to be possible to move 75,000 head from points in the North-West to the seaboard at Montreal or St. John in less than three weeks. The people of Ontario and of the Eastern Provinces generally do not understand the conditions of the West, perhaps, as we would like them to understand these conditions. They have not before them in their daily local papers, as we have, complaints of delay in shipping and inadequate car supply; and they think that, when we come here and make a kick, we are kicking about something that does not amount to much. But, if we could show them that many of our small shippers are prevented from shipping by the lack of cars, and that from this cause they lose a great deal of money, we should be more likely to have their sympathy in the movement we undertake.

"There are many of our people in the West who have attempted to ship, and, if you could read their reports, you would understand our conditions better. In the efforts which our Government is making to bring immigration into the country, it has not paid sufficient attention to the details of transportation. Nowhere in the world, perhaps, are we afforded better facilities for the transaction of business from town to town. Nowhere in the world, perhaps, can a man undertake a local journey with a greater certainty of arriving at a certain point at a certain time than in our West; but when it comes to shipping stock, the case is quite different. One has but to read the papers published in the ever-multiplying towns of our great West in order to recognize the fact that production has exceeded the means of transportation, and this condition has been responsible for the unsatisfactory and widely fluctuating market. The industrious immigrant to our Western heritage will be no better off than on the steppes of Russia or the downs of England, if the transportation facilities accorded him are not on a par with his activity. What matters it if the markets of Chicago, New York, Montreal, Liverpool and Manchester are high or low, if we cannot put our products on the market at a time when, in our opinion, they are finished? I venture to say that no other question is quite so irritating to the stock grower and grain grower of the West as this question of transportation; and if we look into it, we shall find that the complaint of the stock grower is greater than that of the grain grower, for while the grain grower has the agency of the wheat pits and grain exchange working on his behalf, where the bulls are boosting the prices while the bears are pulling them down, we have but one agency, and that is the agency of the bear.

"I suppose that all stock associations, and also all our associations which have for their aim the betterment of all social and financial conditions, find in their immediate vicinities peculiar conditions which defy their efforts to overcome. But there are two points on which we ought all to be able to agree, and these are: Better transportation of our products and the placing of our products on the market in a more perfect condition. As the traveller who has purchased his ticket from any transportation company would expect to be transported to his destination safely and in a reasonable length of time, so one who has ordered facilities for transporting his product has a right to expect that that product will be carried forward in a reasonable length of time. If this be not his right, then we, as producers, are at the mercy of every combination which men can devise to enable them to live at the expense of the toiler.

"Mr. McMullen told us that the season of 1907 offered more stock for transportation than did the season of 1906."

"I don't think I said that," corrected Mr. McMullen.

"I understood Mr. McMullen to say that more cattle were shipped in 1907 than in 1906," explained Mr. Whiteside.

"In the month of October," said Mr. McMullen.

"I was going to say that it was reported in the papers that both Mr. McMullen and Mr. Lanigan stated that there would not be within 25,000 or 30,000 head of as many shipped in 1907 as in 1906," continued Mr. Whiteside.

"Had the season of 1907 been as favorable to the grain growers as that of 1906, and had not the severe winter of 1906–7 decimated the herds on the Western ranges almost fifty percent, I venture to say even a greater congestion of traffic would have been found on our railways than was found a year or eighteen months ago, when quite one-third of the grain was stored in elevators and granaries, and thousands of cattle were turned back on the ranges to drift and die in the blizzards. The evolution of the cattle industry in the West undoubtedly tends towards a reduction of numbers in the individual bunches, but this does not by any means tend toward a reduced output. All precedents point otherwise, and as Alberta's stock has in the past given her the premier place among the Provinces as a grazing country, so her stock will in the future maintain her in that position, both as regards quality and numbers, so that if we are to avoid those conditions which, for the past few years, have made the individual shipment of stock a hazardous and unsatisfactory undertaking, we must have some guarantee that our transportation companies will bear us up in our efforts to establish satisfactory markets, both at home and abroad, and I firmly believe this can be most quickly and permanently arranged by a system of reciprocal demurrage wherein the delayed shipper will receive some recompense for his losses through shrinkage and a falling market.

"There are certain facts in connection with the shipment of stock from the West which were fully brought out in the investigation of the Beef Commission. That Commission reported that the facilities for transportation from the West are outrageous. I have their exact words here:

"'In many instances, we consider that the time occupied in shipping cattle from Alberta to Montreal and the treatment that the rancher receives at the hands of the Canadian Pacific Railway must be expressed in no milder terms than outrageous. That delay in transit is a matter which benefits no one, and, in many cases, portends ruin and disaster to the Western rancher.'

"There are, at the present time, only two markets to which we can look with any degree of success, and that is to the market to the south of us and to the market in the Old Country. The market to the south, under present circumstances, I think offers the best advantage to the small shipper. Even with an import duty of 27½ percent, it has been proven that average beef cattle can be marketed on the Chicago markets in September and October to compete favorably with the American rancher's cattle; in fact, they will outdo them. Some members of our Association, I think, had the honor to make the first shipment to Chicago last season. Four of our men undertook the shipment of 350 to 360 head. They received the best of treatment on the first shipment with the exception that, when they were coming back, though they had received transportation over the

American lines, when they came to the Canadian side they were charged one cent a mile and were compelled to ride in the colonist coaches—not permitted to go into the first-class coaches. This should not be. These people had paid freight to the Canadian Pacific Railway to the extent of something like $4,000, and yet they had to ride in colonist coaches and pay transportation at the rate of one cent per mile. This shipment resulted so favorably, and reports of its success were so widespread, that a great many ranchers, both great and small, thought they would go and do likewise. Mr. Reid, of Baxter, Reid & Co., of Olds, wrote to the Company at Calgary, and undertook a shipment of something like 350 head. But, instead of loading them at Olds, he drove them over eighty miles to Strathmore on the main line of the Canadian Pacific Railway. He had ordered cars from Calgary and sent his son with the cattle. His son arrived at Strathmore with the cattle. The cars were not there, and the agent said that they could not furnish cars to ship stock to American markets. Young Reid did not understand the conditions, so telegraphed to Calgary to his father. Mr. Reid, senior, employed an attorney and went to Strathmore, and there was no more difficulty in getting cars. He got shipment through to Chicago. This shipment was followed up by one from Medicine Hat by a man named Pruitt.

"Mr. Pruitt was going to drive his cattle across the line. The Canadian Pacific Railway asked him to send the stock over their line instead, and promised to have sixty cars for him—Mr. McMullen gave his written promise—on October 23rd. But, on that date, instead of the cars, he got a telegram that the cars would not be forthcoming. He had acted in good faith on the written promise of the Canadian Pacific Railway. The morning after he arrived at the point, one of Mr. Gordon's men came and offered him $27.50 for his bunch of steers. Pruitt thought he was in a box, but still he refused this offer. But that afternoon another buyer came and offered him $40. Still he refused. He decided to drive his cattle away, when another buyer came and offered him $42.50, and, rather than undergo further delay, it is reported he took this offer.

"Another man named Umstall shipped to Portal, and was held up for 27½ percent, duty on a valuation of $25, $35 and $45. The officer telegraphed to Mr. Gordon at Winnipeg, asking him the price of Alberta steers, but he telegraphed back a valuation of $35.00. Therefore, Mr. Umstall had to pay a 27½ percent, on $35 and $45, and pay a fine of 33½ percent, for swearing out, as they said, a false affidavit. He paid it, and, in Chicago, placed it in the hands of attorneys, and they took it to the Supreme Court of the United States, and I have here the decision that the customs officer must abide by the schedule rate, which is $20, $30 and $40, so he received back his fine and the excess duty paid.

"Now, our men who shipped from Red Deer in September undertook another shipment in October. While in Chicago, they had proved the market was good enough for Canadian ranch cattle. They wanted to prove to us that we could put our cattle on the Liverpool market profitably, and to better advantage than handling in the ordinary way. So they made arrangements with Armour & Co. to have a shipment sent through and handled by the Armour men in Liverpool. Mr. Coughlin, of Winnipeg, was to act for the Armours. They were to route the cattle on the 19th of October, and were told by the Canadian Pacific Railway agent that the cars could be supplied any time after the 14th October. They had their cattle in the corrals on the morning of the 19th, but there were no cars there. On the 21st, I think it was, they received a telegram from Mr. Coughlin that Armour's man would pay four cents on prime exporters and 3½ for second grade. Remember, cars were not available. On the 22nd a second telegram came, stating that the market had dropped and the best he could do was $3.75 and $3.85, for immediate shipment. The cars had not yet arrived. In the meantime, the boys had driven the cattle outside the town, and fed them with hay at $6.00 a ton. On the 26th, the cars were delivered. At that time, the best they could get was 3½ cents. If you will figure out the loss on the drop of the market which these men had to stand, you will see plainly that it is not in the interests of the small rancher that these delays should take place. Their bill for feed was $300. The first drop in the market cost them a loss of $1,500 to $1,600. Finally, when they found that the offers from Winnipeg were so low, they decided to take the matter in hand and ship again to Chicago. They loaded the cars in Red Deer on October 26th. In the meantime, something else happened.

"The capacity of a car is 20,000 pounds—that is the usual rate charged by the Canadian Pacific Railway. But when these cars got to Portal, the rate was raised from 20,000 pounds to 24,000 pounds. Though no more cattle had been put in the car, the freight rate was raised six cents per hundred above the rate charged them on shipment in September. This would mean a loss of upwards of $400 from this change of capacity and the extra freight. At Portal, also, they had to pay the increase in the schedule rate to the Customs Offices. They made no kick, thinking that it would be impossible to get the money back. So they paid on the valuation of $25, $35 and $45. There were a great many other expenses which added still further to the loss.

"I think that it is up to this Association to take steps to have better transportation provided by the transportation companies. Otherwise, we shall not be able to take advantage of the market conditions as we should."

"What is your point, Mr. Whiteside?" asked the President. "Is it that small shippers do not get a fair chance with the large ones?"

"No," said Mr. Whiteside. "My point is that this convention should bring in a resolution that transportation companies should pay reciprocal demurrage. If we hold their cars unloaded a certain length of time we have to pay demurrage. When shippers lose from $300 to $1,000 on a single shipment of stock, it is time that the railway companies should be called upon to pay some compensation; especially should this be so when they give written promises to provide cars, and then do not carry those promises into effect."

"What has Mr. McMullen to say?" asked the chair.

"I have nothing to say," remarked Mr. McMullen, "except that I am to be found in my office in Calgary. If there is a complaint to be made, I am there."

"Do I understand that the shipper makes an appointment through the agent of the railway company for cars, and when he drives the cattle to the point of shipment the railway company[s] fail to provide cars?" queried a member.

"That is the point. I can prove that such cases have arisen again and again."

"Mr. McMullen has addressed the meeting," remarked Col. Campbell, "and has been listened to attentively. It seems to me that when he is asked for an explanation it is not an answer satisfactory to this meeting for him to say that his office is in Calgary, and that if anyone has a complaint to make that person can go there. I claim that Mr. McMullen's answer is not a fair one in this meeting."

"Here is a letter I desire to read," resumed Mr. Whiteside. "It arrived the night before I left for this meeting. The man who writes this letter is ready at any time to make an affidavit to the facts of his experience:

'Velva, N.D., January 22, 1908.
'Mr. Whiteside:
    'Dear Sir,—In reply to yours relative to our shipment of cattle from Penhold, Alta., on October 29, 1907. We bought 187 head of beef cattle about 35 miles east of Penhold. We ordered stock cars of the Canadian Pacific Railway agent at Penhold on October 5th for shipment on the 15th'—shipment from Penhold on the 15th—'for Chicago, U.S.A.
    'There was plenty of time for the agent to see that these cars were supplied.
    'There was nothing said at time of ordering cars relative to Canadian Pacific Railway being unable to furnish stock cars for shipment on that date. We ordered our cattle to be delivered at Penhold on the 14th. When we got to Penhold with our cattle on the 14th, the Canadian Pacific Railway agent informed us that the Canadian Pacific Railway would not be able to furnish us with cars, neither could he say at what date the

Canadian Pacific Railway would furnish us with cars for cattle shipments to the States. We went to Calgary, Alta., and took it up with the car distributor in person—the cars for shipment to the United States.

'What is the reason for that? They say it is because the cars are held across the International border. That is a poor reason, it seems to me, for refusing fair play to the people of Canada. They call themselves a national transportation company, and they certainly have received great favors from the nation—$25,000,000 in cash, 25,000,000 acres of land, and many millions of dollars' worth of completed road which the Canadian people gave in order that they might have a national transportation line. Well, from that national transportation line we ought to receive better accommodation.

'The Canadian Pacific Railway would not furnish stock cars for shipment to the States, but would try and get foreign cars for us. The only way we got cars was by billing to North Portal and loading out in boxcars, which we did on 29th October.

'There is a state of affairs—cars ordered for the 15th October, and the cattle held in corrals at Penhold until 29th October.

'After begging of Canadian Pacific Railway repeatedly, they did furnish us with boxcars after we had cattle at Penhold station and 14 days after we had ordered cars, with an extra expense of over $800.00:

> 'Shrink of 100 Pounds per Head on 187, at
> $3.00 per 100 .... $561.00
> Hay—
> Four Tons per Day for 14 Days, at $4.00 .... $224.00
> Three Men and Extra Horses $98.00
> Total $883.00

'We lost an additional $1,000.00 on account of drop of over 50 cents per hundred in cattle at time we could have had them on the market and time we did sell.

'Yours very truly,
'Willis & Downing.
    'Per G. Downing.'

"There is $1,883 lost by such a deal as that. That is only one instance added to the others I have given, and I could quote a dozen instances. I contend that it is up to this convention to declare that we should obtain, either through the Railway Commission, or through legislation, the establishment of a system of reciprocal demurrage."

"I was interested in about half a dozen shipments last fall," began A.E. Cross of Calgary. "They all went to the Old Country and we had no difficulty in getting cars. The first shipments referred to by Mr. Whiteside went to Chicago, and, as American railway systems were short of cars, they simply monopolized Canadian Pacific Railway cars and would not send them back. The consequence was that the Canadian Pacific Railway lost the use of the cars in the busy end of the season. That is the explanation of a good deal of the difficulty. They were naturally more favorably inclined toward shipments that they could take over their own system and have control of the cars. That is, they favored shipments to Montreal for export to the Old Country rather than shipments to Chicago, where the American roads took possession of the cars and would not send them back. That would explain the matter to a very large extent, where the action of the Canadian Pacific Railway was, in many cases, seemingly unjust to the shipper.

"The other question is as to the mange. That is a very important question to us in the West, and I am sorry that the Government has not been strong enough to stamp out this disease. They have, unfortunately, removed for one year the restrictions with regard to having the cattle dipped. The consequence was that the mange increased to tremendous proportions, and this was all the worse as it followed a very severe winter. I lost, I suppose, $25,000 last winter, and I should not have lost nearly so much had it not been for the cattle being infected with the mange so that they went into the winter in a bad state and the hard winter completed the destruction. We ought to do all we can to stiffen the back of the Department of Agriculture and help out Dr. Rutherford, because he needs every assistance possible. The difficulty is with the small man in districts where they have the mange. He seems to think it a hardship to be obliged to obey the regulations. A man with forty or fifty head, who has not had any mange among his cattle, and perhaps there may be no mange for twenty miles around, considers it unreasonable to be compelled to dip cattle. And every one of these men will go to the local member and say: 'We are not going to vote for you unless you get this mange business knocked off.' I know they have done that—and the member goes and says to the Government: 'I am not going to get elected unless you get the mange business knocked off.'"

"I did not know you had politics out on the range," remarked the chair.

"We have. And we have politicians there as well as in Ontario. But there is the unfortunate position. We will never get rid of the mange unless we look at the question in a broad way and insist on cattle being dipped at least one more year. Last year they were dipped, and it did a tremendous lot of good. It nearly wiped out the mange. But, unfortunately, some have escaped the inspector's eye, and it is very difficult to get anybody to enforce the law. So, I would urge this convention to strengthen Dr. Rutherford's hands, and lay before the Government the importance of having compulsory dipping at least for another year."

"I thoroughly agree with what Mr. Cross has said about the mange," said John Turner, of Calgary. "But there is one point on which equal treatment is not meted out to every one in that country. If you can prove to the Department that your animals are inside your fences, and are free from mange, you are not forced to dip. Those that are running on the open range ought to be dipped."

"I want to endorse what Mr. Cross and Mr. Turner have said about the enforcement of the mange regulations," said Mr. Whiteside. "I happen to be in one of those districts where we petitioned Dr. Rutherford to have part of our district examined, and he will bear me out that the inspector reported that no animal in that district that we asked to have examined had the mange. Last year, we dipped the cattle even when the snow was on the ground and a heavy storm raging. But we did not think it was necessary this year. At one of our last meetings we drew up a resolution—and I think Dr. Rutherford has it in his office—asking that our district be examined during the present winter, as the inspectors told us that the mange could be more clearly seen in the cold weather. Of course, if we have the mange, we want to dip, and will be only too glad to do so."

"The compulsory mange dipping order was not enforced in 1906. Mr. Cross told you that, but he did not tell you why," remarked Dr. Rutherford. "He is a member of the Western Stock Growers' Association, and it was by resolution of that body, who were satisfied with the work done in 1904 and 1905, that it was decided that it was quite unnecessary to have the order in 1906. And, knowing that without their assistance and co-operation, which had been invaluable to me in the two preceding seasons, the work could not be well done, I very reluctantly allowed myself to forego having a compulsory order in 1906. I think it is only fair that I should disclaim responsibility and place it on the proper shoulders, namely on those of the Western Stock Growers' Association.

"Mr. Whiteside has mentioned that they asked that an inspection should be made of a district which, if I recollect right, is twenty-four miles wide and running fifty miles from east to west, and this at a time when every inspector we had was busy as he could be and we were spending thousands of dollars all over the country in trying to enforce the order. While this work was going on, the deputation came and asked in the most casual way to make a thorough inspection of

Cattle swim through a mange dip at a dipping station in the Hand Hills in this pre-1908 photo.

every animal in that district. Our inspectors, in many places, could not find a place to sleep over night and had to keep going both night and day, or camp on the prairie when night overtook them. We made that inspection, and found that in a part of the territory there was mange—east of range 5, west of the 4th. West of range 5 we did not have any mange. But where you are dealing with an enormous area, extending from Wood Mountain to the Rocky Mountains, and from the International Boundary to where Mr. Whiteside lives, you can understand what a difficult thing it is to make these inspections that these people talk about so smoothly. But we have made some progress. There were only two vats in the whole of the North-West in the spring of 1904, both of them out of order so that they would not hold dip. By the end of 1904, we had one hundred and ninety-six vats. We have done a good deal of educational work. A great many intelligent people have learned that, if they are going to rid the country of mange and avoid the enormous losses that it causes, as shown by Mr. McMullen and others, they have to take told of the problem in earnest and help us out. This year, between $50,000 and $60,000 of the public money is being spent in trying to compel the people in the West to clean up their cattle.

"Just to show you the difficulties that confront us, let me give you a few facts. We have the privilege reserved to us under our order to exempt cattle which, on careful inspection and examination of the animals and circumstances, are found by our officers to have been kept inside fences and to be reasonably free from any possibility of having been exposed to infection. But some of our friends translate that to mean that we must exempt all animals that are kept under fence. The other night we had a paper which spoke of making a certain municipality into grazing lands, and the necessity was pointed out of having men riding that fence all the time in order to make sure that it was not cut or broken down. Anyone who knows the West knows that two or three strands of barbed wire with a post every half mile is not always an absolutely impervious barrier against the mange. We all know that it is nothing unusual to find— and I make no reflections on the individual who owns the cattle inside the fence—a considerable number of cattle with a different brand inside some of these fences. And in the same way that these cattle got inside the fence, they can go back again. Even a good many prominent people have insisted that in every case where there is a fence, or even an apology for a fence, all the animals inside should be exempt. Of course, that is impossible. We must use judgment, and must make our decision in each case individually. I do not want to take up your time, but I thought I would add a few words to what Mr. Cross and Mr. Whiteside have said on this mange business."

"It is to the interest of the cattlemen to help out Dr. Rutherford in every way," declared George Lane of Calgary. "Dr. Rutherford shows what he has done for that country and how, with the co-operation of many of the cattlemen there, he has tried to stamp out the mange in that country. But how about it where one man complies with the order, pays out his money to clean up his cattle, while a man just on the other side does nothing at all, and by his neglect scatters the mange through the cattle again?"

"He ought to be run off to the North Pole," interjected the chair.

"We have to confront that kind of thing every day," continued Mr. Lane. "What is the good of making one man comply with the law and leaving another man free? Now, there is just one word I would like to say concerning Mr. McMullen as Live Stock Agent of the Canadian Pacific Railway. I have been shipping on the Canadian Pacific Railway since 1887, and I am free to say that Mr. McMullen, since he was appointed as Live stock Agent, has done more good to the cattle business, has done more to protect the interests of the shipper, than any other man we have in that country. He is not responsible for not furnishing cars; he is not responsible for the cattle before they are loaded or after. All he is able to do is to give information and to help the shipper when the operating department supplies the cars. He has done that and has given all the information that was in his power. I am satisfied as to that. But still, I must say that the company is not looking after the business as it ought to do. Just as Mr. McMullen says, they are hurting us and themselves, too. But, at the same time, they are making more money out of it than we are. But they have too many other irons in the fire. They should consider that they are not carrying the stuff as fast as they should; they are not getting to the market as quickly as they should. In that, I think, there might be an improvement."

"I have only a few words to say," said H.C. McMullen. "It was in deference to this meeting that I refrained from entering into any controversy as to the specific complaints. We were brought here, as I understand it, to find out how we can get along together and make money out of our business. If there is a specific complaint, as Live stock Agent of the Company I am paid to investigate that complaint and locate the trouble. As to the complaints that we have heard here today, I never heard of them before except in the most general way. They have never been made known at my office. But I will be pleased to meet the gentleman who made the complaints and take them up with him."

"You know that complaints have been made," remarked A.B. Macdonald.

"But, if I understand the position, this is not the place to discuss them. Out of deference to the gentlemen who invited me here, I did not go into these questions.

I thought I had talked long enough, and I think that some of you gentlemen will be inclined to agree with me."

"We have heard a great deal about eradicating disease in the North-West. Are the Provincial authorities putting forth efforts to eradicate disease in their several jurisdictions?" asked a member.

"Do you refer to the work done in any particular Province?" said Dr. Rutherford.

"Speaking generally of the North-West."

"No; the Provincial authorities do nothing with regard to contagious diseases. They used to do it in Manitoba, and, at one time, in the North-West Territories. They finally transferred the responsibility to the Dominion Government."

"Which means that they are putting forth no effort at all?"

"They take no action."

"They have nothing to do with it under the arrangement. I do not know that I am right, but, reading between the lines, I should be inclined to say that you are bothered with this mange question because the mange is mixed with too much politics. I can see that the small man who won't clean up his cattle and so places others under the risk of spreading the disease does this because he thinks he is protected by the man who represents the constituency. When a man with a few hundred cattle allows those cattle to get into a condition that makes them a menace to the whole country, he ought to be run out of the business somehow, by some power or other," summed up the President.

CHAPTER II

A Resume

ALL through the twenty years immediately following the disappearance of the buffalo in 1879, the Indians despoiled the herds of the ranchers, creating heavy losses in some years. At first they killed quite boldly, but as time and police punishment taught them the results of such doings, they resorted more to craft and cunning. Indians were bold horse thieves, good riders, and improvident providers. Sufficient unto the day was the meat thereof, and they killed and wasted many beeves. Winter time was the hardest for the red men, as their hunting was going, the buffalo gone, and they had not learned to save when they had plenty. Consequently they must provide food, and the police had in many ways taught them they must exercise their brains in order to escape the hated confinement of stone walls and iron bars. One very clever method that they used was in the sense of a winter sport. In the winter the springs seldom freeze, continuing running until broad sheets of smooth ice form where the water gushes from the grounds. These "glare" ice sheets are very dangerous for cattle, as they slip and sprawl, and if they should "spread"—go down with all four feet radiating in different directions—they are no longer any good, only fit for death. Ligaments and muscles are torn out by this "spreading," and there is no recovery for a horned beast that has been well stretched out in this manner.

Knowing this, a party of the Indians would ride until they came to a bunch of stock, herd gently until they had them headed toward some particularly slippery sheet of ice, then stampede them with quirt and voice. Usually some unfortunate animal would "spread," and the natives would ride eagerly to the owner's ranch, telling him of the calamity which they had "just found." As a rule the owner would let them kill it and keep the beef.

The half-breeds of Alberta were usually a shiftless, useless, lawless class, this being generally caused by their environment. These people were good riders, good stockmen, but remarkably lazy, exceptions to these rules being marked by their rarity. Some half-breeds made splendid men, but assuredly not all. If a half-breed were removed from the prairies and hills, from association with the Indians, from the shiftless ways of the tribes, he would be a much more creditable addition to the human race, for educated half-breeds who have left the red men and consorted only with whites have often become men of mark. They frequently have active, strong minds, and half the mischief done by the red men of latter days has been directed, or suggested, or managed by half-breeds.

A half-breed boy who was born in the Macleod district, educated at Port Hope and given every advantage, became a fine young man and a gentleman. He consorted with educated people, took part in their pastimes and was a white man in every sense of the word. When the Boer War broke out he went to South Africa and became a lieutenant in the army there. He was reckless, brave, and had a capable, masterful brain. One day he and four or five others were carrying despatches when a party of about a dozen Boers spied them and pursued. The Canadian soldiers knew that if they gave up they would not be hurt, so they discussed the advisability of doing it—hiding the despatches and surrendering. The half-breed was carrying the papers, and he, after silence, declared he was going on.

"If they catch me they'll shoot me for a Cape boy," he said, referring to his own dark coloring and the pleasant habit the Boers had of shooting all half-breed Negroes whom they found assisting the British.

"We stay with you," decided the white men, and they rode on, the half-breed planning a clever ruse. He said they would ride in a wide circle, one man dropping off every now and then and concealing himself in the rocks, leaving his horse to go on with the others, and picking it up on the completion of the circle again. The dismounted and hidden fugitive could then pick off one or more pursuers. This plan worked splendidly, and before long the Canadians had incapacitated every Boer but one. Then the half-breed suggested his grand finale.

"I know how we can catch that fellow," he said. "Just over this hill there is a wide slough with a corduroy bridge across it. You fellows herd that Boer into it, and I'll wait on the other side, and we'll get him." A fine scheme and carried out to perfection, the fleeing Boer dashing down to the bridge and seeing the half-breed lieutenant standing at the far end with rifle levelled.

"Surrender!" ordered the half-Indian, but the Boer threw up his rifle and fired, leaping then off the bridge into the deep mud, while the exultant pursuers thundered down from behind.

"Surrender!" ordered the half-breed again, but the stubborn enemy only replied with his rifle. Sudden anger flamed within the Macleod half-breed; he threw his rifle down and plunged blindly toward the Boer, the lust to kill coming and overwhelming him. Absolutely regardless of the cornered enemy's bullets which cut his uniform in ribbons, he walked coolly closer, his service revolver ready. Finally the Boer's gun clicked, his ammunition was gone, and then he threw the useless rifle down and cried: "I surrender."

But the Indian blood was up, the stubborn, senseless opposition, the whistling bullets, the smell of smoke, the heat of the pursuit had resulted in a blood-frenzy, and the half-breed gritted as he shot steady and true:

"I gave you your chance and you wouldn't take it."

This man went to his home country after the war, away from the East, away from his early companions, back to the range where the red men wandered and the tepees of his mother called silently. He went, and he became—"just a half-breed," shiftless, unmoral, whisky-sodden.

Throughout the early part of the ranching history, I.G. Baker & Company bought all animals that the ranchers had to sell up to the time the railroad came in the first eighties. They paid cash or trade, as the vendors willed, then they placed their own "Figure 3" brand on the animals and turned them on the range. I.G. Baker & Company were the first "beef" ranchers in the province, their markets being the Indians, the Mounted Police stations, and latterly the construction camps of the railroad. They gave fair prices, ranging from thirty-five to forty-five dollars, and they bought indiscriminately, for beef was beef in the early days, and old bulls were as toothsome to Indians as the best-fed three-year-old.

"Beef" ranchers who succeeded I.G. Baker & Company were P. Burns, later P. Burns & Company, J.H. Wallace, Gordon & Ironsides, and J.D. McGregor, though McGregor always had a breeding herd too, and Gordon & Ironsides and their associates in later years did extensive breeding. Certain ranchers declare that P. Burns and other buyers—but particularly Burns—harmed the ranching industry when a demand was made on the ranchers for fat cows, thus cutting down the breeding strength of the herds. Steers were always considered the only "right" stock to market, but Burns opened a wide market for rougher meats, supplying construction camps and lumber outfits in eastern British Columbia, selling them cheap beef and calling on the ranchers of Alberta to supply it. Thirty dollars for a fat cow was as good as forty dollars for a fat steer, ready money was always welcome, and many ranchers cut cruelly into their herds to sell their fat animals, cows and steers alike. Combinations of buyers also hurt the business, holding the price low, and making it difficult for a rancher to go to the expense of putting up hay and feeding his stock, also keeping them from being in position to purchase enough grazing land to allow them to expand. The time of the twenty-one year closed leases did not last long, the leases that many ranchers obtained being those with the two-year clause enabling the Government to close out the lease with that period of notice. Then came Hon. Frank Oliver's strong policy of encouraging the settling of farmers, and broad leases were thrown open to the settler, much to the disadvantage of the ranchers, who were crowded back and ever back. Ranchmen were forced from business, their range gone, their cattle starving on overcrowded pastures, and many a wealthy stockman now sits with folded hands and longs for the action of ranch life again. Tony Day, a ranchman who ran huge herds south of Medicine Hat some years ago, is one of these, and when asked if he would resume his business could he but get an assured range, he answered, "I'd fly at it."

When settlement began to crowd, and the old ranchmen began to worry about their range, many feeling greatly restricted if there was a fence anywhere between themselves and the horizon, P. Burns commenced to buy great tracts, purchasing a choice spot here, another there, until now he and his company own the "Home" ranch, twelve miles south of Calgary, the Mackie Ranch of one hundred and fifty thousand acres on Milk River, the Quirk Ranch of six thousand acres at High River, the Imperial Ranch of five thousand acres north of the Red Deer River, and two ranches, five thousand and three thousand acres respectively, twelve miles east of Olds, with leased lands adjoining.

George Lane, foreman of the Bar U Ranch in 1885, is another stockman who has become a king of cattlemen, owning the Flying E Ranch, the YT Ranch, great tracts in the Bow Valley country near Bassano, and being associated with Gordon, Ironsides & Fares in the ownership of the Bar U.

E.H. Maunsell, the only absolutely independent "big" cattleman, who owns his herds with no associates, has great numbers yet, nearly all on leases. He holds a lease of the great Piegan Indian reserve, and huge leases in the Grassy Lake and Pot Hole countries, where many thousand of his "IV" cattle brands are seen on the hills and flats. These three men, of all the old-timers, have grown and strengthened, seeing the future of ranching and preparing for it. The future of it, as seen by Lane and Maunsell in particular, is a gigantic mixed farm, stock fed throughout the winter, happy relations with farmers who take stock to feed, and a market that can never grow less.

The ranchers of the first few years in Alberta showed certain defects of judgment and inexperience—the overcrowding of choice ranges, the promiscuous mixing of different breeds of stock, the careless dependence on the chinook winds to bring the cattle safely through the winter months. Cattle, when turned on the range, will "locate" in a short time, selecting that section where the most feed is, where the water is handiest, where the shelter is best. The turning of the stock into the open range with no hindrance led to the choicest, most sheltered districts being picked by all herds, and a consequent summer destruction of the finest

Mounted on their horses, left to right, are independent rancher Edward H. Maunsell of MacLeod and dairy rancher Joe McFarland, *circa* 1882.

winter ranges, the fall finding the pasturage gone and the lack of grass driving the herds to less favored regions. Another weakness was the fact that nearly all ranchers turned their bulls loose—Shorthorns, Herefords, Polled Angus, Galloways, Devons, and all sorts of scrubs eventually getting together on the same range and causing a great slump in the fine grading of stock. The Cochrane Ranch, after it located in the Waterton Lakes district of extreme southwestern Alberta, was more fortunate than many others in this respect, they having a range so far removed that stray bulls seldom reached there, the result being that the Cochrane cattle were for some years the finest grades that went from the ranches of the Province.

The system of leasing was introduced by the Government in 1880, though the cattle did not begin to come in to stock these leases until the next year, when the great Cochrane herds trailed across from Montana. Practically no effort was made to fence the leases, they being left open for any stock to wander over, one of the chief reasons for taking the early leases being that the securing of a lease enabled the lessee to bring cattle from the United States without payment of duty. Big companies secured the finest leases, the choicest districts, and the small man was forced to content himself with what was left. There is no doubt but that the southern foothill country and adjacent plains is the finest pasture and grazing land, winter and summer, in all of North America, and the choice spots in it are select indeed. The Waldron Ranch took many thousand acres, the North-West Cattle Company many more, the Oxley and the Cochrane picked others. What was left went to the smaller men, and the open range was for all, there being no fences anywhere. Consequently cattle drifted to the choice spots, roundups were held, and the small rancher must be on the spot to get his share of the brand-crop. This expense, not particularly felt by the big ranches, was an added burden to the small man.

Time passed, and the ranchers began to realize by their spring calf loss that the winters, though mild enough for a strong and sturdy steer, were not sufficiently mild to enable a weak cow to suckle her calf and survive the spring, with its general storm. It was found that it took a goodly number of cows to produce one calf, and some ranchers took to feeding their weak cows and all calves in the fall, thus strengthening them for the winter and spring, while others entered into the business of importing stockers, using the range for a feeding and fattening district, and cutting out much of the breeding. This paid fairly well for a time until it was found that the class of stockers brought in, the yearling and two-year-old steers of Ontario and Manitoba, were not adapted to rustling. These dogie stock could eat grass all right in summer, but they could not find any in winter, the comfortable barnyards and stacks of Ontario and Manitoba being a poor education to give a beast that was going to be left to carry himself through an Alberta range winter. Considerable loss was experienced among these first herds of stockers before the ranchers learned how best to guard against the drifting, helpless spirit of the tenderfoot, pilgrim stock. Ranchmen looked for more adaptable stockers, and found them in Mexico, a good distance away, but an animal which, when cheap, proved more successful than the Ontario and Manitoba beast, his education being assuredly Mexican, and his nature being to rustle and take care of himself. It is known among cattlemen that range cattle put on additional weight as they come north, the colder climate, more luxuriant, suitable feed, plenty of water and little wandering tending to assist the fattening process. A hundred pounds is an average increase for every three hundred miles north the animal comes; consequently, when Mexican steers on the home range will ship at eight or nine hundred pounds, the calves from those same herds when fattened for three years on the Alberta range would weigh, when prime, some twelve to fourteen hundred pounds. The chief difficulty experienced with the narrow-hipped, poorly formed Mexican dogies was that their conformation was such that they would not be suitable for export, the home market being best adapted to them. The first Mexicans were shipped to Alberta in 1902, the shipment consisting of eighteen thousand head of yearlings and two-year-olds, billed to the Bar U, to Waldeck, Dunmore and Grassy Lake. Only ten thousand head of these animals were delivered, the majority going to the Bar U, for, owing to the long haul, various errors of shipment and losses en route, the Canadian Pacific Railway Company lost heavily in damages, thousands of head of the Mexicans dying on their hands.

The export beef of the ranges had another hard condition to contend with when ready for market. The fat animals of the Alberta prairies are wonderfully fit when the grass of the fall has matured and the animals have "trimmed to solid meat and marbled flesh," but the long haul, the handling, transferring, crowding in cars and jamming up chutes and into the strange interiors of vessels, the pitching, tossing voyage, all count to worry the wild, nervous range stock. Bruised, battered, worried out of hundreds of pounds of the fat they carried when crowded up the first loading chute, they landed in England, where almost immediate butchery awaited them, and the bruises, cuts and jams showed all over the butchered meats. This hurt their sale, and the reduction in weight during shipment subtracted considerably from the pockets of the owners. An average shrinkage of nine to ten percent, was the usual result of the long journey to the British marts.

Farmers, ploughs and cultivation in general did away with millions of acres of the unrivalled grasses of the prairies. There is no pasturage to equal the wild grass of the open range, no grass so strong and nutritious for summer pasture, none other that cures into hay as it stands and furnishes a nourishing, strengthening food for winter feeding. The weakness of the prairie hay was that it took too many acres of it to support a beast the year around, and too much territory was required once the farmers came in. It requires at least thirty acres of range to carry one "critter" through any kind of a twelve-month, and winter grazing had to go, especially after it was discovered that no tame grass could be used for rustling stock, that no known domestic grass would cure as it stood, like the native grasses of the West. One ploughing kills forever the nutritious natural food supplied by the prairies, the breaking and turning of the sod being enough to spoil that part broken and turned. A plot of ground that was ploughed in the Macleod district in 1879, and has lain idle since 1883, has never recovered, though vegetation grew again within a year. But the new grass was no good for winter. It was a strong, good summer feed, palatable and nourishing when green and growing, but with the first touch of frost it withered, died, and the nourishment left it entirely.

So the farmers killed much winter range, and the herds suffered in time. The plough went everywhere, in the choice districts, in the poor districts—the Government wanted settlement and allowed them to rush where they willed. There are districts in the south country of Alberta where wheat is now grown that would have done much better as perpetual range and grazing districts. The first crops were fine, the second a little worse, and ensuing crops lessened to mediocre. This condition is due to the soil and the winds, the ground when broken powdering to a fine dust, the high, dry winds then blowing it away in clouds. The soil has drifted away, the plough has killed the native grass, and the particular districts referred to are not nearly so desirable as they were when the first farmer went on them. Alberta will always have vast tracts of superb grain districts, soil that will grow the finest grain in the world, and grow it in profusion, but it is foolish to think that because some sections are thus favored, all should be; just as foolish as to declare every foot of the Province is grazing land and nothing more. Then, too, there are splendid grazing tracts where the soil is good and the growth of grain is promising, but where the summer frosts are almost as reliable as the seasons; these districts have also been given over to the farmer, the plough has killed the range—and the farmer is often hard up.

Some years ago it was the intention of the Provincial Government of Alberta, if they obtained control of the Dominion lands in the Province, to set aside certain portions of the country for a grazing reservation, and this would have assisted the ranching industry tremendously; but the Dominion kept control, and this opportunity was lost.

When times were hard and money tight, the settlers often resorted to various shifts to add to their incomes, especially from the Government. Governments are always looked upon as proper prey, if preying is possible, and the tricks that were played were often very ingenuous as well as ingenious. Dug Elison was a farmer of the very early days, and he and his partner at one time sold considerable oats to the Mounted Police. Many sacks were brought in by these energetic sons of the soil, piled high in the storeroom of the police force, and left to be used as the horses required. Nothing was ever measured, everything being sold and purchased by weight, and in selling a wagon-load of oats, the simple process of weighing the sacks and totalling the tally was used. Consequently the farmers received their pay orders upon delivery, and some time had elapsed before the police learned of a simple subterfuge that Elison and his partner had adopted to swell their income. Two policemen were rearranging the sacks; one was rolling them from the top of the pile, the other stood on the ground and caught them. No matter how strong a man is, if he is catching known weights, and is set to stop just about that much, he is disconcertingly jarred when any weight exceeding that to any extent comes upon him; so when a sack dropped like a plummet and nearly drove the sturdy catcher into the ground, it produced investigations that brought to light a rock weighing one hundred and forty-two pounds, neatly sacked and trimmed with a few oats.

All sorts and qualities of men came into Alberta from the first year of its settlement, and the Mounted Police were frequently hard-pressed to keep them in order. The men were sometimes thieves, sometimes murderers, and many were caught and handed back to the American authorities after hard pursuits. One case where the Mounted Police failed was interesting enough to the layman, and was not the fault of the Canadian police at all.

It was somewhere around the year 1900 that a gunman named Gibbons drifted into the Red Deer River country and secured a job as a cowboy, a position he was eminently fitted to fill. He was a capable stockman and an able person with a revolver, quick and deadly on the draw, being wanted in the States for murder, which he had committed in the heat of an argument. Under the name of Elgin he secured work in Canada, proving to be a cheerful, likeable, devil-may-care sort of fellow, who soon became popular with all who knew him. His guns were always sagging ready by his hips, but he never made any sign of using them. In the meantime the Mounted Police had been notified that the murderer Gibbons was in Alberta, and would they kindly locate him for the American authorities? Sure they would; that was their duty; so a constable inquired a bit and finally located the desired party by the description, the police authorities then communicating with the United States powers and telling them to send a man to identify and carry the suspect back if he proved to be Gibbons. A sheriff came, and he went to the ranch in question, accompanied by the Mounted Police constable who had recognized the quarry. Gibbons saw them coming, arose and wandered carelessly from the house, out into the open, where a man would have more chance, guns swinging, ability in every curve of his careless body. The American sheriff looked at him, saw the serviceable guns, remembered their effectiveness, thought of his own hide, and said: "Nope, that ain't Gibbons!" while the ready Gibbons strolled past, with no sign of recognition, mounted a horse and rode away, ostensibly on some ranch duty. The American sheriff returned to his home, the gunman remained at work on the Red Deer, a good cowboy, a quiet, reliable hand. But a higher power was watching, for the following spring he was drowned in the waters of the river while crossing the stream with two led horses behind his wagon.

CHAPTER III

The Indians

THE Indians of the western prairies belong with the story of the ranching industry; their later history is bound up with that of the stockmen; they were one of the original dangers, sources of loss, and worry. The first Indians were wild, lawless and bold, they stole cattle when they willed, and they to a certain extent had control of the ranges even for a short time after the Mounted Police came in. From restless, wandering tribes the red men of Alberta have become fairly respectable units of the great mixed farm of Alberta; they have been taught to plough, to harvest and thresh; they have learned the advantages of improving themselves, their herds and their prospects. From horse thieves, they have become breeders of livestock, owners of goodly herds, and are frequently self-supporting. Several great tribes of Indians lived in the southern half of Alberta when the white men came—Crees, Stonies, Sarcees, Kootenays, Blackfeet, Piegans, Bloods—of whom today large numbers remain, some creditable, some disreputable. The Kootenays, mountaineers and hunters, are comparatively worthy today, the Crees are sometimes hard-working but often shiftless, the Stonies are essentially a mixed-farming community of energetic hunters. Of the Blackfoot Nation, the three great tribes of Bloods, Piegans, and Blackfeet, the Piegans are the most quiet and industrious, the warlike Blackfeet are mining coal and raising cattle, the Bloods are stockmen, cowboys, ranchers, and good workmen, when they work.

The story of these Indian tribes if prepared at length would present a most interesting tale of war and murder, theft and raid. It would tell of wild men and wild animals, free life and underhand fighting. The source whence they all sprang has been traced back until today the ancestors of the great Western tribes are pretty well assured for two hundred and fifty years.

The Cree tribe, the most numerous native division in Canada, is said to have descended from the Ojibways of the northern great lakes, the Blackfoot Nation is sometimes asserted to be also of Algonquin stock, though some say their correct father tribe was the northern Slaves of the great north inland seas. Crees inhabit the entire land from the Hudson Bay to the mountains, Blackfeet are essentially a race of the plains, Bloods of the hills and levels, Piegans of mountains and hills. These last three tribes are all one in reality, springing from one parent stock. From the rough estimates of Indian traditions, it would seem that it was somewhere around the year 1650 AD that a band of Slave Indians ventured south, hundreds of miles out of their hunting grounds, and into the unknown prairies on the great Saskatchewan, to the very banks of the Missouri where the hostile Minnetarees, or Sioux, and their allies of the Crow, Cheyenne and other tribes, turned them back toward the north again. Starting the long return trip, these Slaves discovered they had been cut off in the meantime by Crees and Assiniboines who had gone in large numbers into the country along the north Saskatchewan. Thus they were hemmed in on the south by the Sioux, on the north by hereditary enemies, the allied Crees and Assiniboines, and the only thing left to do was to locate on the plains of the Bow and South Saskatchewan rivers and make as vigorous a stand as they could. Crees, Sioux and Assiniboines hesitated to beard the savage Slaves in their new-found hunting-ground, and the tribe waxed powerful, rich in ponies, brave in war, skilful on the hunting trail. As late as 1850 the Blackfoot Nation was generally called Slaves, though the name is now forgotten. As time passed and success and prosperity came to the arms of the new prairie nation, there arose of course various internal differences, conflicting ambitions, jealousies, and other causes of agitation. Young and ambitious chieftains would arise and feel trammeled by the rule of the leaders. They would assert an independence, collect a band of reckless young men, take squaws and go to a new district, forming a new tribe from the old stock. Thus there sprang from the band of Slaves the three big tribes, the Piegans of the muddy waters that arise in the hills, the Blackfeet of the burned prairies, the Bloods in the edge of the hills and out into the plains below, where the Blackfeet had chosen their special range. In early times they were all indifferently known as Blackfeet, and as Blackfeet they fought when Cree, Sioux, Crow or other enemies swooped into the land.

The original Slaves of northern Alberta were an outstanding tribe insofar as they treated their women with deeper respect than that shown by other tribes toward their squaws. The Slaves braves did much camp drudgery, which was something no other tribes did. When the Blackfoot Nation came into the southern sections, they apparently did away with this, and when white men first went among them, the squaws were the same hard-worked menials that they were among the Crees, Beavers, Assiniboines and other tribes.

After the new nation had gained sufficiently in strength and reputation, they became the scourge of the southern prairies, carrying their depredations even into the great country of the Snakes, across the mountains, deep into Oregon territory. During the early years of the nineteenth century they were the most feared raiders, next to the Sioux, on the North-Western prairies.

Some authorities say the Blackfoot Nation are of the Algonquin family, the same as the Crees, basing this declaration upon the fact that they belong to the same linguistic stock as the Crees. Slaves of the north belong to the same linguistic family as the Beavers and other northern tribes, differing considerably from Blackfoot and Cree. Alexander Henry found in 1810 that there were natives among the Piegans who had fair skin, grey eyes and light hair. The Blackfoot Nation during the early nineteenth century claimed all of that territory between Fort Vermilion along the South Saskatchewan to the mountains, thence north to the North Saskatchewan and down that stream to Vermilion River. Henry states that when he was among the Blackfeet they were rich in ponies and strong in war, that few warriors could not count ten scalps, and few had less than fifty horses, while some had as high as three hundred. He makes also the first mention of mules, the Piegans having one at least. A huge chief of that tribe who measured sixty-seven inches around the shoulders and seventy-six inches around the waist rode a large white mule, ordinary horses being too weak to carry his bulk.

The Stony Indians are easily traced from the Assiniboines or Stone Indians who lived in great numbers on the plains of Manitoba when the first white men ventured into that land. The Assiniboines are a branch of the American Sioux, the Alberta Stonies are a division of the Assiniboines who found the hunting good when they came to the foothills and decided to remain, despite vigorous opposition of Blackfeet, Kootenays and Sarcees. Though their race had been on the flat plains for many years, these Stonies adapted themselves so well to their new surroundings that they became mountaineers and goat and sheep hunters of such skill that the natural mountain tribe, the original holders of the peaks, took great umbrage and frequently killed and scalped to force the interlopers to go. But the Indian seems to fight best when cut from his base, as is instanced by the Slaves and the Sarcees, so the Kootenays failed to dislodge the Stonies and in time came to recognize them as having some share in the hunting grounds of the eastern slopes at least.

The Kootenays were themselves a branch of the Beaver tribe, a northern Indian race that controlled a vast territory along the Peace River until the Crees came and disputed their authority. Doubtless the ordinary tribal difficulties arose and a band took to the mountains, where they established themselves and soon held independent control over a new garden of hunting; the Sarcees of the plains, also a branch of the Beavers, coming down later through the open plains and locating in the Battle River country and thence down until they touched the hunting grounds of the Blackfeet, with whom they often made common war against the Crees.

When the white men came to trade with the natives, they found the Blackfeet a warlike race of magnificent horsemen, trappers of beaver, hunters of buffalo, living handsomely on the spoils of chase and war. They found them already engaged in almost incessant war with the Assiniboines and Crees—both of the latter being still without horses; they found them treacherous, reckless, brave, underhanded as the occasion required, and quite open to trade for white men's blankets, guns and whisky. The history of the white traders among the red Western tribes commences with the hardy French explorers, swings to the Hudson's Bay and North-West companies, stays with these great companies until amalgamation, and continues on to and under the Dominion Government. French traders were operating in the countries of the great lakes before the Hudson's Bay Company started, yet they did not get into the rich Western trade until fifty or sixty years afterward. The Hudson's Bay Company, starting in a new land, had a virgin trade to draw from and did not worry about extending, because the profits made with their one post were enormous. But when the French pressed in to Lake Manitoba and beyond, the canny merchants of Montreal under the name of the North-West Fur Company ventured into the new West, and after them came the Hudson's Bay, outstripping the French and racing neck and neck with the aggressive Montreal organization, opening new fields, braving untold dangers, seeking always for new territories where furs would be more plentiful and cheaper. The impetus to these explorations was really given by the French traders, who taught the British and Scotch that the best way to get trade was to go out and seek it. Having learned this, the British companies, assisted by government and money, soon outstripped the energetic Latins, though the latter have assuredly left their traces among the tribes.

In the year 1668 the Hudson's Bay Company of Adventurers built their first post, Charles Fort, on James Bay. Their charter was granted to them for the purpose of trade and exploration, but during the first couple of decades they devoted themselves to the more profitable of the two followings. About the year 1690, criticisms of this policy made it expedient to show some results from explorations, and so in 1692 Henry Kellsey's report of his journey to the land of the Assinae Poets (Assiniboines) was published. The Hudson's Bay factor in charge of the post whence Kellsey went took upon the company the credit of sending out the expedition, but independent information from a contemporary of the time tends to give the impression that Kellsey's journey was simply used to Hudson's Bay advantage after it had been made.

Kellsey's account of his journey is itself a hazy one, though there is little doubt that he made it. According to the independent historian, the Hudson's Bay governor at that time, following a policy of the company, strongly discouraged all intercourse with the natives excepting on company business. Kellsey, a young man who had a liking for Indian company, resented this and finally went away with a party of red men, returning a year or two later. Before he returned he forwarded a letter to the governor, written with charcoal on birchbark and asking for pardon. It was granted, and he came in, accompanied by his Indian wife.

Kellsey's story indicates that he reached either the forks of the Saskatchewan or the foothills of the Rockies. He is said to have met with grizzly bears, and this is used as an argument in favor of the journey reaching the foothills, but grizzlies covered a vast country in early days, reaching as far east as the Red River even as late as a hundred years ago.

In July 1691, Kellsey left the Hudson's Bay fort and plunged into the unknown wilds, travelling on foot and by river and canoe. On July 24 he arrived in the country of the Stone Indians, or Assiniboines. With him were a number of Nayhaythaways, or Crees, who were very much inclined to take the warpath against the Naywatamee Poets because the latter had killed three Cree squaws that spring. Kellsey, after urgent appeals, succeeded in persuading the Crees to abstain from fighting, though it took them a long time to understand why they should make their tribal affairs secondary to the interests of the white men's trade. Kellsey's arguments won the day, however, and the Crees stayed with him, pushing onward. But the thoughts of unpunished Naywatamee Poets rankled, and when one of the Crees took sick and died, the Indians swore it was caused by the anger of the gods, who were offended with the party because they had not sought to bring punishment to the Naywatamees. Again the white man must use his persuasive talents, and again he met with success.

On August 19, 1691, the party sighted buffalo in abundance, "not like those to the northward," says Kellsey, probably referring to musk ox, "their horns growing like an English ox, but black and sharp." Pressing forward, they crossed brush country and open plains, gradually drawing nearer and nearer to the land of the Naywatamees until, on September 8, they met four of that tribe, who reconnoitered and after satisfactory conference hastened westward to notify the main camp. Three days later Kellsey and his companions arrived at the camp, where the white man immediately commenced to pave his way with presents and gifts. Apparently, if Kellsey was on a freelance journey, he was very wisely arranging for a welcome return to the white people when he wanted to go back, for he urged upon the chief of the Naywatamees the desirability of trading with the Hudson's Bay posts on the coast, and drew from that potentate an expressed intention of trade in the spring. This promise the chief kept as far as he was able, for he started eastward the following spring with a party of young men and a rich load of furs, but was intercepted by the Crees, who had a little matter to settle, and when the discussion over the slaying of the Cree squaws was finished, the Naywatamees were going hastily back to their own country. Yet it is fairly reasonable to suppose that if the Crees captured the camp of the trading Naywatamees in that spring of 1693, the rich peltries eventually came into the possession of the Hudson's Bay Company anyway.

One day, while on the western end of his journey, Kellsey and an Indian were surprised by two "grizzled" bears, who charged without warning. The Indian promptly shinned up a tree, while Kellsey took refuge in some thick willows, from which vantage point he shot and killed both bears. This deed won for him the Indian name of "Little Giant."

Kellsey was the first white man to hunt buffalo on the plains of Western Canada; he was the first white man to reach the prairies. His account of the Naywatamee Poets is indefinite and unsatisfactory, and, as no further knowledge has been given of them, it is impossible to say which tribe they were. Naywatamees suggest Nodwayes or Sioux, but the Sioux ranged too far south then to have been reached by Kellsey with any degree of probability.

The first fully authenticated journey into the West by a Hudson's Bay Company's officer was that made in 1754–5 by Anthony Hendry. Hendry was born in the Isle of Wight, and had been outlawed in 1748 by the British Government, smuggling being the direct cause. Enlisting in the Hudson's Bay service, he was shipped to America, though the old Company officials declare they did not know the man was an outlaw when they took him in. At any rate he proved to be fearless, energetic and valuable, and brought out more real knowledge of the great country of the West than had ever before been gathered.

On June 26, 1754, he, with a party of Indians, left York Factory. Journeying westward, they reached the Saskatchewan, and eventually came to a French fort or trading post that had been built by La Corne some time previous. The French, courteous and affable, endeavored to bluff Hendry into giving up his intention of proceeding westward, but the white man was made of different timbre, and pushed onward after spending a pleasant night with his French hosts. On July 30 they reached the plains and killed a moose, later meeting some Assiniboines and urging them to trade with the Hudson's Bay posts on the coast. The reply was that the Indians were well supplied with white men's goods by the French.

August 13, 1754, saw the adventurous Hendry on the edge of the big plains, expecting every moment to sight the Archithinue Indians hunting on horseback. He had heard of these people, and was anxiously looking forward to the meeting. Kellsey, sixty years before, penetrated deeply into the plains, but he says nothing of horses. Perhaps in the interim the horses first introduced into western North America by the Spaniards in Mexico had spread by trading and theft to the tribes of the Saskatchewan and Belly river plains. But, from Hendry's description of the finished horsemen he encountered, it must have taken many years of acquaintance with the animals to reach the described stage of perfection.

The name which Hendry uses to signify the equestrian Indians seems to mean several tribes, for in 1772 Cocking says that the Archithinue people embrace five tribes: the Powestic-Athinuewuck, or Waterfall, Indians; the Mithco-Athinuewuck, or Bloody, Indians; the Koskitow-Wathesitock, or Blackfooted, Indians; the Pegonow, or Muddy Water, Indians; and the Sassewuck, or Woody, Indians. Thus the Bloods, the Blackfeet, and the Piegans were the early horsemen of Western Canada. The Waterfall Indians were the Gros Ventres, who then lived in the hills between the North and South Saskatchewan, and a few years later were reduced by smallpox to about a hundred lodges on the Bow River. There is a tribe up around the mountain country north of the Yellowhead Pass that is known as the Wood Indians, and a family of Crees has for long years been known as Wood Crees. Barring out the Woody and Waterfall tribes, the Blackfeet, Piegans and Bloods were horsemen a hundred and sixty years ago, while the Assiniboines of the Manitoba plains still travelled on foot and with dog-trains.

Prosecuting his journey, Hendry met various bands of horsemen, members of the Blackfoot confederation, and by the middle of September was in the midst of countless herds of buffalo. Crossing the South Saskatchewan near what is known as Clark's Crossing, he met a band of Eagle Indians in the country near the Red Deer River, perhaps a little west of that stream's junction with the South Saskatchewan. These Eagle Indians rode horses, and were most easily discerned from the other tribes owing to their habit of wearing no clothing at all (traders later found that the Blackfeet often rode thus in summer). Hendry sought to lead them to trade at York Factory; they promised to do so, and this proved one of the very few instances where such promises made to Hendry on this trip were kept.

Great herds of buffalo covered the prairies, everywhere the Indians were slaughtering, taking only the choice pieces, tongues, etc., and leaving the greater portion of each slain body to the wolves who followed the herds and the hunting parties in large bands. In September, somewhere near the Rosebud country, a grizzly bear horribly mangled a couple of Indians who belonged to a party that Hendry encountered. On October 11, 1754, Hendry crossed the Red Deer River, and, bearing southwest, left the river probably between the mouths of Knee Hill and Three Hill creeks, reaching the head camp of the Blackfeet in three days. The tribe's headquarters consisted of a fair-sized portable town of two hundred lodges placed in two parallel rows; the horses were picketed with buffalo thongs and hair halters, the saddles were of buffalo-skin pads, while the stirrups were of the same material. Strict discipline was kept in camp, guards being always on the alert, while when on the trail the tribe travelled with a main body guarded by clouds of mounted scouts scouring the advance and the flanks. Hendry was greeted with great courtesy; the head chief, awaiting his arrival, was seated on a sacred white buffalo hide; the peace calumet was presented, accepted and smoked, then the white man told of his mission of securing trade. The Blackfoot chief, apparently a red Solomon with the gift of judgment and tactful expression, turned the trader down decidedly, but with the greatest courtesy. He gave the matter mature deliberation, and he spoke thus:

"Your forts," he said, "are too far away, and my people could not live without buffalo meat. My young men do not understand canoes, they only know horses, they cannot live on fish. Why go to the white posts anyway? There is much food on these plains and with bows and arrows we secure all that is needed. Also," concluded the wise leader, "I have been told that the Indians who frequent the white settlements often starve on their journeys."

Hendry spent two or three days with the Blackfeet, but the chief had spoken, and hopes of arousing the trading desire were blighted. He left the camp and travelled north again, bearing slightly westward to the country about due north of Calgary. Three days later he pitched camp. It was now November, and by the middle of the month the squaws commenced to prepare skins for winter use. Beaver and otter swarmed in the creeks and rivers. The party wintered in the Blackfoot country "west of the Red Deer River," though there was little winter until December. On the first day of that month, Hendry notes that he is still wearing his summer clothing, and the weather is the finest he ever experienced at that time of the year. But the next day a snowstorm swooped down and buried the land, and then the Indians, improvident as always, bestirred themselves and commenced trapping and hunting to get a sufficient supply of skins for winter clothing. When they had what they deemed enough, they resumed loafing, though Hendry reproached them and set the example of trapping. His

Indians laughed, saying, "We'll get all the packs we want from the Archithinue in the spring."

On February 27, 1755, Hendry came to Devil's Pine Lake, east of the present town of Innisfail. Buffalo were plentiful and hunting was good, various parties from different tribes being out. A number of unknown Indians visited Hendry's camp and, when leaving, offered him the present of a Blackfoot boy and girl whom they had taken in a raid on that tribe. Hendry says he refused with difficult diplomacy. Pressing northward again, he came to the Red Deer River, just above Tail Creek, and then retraced his steps and waited for spring, when he intended returning to York Factory. The ice breaking up in the rivers on April 24, Hendry commenced his long homeward trip on the 28th of that month. They went down the river (the Red Deer), and on May 23 reached the French fort, where De La Corne in person greeted them, attired in the uniform of an officer in the French army. His men wore only thin drawers and striped cotton shirts, and the weather must have been mild even at that early season. The French "fort" was a stout log structure, twenty-six by twelve feet, with walls nine feet high. Hendry reached the York post on June 20, 1755, a year, lacking six days, after he set out on his hazardous undertaking. The result of his explorations and investigations was the conviction of the Hudson's Bay Company that they must build posts inland in order to gain the trade of the red men. Hendry had explored the vast country between the North and South Saskatchewan rivers, as far west as Calgary. He was the first white man to see the Red Deer River, and the first Hudson's Bay officer to visit southern and central Alberta, though the French explorer De Niverville in 1751 possibly went as far west as Hendry did. Hendry found the French were on the ground, that the Hudson's Bay Company must be there too in order to get the business, and his recommendations were tardily followed, even though he was considered a most colossal liar with regard to Indians riding horses. Hendry was thirteen years ahead of the first English traders of Montreal in getting into the upper Saskatchewan country.

In 1768, exactly one century after the Hudson's Bay Company first located on the North American continent, the North-West Company, composed of English merchants of Montreal, canny men and good traders, built posts in the great inland and soon controlled a huge native trade. Six years later, Samuel Hearne of the slow-moving Hudson's Bay Company built the first inland fort for that organization. In 1774 Hearne erected Cumberland House on Pine Island Lake on the lower Saskatchewan. Here, for many years, the policies of the company were directed. The North-West Company, having more energetic explorers, had located before them with strong posts on Lake Winnipeg and the Red River, and were in high favor with the Indians. It was here that the rivalry of the two great trading companies commenced, a rivalry which led to bloodshed, robbery and war, and was never finished until the powers of the two organizations, seeing plainly that they would destroy each other if the struggle continued, sought a remedy and found it in an amalgamation that took place early in the nineteenth century (1821), when the North-West Company was merged into the Hudson's Bay Company, all under the latter name. But before this took place, the toll of lives and money had been enormous, considering the small population in the West.

The next white man of the Hudson's Bay Company of whom we hear in the West and in the foothill country is Peter Fidler, who left Fort George on the lower Saskatchewan in 1792, went southwest across the Battle, Red Deer, and Bow rivers to the Little Bow, and then, turning back, crossed the Red Deer at the mouth of Rosebud Creek, which he named "Edge Coal Creek," this being the first record of the discovery of coal in southern Alberta.

Fidler wandered throughout the West for many years after this, drawing more or less authentic maps, and left many valuable notes and writings. David Thompson, a Nor'-Wester employee, was a contemporaneous explorer with Fidler, their trails crossing many times during the last decade of the eighteenth and during the first few years of the nineteenth century. Thompson reached points further west than Fidler did, penetrating into British Columbia and discovering the source of the Columbia River.

Much honor is due the North-West Company and Hudson's Bay Company for their work in exploring the western part of Canada, but the palm belongs to the French, who as explorers and traders pressed far into the wilds long before the English companies ever thought of establishing posts beyond Hudson Bay. They came up the St. Lawrence into the great lakes, building posts, trading and fighting with the natives. Jacques de Noyon in 1716 met a party of Assiniboines in the then far West, who told him they were going on a war trail to the western sea to fight a race of dwarfs—stout, wicked warriors whose height was between three and four feet. The red Munchausens further told the credulous Frenchmen of fortified towns and villages where white men, bearded and brave, rode on horseback with their women behind. Doubtless they were simply filling de Noyon with tales based solely on imagination.

The great explorer, the first white man to reach the North-Western plains, was undoubtedly Pierre Gaultier de la Verendrye, who covered a vast land in the far West during the first half of the eighteenth century. La Verendrye and his sons visited the country of the Assiniboines, the Mandans, the Minnitarees, and saw the Rocky Mountains, "the Mountains of Bright Stones." In 1737 they were among the Assiniboines, whom they found to be still horseless, using dogs and canoes for travel; they visited the Mandan villages on the Missouri, finding them to be a rather superior race of agricultural Indians who lived in villages well-protected with palisade walls and ditches, clean, neat streets, and wide defensive ramparts on the walls. One of La Verendrye's sons discovered the Saskatchewan between the years 1732 and 1743, exploring it to the forks some time before 1750. La Verendrye and his sons are unquestionably the true discoverers of the great North-West.

In the spring of 1751, Legardeur de Saint-Pierre sent De Niverville into the West from a post the French had established on the Saskatchewan near the forks. De Niverville, being ill, sent ten men in canoes, and they are supposed to have gone up the Saskatchewan as far as the Rocky mountains. De Niverville followed later, and a fort was built against the base of the mountains, just where, it is impossible to say. The indications given by De Niverville and Saint-Pierre lead to the supposition that it was built on the South Saskatchewan, at or near the present city of Calgary, though some historians hold that the party went up the north fork. From the description of the country given by De Niverville, it is most probable that they were in the prairie and foothill portion of southern and central Alberta. They named the fort La Jonquiere, in honor of the then Governor of the French colonies in Canada, and they found Assiniboines in the vicinity. It is possible that these Indians were the forerunners of the present tribe of Stony Indians who live just west of Calgary in the edge of the foothills.

Beyond the reports and accounts of Saint-Pierre and De Niverville, there is no further mention of Fort La Jonquiere, though Hendry in 1755 must have been within a couple of days' journey of it. Perhaps the French relinquished it after the first year, and it was destroyed by time and red men.

As has been already said, Hendry found equestrian Indians on the plains of Alberta, though De Niverville does not speak of them. In fact the French do not mention horses for many years afterwards, though they must have met with horsemen in their wide journeys. They speak in 1776 of the Assiniboines' methods of transportation, saying they are yet without mounts and still use dog teams and canoes, hunting buffalo in winter by forming great pounds where the entrapped animals were slaughtered at ease and in great numbers.

From this brief resumé of French explorations in the West, we will go back to the journeys of the Hudson's Bay and North-West explorers. David Thompson, the famous Nor'-Wester, left the mouth of the Souris on November 28, 1797, with the Mandans and the great western plains as his goal. He was accompanied by a guide named René Jussaume, a famous trader who had been much in the Mandan country. Thompson found that already, white scoundrels, offscourings of civilization, outlaws, and whisky dealers, were among the red tribes and were commencing to demoralize them. He found the Mandans and their villages, the tribe being somewhat reduced in numbers since La Verendrye's time, living in company with some Willow or Fall Indians, these latter being possibly the Gros Ventres of the French. Thompson describes the Mandans as being a superior race who tilled the soil, grew corn, beans and other foods, and had villages and houses that were strongly built and well fortified. Their hair was of fine texture and sometimes was brown, though some of the children had white hair. Their knowledge of mortising logs and building houses indicated a high intelligence or a basic education from some better educated race than the Indian tribes surrounding them. No one knows from whence came the Mandans; they are now practically extinct, but the mystery of their foundation remains unsolved. I will repeat a tradition which was given me by a man who traded among them forty years ago, and which he seemed inclined to think might be a possible solution:

In the sixteenth century, a party of Welsh sailors were said to have been lost in the country to the east, just where and when the tradition does not say, though it does insist that they were lost, and the Black Hills in the country west of the Mississippi along the Missouri are thought to have been the spot where they finally disappeared. My informant tells of meeting a Welshman on one of his trading trips, and persuading the man to accompany him into the Mandan country. There they met the Mandans, and in conversation the Welshman recognized many Welsh words that the Indians were using in their language. Intonation and meaning were exactly correct, and the Welshman made further inquiries. He was told by the head chiefs that the Mandans came from the Black Hills many years before, being driven out by stronger tribes, and, braving the Missouri in boats made of beaver hides stretched over willow frames, went to the spot they then held on the river. They could not tell where they learned to build their palisaded towns and their houses; they could not tell of any traditions of white men living among the tribe—and that is the tale told by the old trader. Whether it is another Western myth is of course unknown.

After the white men commenced making frequent journeys into the West, they grew to know more of the red men and the different tribes. They knew the Blackfeet, Bloods, Crees, Piegans and Sioux, the latter a great nation of horsemen who lived along the Missouri and made frequent forays into Blackfoot country in retaliation for similar visits from the Blackfeet and their confederate tribes. It was seldom that they bothered white men, and the first known mention of such is in 1895, when Charles McKenzie and F.A. Larocque saw Minnetarees (Sioux) on the Missouri attired in the clothing of Canadian voyageurs. The explanation of this seems to fit well into an experience of John McDonald of Garth on the South Saskatchewan in that same year. McDonald, while at a small fort built at the forks of the Red Deer and the Saskatchewan, was trading with a large number of Blackfeet when he saw a splendidly mounted Indian, gay in trappings and fine skin clothing, dash through the ranks of Blackfeet and ride right to the door of the fort, where, dismounting and throwing his reins to a man standing near, he approached McDonald, and proved to be a Minnetaree chief. McDonald mentions that he rode a superb horse. The Indian, scornfully turning his back on the scowling Blackfeet, told the white man that his tribe was at peace with the whites, but was at war with the Blackfeet.

"I am a member of a small party of Minnetarees," he said, "and we would trade with you. If you will receive us as friends we will fight our way in." McDonald thought he meant to secure the services of the whites in this puncturing of the Blackfoot lines, but the trader would not do that, simply saying he would remain neutral, and that he would trade with the Minnetarees if they came. The chief sprang to his horse and dashed away, again safely passing the numerous Blackfeet. The next day he attacked, but the entire party was surrounded and captured, escaping again that night, taking with them their horses, their wounded and their dead.

Some little time later McDonald ordered his guide, Bouche, to wait for him at a certain point on the river. Bouche waited, but McDonald did not come. Bouche's men saw hostile Indians, and urged the guide to take to the canoes, but he would not, saying he had been told to wait. That night the Minnetarees attacked the camp, killing Bouche and two men, the others escaping to the canoes and holding the red men off. Doubtless this was the party that McKenzie and Larocque saw on the Missouri attired in the Canadian garments.

With the people of the West in the early nineteenth century there was no law; the traders were present on sufferance, the Indians allowing them simply because they wanted their whisky and their arms. Yet the reckless whites pressed on and ever on, though many prairie tribes objected to the traders venturing into new territories, because the western tribes would therefore gain the advantage of firearms also, and would become more insolent than ever. The Blackfeet, Piegans and other extreme western Indians seem to have been held in respectful awe by the tribes further east. White traders must wink at lawlessness, must ignore murder and brutality, for they were present to trade and that was all. Alexander

Henry, who was in charge of a post on the Pembina in 1803, unconsciously tells in a few words the condition of affairs among the natives, his diary, written in all seriousness and simply noting the happenings of an ordinary day, reading: "Le Bœuf stabbed his young wife in the arm, Little Shell almost beat his old mother's brains out with a club, and there was terrible fighting among them. I sowed garden seeds." A plain, simple and forceful word-picture of conditions of the white men among the western tribes in the very early times.

By 1809 the energetic and warlike Blackfeet had discovered a new and enticing method of life. Every spring they went on long pilgrimages to the south, where they raided among the Sioux and white traders, murdering right and left, stealing horses, fur packs and arms. The furs they brought back to Canada and turned in to the coffers of the Hudson's Bay and North-West posts at very reasonable prices. This entirely eradicated the cold, disagreeable drudgery of winter trapping, supplied a pleasant season, and brought as large results as if they had trapped the hides themselves. The companies' officials, being simply in search of furs, never prosecuted inquiry as to source.

Whisky and rum were already getting their grim grip on the prairie and foothill races. Every trader carried the stuff, every trader used it as a strong incentive to trade. In fact, it was so important that a sort of prescription as to how little the different tribes would take was drawn up by Alexander Henry in 1809. Spirits cost money and had to be packed long distances, consequently they were heavily diluted when retailed to the Indians. Four or five quarts of high wine, emptied into a nine-gallon keg of water, was of sufficient strength to pass satisfactorily among the Blackfeet, and was known as "Blackfoot Rum." The Crees and Assiniboines, having had a longer education in white ways, and having appetites and capacities better developed, insisted on six quarts to the keg, while the grasping and thirsty Saultaux, hardened topers, would not accept anything weaker than eight or nine quarts to the nine gallons. To the traders, the Saultaux were not nearly as welcome customers as the Blackfeet.

Free traders, those rough, fearless opposition members of the commercial world who looked upon the land as being theirs as much as the Hudson's Bay Company's, were proving themselves painful thorns in the sides of the two established organizations. They delved into new territory, paid larger prices than the companies wanted to recognize, treated the world and the dignified HBC with a flippant disregard, and in every way annoyed the factors. Their over-bidding of the Hudson's Bay Company prices was one of their greatest sins, for it tended to so badly "demoralize" the native idea of values. Henry, of the North-West Company, was also very bitter against them, and in 1810 expressed himself in biting words. Writing at his new post on the White Mud River, west of Edmonton, he says:

"Five young vagabonds came to us, on their way, they said, to the Columbia to trade for beaver. I took much trouble in trying to make a division among them to prevent their crossing the mountains, where they will be even greater nuisances than they are here."

Evidently his thoughtful efforts in behalf of these young adventurers were futile, for on the following day his irritation breaks forth again:

"Troubled with these mongrel freemen all day. There is no dependence to be placed on them. They have neither principle, nor honor, nor honesty, nor a wish to do well; their aim is all folly, caprice and extravagance; they make more mischief than the most savage Blackfeet on the plains."

During this same year, the North-West Company and the HBC were being interfered with considerably in explorations by the Piegans, who ranged the foothill country from the United States to Rocky Mountain House, a North-West Company post that had been built by John McDonald of Garth in 1802. The Flatheads having badly defeated the Piegans the preceding season, the latter were determined that no traders should take firearms and ammunition across the mountains. So they were on the *qui vive* for passing parties. Alexander Henry, who accompanied a party from his post to the Mountain House shortly after the young free traders visited him, found things blocked by the Piegans. His party's aim was to cross into the Columbia country. But Henry was not stopped in the least. He simply filled the guarding Piegans with spirits, and, when they were unconscious, sent his party westward, a very simple and effectual expedient.

Rocky Mountain House was the farthest west and the farthest south of any North-West Company post. It was for years the outfitting point for southern, northern and westward exploration and trade. No posts were built south of that point, though in November 1800, David Thompson came down the Red Deer River, crossed over to where Calgary now is, and went south to the stream now called High River, where he met some Piegans on a hunting trip. Some eighty years ago, possibly shortly after the merging of the North-West and Hudson's Bay companies, the Hudson's Bay people endeavored to establish a trading post on the Bow River at Calgary. They built a strong post and put in a strong garrison, but were frightened out by the overbearing manners of the Blackfeet and Bloods, who resented their presence. From the time the first white man set foot on the prairies to the day the Mounted Police arrived at the foot of the mountains, the white men in the southwest of the present Province of Alberta were living simply because it was the pleasure of the Indians to allow them to do so. Any undue interference with their methods or their habits would have been the signal for the wiping out of every white soul west of Lake Winnipeg, and the traders knew it. But the longer they traded, the more essential their presence became. They supplied blankets and arms and ammunition, they introduced the insidious alcohol, they gradually—slowly but most surely—wrought demoralization among the reds. But, as yet, none dare interfere with the religious ceremonies or general policies of the tribes.

Little mention is made by early explorers of the Sarcees or Surcees, as they were often called in later days. They seem to have been so small a tribe or so retiring that they were beneath notice. Yet they carved a corner for themselves in the Indian history of the Province.

Somewhere around the time that Peter Fidler was wandering over the south part of Alberta, there was internal trouble in the Beaver tribe along the Peace River. The Beavers were numerous, brave and warlike; they fought with the Kootenays, a branch of their own family, on the west, and with the Crees on the east, and rendered a splendid account of themselves. Finally the peace pipe was smoked between them and the Crees, the ceremony taking place on the banks of the great north waterway which ever afterward bore the name of Peace. After this pact had been consummated, the Beavers hunted and fished through the north, and one day, when camped on the Spirit River prairies, had troubles that resulted in a split and the foundation of the Sarcee tribe. The Beaver historians tell how a small portion of the race desired to eat a fat dog, the majority objecting and sustaining that objection so vigorously that the dog eaters fled, shattered and beaten, never stopping until they were safe south of the Saskatchewan. The victors did not necessarily object to eating dogs, but they simply objected to the devouring of that particular fat canine. Thus, according to the Beavers, the Sarcee nation was formed from a defeated section of their great tribe.

On the other hand the Sarcees, although admitting relationship to the parent Beavers, tell a more flowery and romantic story of the separation. They tell of a winter day in the northland when the tribe, then undivided, travelled over the country and encountered a small lake, which they decided to cross on the ice. Halfway over, a provident young squaw spied the antlers of an elk sticking from the ice and mistook them for a branch of dry wood. Having in mind the necessity of building a cooking fire, she hastened toward this handy supply of kindling and smote it with an axe. Immediately there was a tremendous upheaval, the injured elk, smarting under this indignity, aroused himself in wrath and, rising, split the ice straight across the lake. A large part of the tribe was on the north side of the split, a smaller portion was south; the latter turned and fled to dry land, while the rest continued onward to the security of the north shore. Horrified at the indications of godlike anger which had been displayed, the southern party kept their noses pointed south until they reached the safety of the prairies south of the Saskatchewan River, remaining there ever afterwards and becoming known as Sarcees.

Whichever explanation is true—and the dog story seems most probable— the Sarcees located on the plains late in the eighteenth century and proceeded to make themselves felt. They were small, dirty, reckless and fearless; they were good warriors and splendid horse thieves. Taken as a whole, they were as blood-thirsty a band of red men as ever killed by treachery, and they held their own nicely. The Blackfeet undertook a few predatory visits and suffered thereby, and the Crees were often victims. The peace pact entered into on the banks of the Peace was not binding to the outcast Beavers, and they trimmed the Crees when-ever opportunity arose. So well did they do it that the Blackfeet formed a sort of friendly alliance with the strangers, and both scalped Crees whenever possible, while the latter tribe was always looking for similar trophies from them. The Sarcees flourished and multiplied, grew rich in horses, and famous in war. At one time they could put a standing army of eight or nine hundred skulking, creeping scourges in the field at a day's notice.

In the early sixties of the last century, they suffered one disastrous defeat, at the hands of the Crees, that would doubtless have been gloriously wiped out, had not the smallpox swept down and shattered the tribe. This battle was fought on the Battle River near where Vermilion Creek enters it. The Sarcees were hunting there in a strong Cree country, and they were not worrying at all on this account; but one night an old man failed to return from a rabbit hunt, and after waiting two days, a party of thirty mounted Sarcees went forth to find him. Seeing a number of horsemen on a hill and thinking they were friendly Blackfeet, they innocently came close to a large war party of Crees. Retreat was in order, and was followed with much energy, though, due to superior steeds, the Crees garnered ten scalps en route. Bull's Head, chief of the Sarcees, rallied his men and led them forth, scattering consternation in the Cree ranks as his men's long knives and effective guns and bows thinned them out. But while Bull's Head and his braves were instilling respect among the ranks of the Cree horsemen, a party of hostile footmen crept around, sacked the camp and, attacking Bull's Head in the rear, turned the tide of battle. Thirty Sarcees and five squaws were lost in this disastrous affair, the men being promptly scalped and the women being taken to the Cree camps, where they were kept prisoners.

The Sarcees, Blackfeet, Piegans and Bloods were devoutly religious in their way, and particularly observed the famous and cruel Sun Dance, which is a Sioux ceremony as well. The Sun Dance, which was forbidden as soon as the Mounted Police had gained sufficient ascendency over the tribes, was carried out in all its horrors up until about twenty-five years ago (about 1890). The dance was the ceremony through which the Indian lad stepped from boyhood to the status of a warrior. It was a shockingly cruel series of tortures the neophyte must pass, giving no sign of pain, else he could not qualify as a brave.

At the ceremony, where the entire tribe gathered to witness the trials, successes and failures, a youth entered the ring, old squaws took sharp knives and cut slits in the lad's breast, medicine men lifted the strips with pincers and, passing rope or hide thongs through beneath the muscles, knotted them there. Then the other end of the thong was fastened to the top of an upright pole fixed firmly in the ground, and the youth must tear himself loose by throwing his weight on the thong tied through the muscles of his chest. Should he fail, showing indications of pain or suffering, he was disgraced. Another method was to cut the flesh of the back, tying leather thongs through these flesh-loops and then fastening buffalo heads to the other ends, the thong being so short as to allow the heavy heads to dangle free of the ground. Then the candidate danced until the skulls tore loose.

Another method was to fasten the long thongs through the muscles in the manner described and then drag the suffering wretch with horses until the tortured flesh gave way. No man was admitted to the standing of a brave until he had passed this test of rope and pain, bearing himself throughout with a stoical dignity that would tend to prove him insensible to suffering. Youths of seventeen and eighteen years of age often graduated with high honors, but woe to the one who failed, for he was a marked man.

The Indian mothers were as anxious for their sons to pass as the sons themselves. The women always attended the dance and watched and criticized, as is the prerogative of the sex. One incident will show how they looked upon it. A young lad was being put through the buffalo head torture, and though he danced with creditable vigor, his bodily strength was not sufficient to enable him to last until the heads had torn out. Finally, tottering, swaying, his face set grim and fixed, he shook one dangling skull loose, but could not tear free of the other. He bent, pitched, and sank to his knees, while the watching tribe stirred and rustled. The lad was going to fail, and already glances of scorn were being directed toward him. Perspiration poured down his face; he struggled manfully to reach his feet, but failed, and pitched forward, just as his mother dashed into the circle on a horse and, seizing the buffalo head, urged the pony away, dragging her son by the thong. Not a whimper passed his lips, not a sign of pain was visible to the critical audience, and eventually the flesh gave out and the lad was a brave, but a very sick one.

Little is known of the country south of the Saskatchewan previous to the beginning of the last century, though doubtless a goodly number of American trappers and traders ventured as far north as the South Saskatchewan. Doubtless, too, many of the American trappers who were murdered by Blackfeet and Bloods in order to supply peltries for trade with the Canadian companies were despatched north of the 49th parallel.

Prospectors, trappers, hunters, free traders and Hudson's Bay men came to the North-West in the early half of the nineteenth century. They journeyed on the northern rivers and explored the unknown prairies; the Hudson's Bay people, devoting themselves to the country north of the Red Deer River, made trails and surveyed roads of general travel. Half-breeds multiplied, all having white fathers. There is no record of Indian men ever treating white women other than with respect. South of the Red Deer River and the South Saskatchewan, the American trappers and free traders grew in numbers, while the foothills were combed by gold prospectors in search of the precious metals. Indians still suffered their presence, though sometimes attacking isolated parties, the most hostile tribe being the Bloods, who were always bent on war, though the Blackfeet were ever in the position of men with chips on their shoulders.

Such was the condition of affairs in 1869, when the smallpox tamed the tribes. The land was lawless, the Indians were lords of it, the white men winked at or encouraged disregard of laws according to their characteristics. Possibly a hundred thousand Indians inhabited the great country between the international boundary and the Arctic Circle, and the eastern slopes of the mountains and the forks of the Saskatchewan. It was the Indians' last kingdom. Then the Blackfeet brought the smallpox from white men on the Missouri, and the dread scourge swept the camps with terrible results, the natives dying in thousands, the tribes becoming crippled, shattered and cowed, the white men taking and holding control, and settlement following the advent of the North West Mounted Police in 1874.

Never was a more terrible revenge carried out by any man than that of a whisky trader named Evans, who, mourning the loss of a partner while trading with Indians in the Cypress Hills, swore to enact an awful payment. The Bible says the Old Testament price was a life for a life, but Evans wanted the repayment ten thousand fold, and few men would have thought of the scheme he carried to success. Some time in the late sixties, Evans and a partner were trading among the Blackfeet and other Indians in southern Alberta and Saskatchewan when they were attacked and the partner was slain, while the horses belonging to the white men were stolen. Upon making his escape, Evans swore revenge, and, hastening to St. Louis, he is said to have purchased bales of blankets that were infected with a most virulent form of smallpox, which had been raging there. Carefully wrapping these bales, he shipped them up the Missouri, and, when in the heart of the Indian country, left them on the banks for the first passerby to take. Of course the red men seized upon this treasure trove with natural avidity, and the smallpox raged through the tribes, sweeping thousands, probably tens of thousands, of natives in the Blackfoot, Blood, Assiniboine, Cree, Stony and Sarcee villages into the happy hunting grounds.

CHAPTER IV

The Traders

IN the ranch country proper, that territory lying between the Red Deer River and the international boundary, the first white men other than explorers were a few wandering Canadian half-breeds and some American traders, prospectors and hunters. The Hudson's Bay Company was bitterly opposed to all free traders who poached on what they considered Company preserves, and they lost no opportunity to discredit them. "Whisky traders" was the general term used to describe them, though in methods of trade there was little difference between the free men and the Company men, with the exception that the free traders used whisky and had few established posts, while the Hudson's Bay Company people used rum and were permanently located. In both cases the necessity of spirits in trade with Indian tribes was recognized, and the commodity was used simply in conjunction with calico, blankets, firearms, ammunition and trinkets. Old-time free traders say that among the Bloods and Blackfeet, though there was much drunkenness prevalent when trading was on, the whisky was not called for until the sober side of traffic had been completed. The "head" squaws, or first wives, of each family would barter first for family requirements, getting cottons, handkerchiefs, blankets, tea, sugar, tobacco and knick-knacks; then, when the squaws had finished, the braves would come in for whisky, and the result was drunken fights and riots. Many Indians and some few white men were killed in these debauches.

An added stigma has been given these free traders by the whisky smugglers of later days, the lawbreakers of the prohibition times that followed the arrival of the Mounted Police. These whisky smugglers were not the same grade as the whisky traders, for with the latter, the commodity was a sideline, while with the smugglers, it was the only thing they dealt with in coming into Canada, though Chinamen were often remunerative articles for smuggling on the return trip across the 49th parallel southward.

In 1864 G.W. Houk, now a resident of Lethbridge, Alberta, accompanied a party of prospectors north along the base of the mountains. They prospected all the rivers and streams, meeting with no opposition from red men, though they often expected it. At a stream some miles west of the present town of Macleod they lost a pair of pinchers, which was a rather notable event, because tools were valuable articles. Little gold was found, though they washed the gravel and sand of every creek and stream they crossed. Stopping at Edmonton, which was then quite the centre of settlement of Alberta, they re-outfitted from the Hudson's Bay store, and went northward and westward into the Peace River country, where they found small-paying sands that netted five and six dollars per day. This was not rich enough to satisfy, so they went westward and back down the hills again. While in the Grande Prairie and Spirit River country of the Peace they saw little, runty cattle, evidently the offspring of work-cattle that the Hudson's Bay Company had used and was still using on the long trail from Winnipeg.

About this time, too, a party of white men camped on the Bow River, east of the present town of High River. Toward evening a band of Bloods surprised them and killed all but one man by shooting through the tent walls before the unsuspecting whites knew there was any danger. The old smooth-bore muskets of the Indians, loaded with slugs and buckshot, created terrible havoc at the short range, and the white man who escaped was very badly riddled. Crawling to the river, while the exultant warriors rushed into the tent and scalped the dead men before plundering the packs, he dropped into the swift current and floated away, using a piece of timber as a support. After drifting some distance he returned to shore and, dragging himself out on the bank, proceeded to crawl across country. He was sorely wounded, weak from loss of blood, without food and with no definite hope of succor, so it is not to be wondered at that when wandering Indians found him creeping painfully along thirty miles from where he started he was babbling nonsense and was utterly out of his mind. Indians have a respect for the mentally afflicted, so this outfit, instead of killing the man, took him with them and turned him over to some white men who were running a small post considerably south of the point where he was picked up. He was given attention and finally recovered.

A man named Mark Lemon, at present living in Montana, tells of a prospecting trip he and a party made into the Porcupine Hills in 1869. They found, he declares, placer gold of such exceeding richness that even their hopes were satisfied, but the Blackfeet or Piegans came down and killed all but Lemon, who, upon escaping, has never returned, though he still asserts he knows the point where the placer deposits can be found. But though the Indian portion of this story might be true, it is improbable that the gold part is, for every foot of the hills has been prospected again and again. No gold has ever been found there, though hundreds of men have panned the hillsides and streams. And yet some of the Piegans say they know where rich washings can be found.

Although the Indians of the present day are absolutely harmless, and are as respectful of law as could well be expected, yet they were very bad at times before the white settlers and the Police arrived. Treachery was the foundation of most of their deeds of bravery, the warrior who slew the most and held the best reputation being usually a man who depended much on surprises. For instance, one can take the case of Little Louie, a small, battered Cree who lived near Calgary until a few years ago, when he perished of starvation. Louie's body was seamed with scars, puckered with the burns of Blackfoot torture fires, broken by bullets. He was crippled and little and skinny, and he used to sell the sight of his mutilated body for the sightseer's quarter of a dollar. Yet Louie had been a warrior with a high reputation among the Indians; he had fought across most of the new Provinces; he had killed many times. In Cree words and with beautifully illustrative gestures he would tell, for money, how he became a brave. It appears that one day in the long ago, when he was young and vigorous, he decided to become one, and taking his knife, bows and arrows, slipped out to the open prairies in search of adventure. There was at that time a brief peace between the Crees and Blackfeet, but Louie did not mind that, for he was freelancing. He marched for days, unfortunate in hunting, with little food, weary, yet filled with his high resolve. Finally one morning he spied smoke, and creeping close, he recognized a lone Blackfoot hunter, who was squatting beside a slain antelope and cutting off generous chunks. Louie's mouth watered, but a very serviceable-looking gun that was lying near the hand of the Blackfoot caused him to move with judgment. Arising and exposing himself fearlessly, he came down with palms outspread, the sign of peace, and the Blackfoot, after reaching for his gun, seemed satisfied, especially as he knew there was then no war between his tribe and Louie's. Accepting a grunted invitation the visitor ate heartily, and then, while gnawing a bone, displayed indications of excitement. Gazing fixedly across the head of the sitting Blackfoot he pointed cautiously and muttered, "Deer."

The Blackfoot reached for his rifle, turned his back fairly on his guest, and prepared to shoot, but at that instant the chivalrous Louie buried his long knife in the other's throat.

Armed with a gun and enriched with a scalp and a horse, Louie rode joyfully afar in search of other worlds to conquer. Deep in the foothills he says he came, hungry and spent, to the cabin of a white hunter. The little Indian's horse had strayed, his luck in hunting was poor, he was faint and almost gone. Crawling on hands and knees he drew close to the hunter's cabin, cached his rifle in the willows and dragged himself to the door, where he knocked. The white man, huge and bearded, responded, and upon seeing who was there, spurned him with his moccasined foot and told him to go and go quickly. Louie went, quite vigorously, energized by anger and hate. Shortly a rifle-ball whizzed through the window and slew the white man as he sat at supper—and Louie then had another hearty meal, garnered another scalp and another rifle, and headed homeward. He was acclaimed a hero when he had told his adventures and successes.

These are instances of the nature of the red men of the plains and hills—treachery and blood. Isolated cases of such happenings occurred in various sections, but as a rule the Indians allowed the whites to stay because they did not want to kill off the source of supply of firearms and firewater.

It was into the country of the Bloods and Blackfeet that the American whisky traders came. Late in the sixties a party of men from Fort Benton, Montana, crossed into Canada to erect a trading post and fort at the junction of the St. Mary's and Belly rivers. Under a trader named Hamilton, they constructed a strong post, with stockade, watchtowers, ramparts, loopholes, and wide gates. Buildings faced the inner three sides of the enclosed square, the fourth being the stockade wall and the wide gate. Strong doors shut the interior of the buildings from the open space enclosed, the storerooms, stables and living quarters of the whites being all connected, so that they could live for days within the structure while the open inside space was crowded with trading Indians. There was no need for white men to enter among the red ones at all. Loopholes commanded the interior of the fort as well as the outside, small openings allowing for the exchange of hides and merchandise. Often the place was filled with drunken, fighting Indians, while the traders placidly waited within the buildings for the riot to die out.

After the completion of the fort, which was officially named after its founder, Hamilton, a German named Wye was sent back to Benton for supplies. He was asked how the traders were getting along in the wild new land, and he replied cheerily: "We're just a-whoopin' on 'em up." Securing his loads of supplies, he headed back northward, and the people of Benton said: "Wye's goin' back to whoop up again." Thus the post was christened "Whoop-Up," and the name stuck.

Whoop-Up was the headquarters for the free traders. From that point, adventurers built posts at Blackfoot Crossing, at Calgary, High River, Kootenay Lakes, and other points. They were hard, rough men, and they used hard methods to hold the Indians in proper place. If, in the interests of commerce, trade, or safety it was thought expedient to shoot a few Indians, it was done. Whisky was freely handled; sometimes the traders held as shocking carousals as the reds.

The United States authorities tried futilely to stop this illicit traffic in whisky, for the traders brought this article in without taking the trouble of paying duty. They considered that there was no law in the land they were going to, they had to protect themselves, and they consequently had no right to pay any tax. Five whisky traders from Montana flitting toward Canada one afternoon were pursued by two United States marshals until they had crossed the boundary line. Then the traders stopped, and, presenting their ready rifles, intimated that the American officials had no right to come further. The latter saw the strength of this argument and returned home, while the traders built a trading shack and named it "Stand-Off," in honor of the discomfiture of the marshals. Stand-Off then became a rather important point for American traders, second only to Fort Whoop-Up.

One night when a number of men were at Stand-Off, the conversation drifted to the discussion of a suitable point to establish another post. One advantageous locality was discussed with much favor, and the men retired to sleep with the expressed intention of going there and building a store. During the night two traders who had originally suggested the spot, and who did not want to see it taken from them, slipped out and hastened away to get early on the ground, start construction, and claim ownership by priority. When the other traders awoke they cursed cheerfully among themselves and christened the new post "Slide-Out," in commemoration of the manner in which the two men had left Stand-Off.

The last great battle between the Crees and Assiniboines, and the Bloods, Blackfeet and Piegans, occurred in the autumn of 1870 in the valley of the Belly River, close to the site of Fort Whoop-Up. Smallpox had ravaged the tribes, being especially severe among the Blackfoot Nation, and their hereditary enemies, the Crees and Assiniboines, judged it was the psychological moment to take a telling and crushing revenge for previous defeats and wrongs. With seven hundred tried and eager warriors, the Crees and their allies crept down into the Blackfoot country, prepared as usual to attack by stealth and win on the surprise. But the Blackfeet, being warned by scouts, perfected a complete and demoralizing counterattack. Some of the southern Indians were armed with splendid single-shot rifles, secured in trade or foray across the line; the best the Crees and Assiniboines had were the short-range smooth-bores of Hudson's Bay Company supply; surprise and long-range rifles took all the aggressiveness out of the invaders, and they retired before the impetuous Blackfoot family. The valley of the Belly was the theatre; the Crees and Assiniboines dropped in scores. When the last scalp had been taken, it was learned that the Crees and their friends had lost three hundred men, while the Blackfeet casualties totalled a scant hundred.

About this time, the Hudson's Bay Company and the missionaries scattered throughout the West, especially in the north, were making vigorous complaints of the whisky traders of the south. The cry was that they were all desperados, outlaws, murderers, thieves, and the scum of civilization. Long communications were forwarded to the Ottawa Government, and newspaper comment grew strong. An article that the old traders say was inspired appeared in the *Sacramento Union,* of Sacramento, California, in 1870, in which it was said that there was a fort called Whoop-Up in Canada that was always garrisoned by two hundred and fifty to five hundred American desperados and outlaws, who were despoiling and butchering the Indians.

The *Ottawa Free Press* of April 9, 1871, spoke regarding the matter as follows: "Latest Saskatchewan advices bring the intelligence of a fight between Cree and Blackfeet Indians in which many of the former were killed at long range by breech-loading rifles. The Crees were not aware that their enemies had been furnished with so deadly a weapon. The rifles had been furnished by American traders. It is a pity this cannot be stopped. No one knows how soon these weapons will be turned against our own people."

Other reports followed. It was averred that the south was overrun with American free-booters, that imposing forts manned by desperados had been constructed, that the Americans defied the law openly and were ready and anxious to resist interference from the Canadian Government. The seed that resulted finally in the establishment of the North West Mounted Police was planted.

In addition to this, and a short time before the Mounted Police came into the West, a grim tragedy was enacted in the country south of Calgary, somewhere between that point and Macleod. A party of German settlers, driving overland from the States, were heading toward Edmonton, where they intended settling. They had good wagons, good horses, and were fairly well armed, though they had little fear of any trouble, as the Indians were supposed to be peaceful. The party consisted of three families, eleven souls in all, counting women and children. One day, while they were sitting at dinner, utterly unsuspicious of danger, a score of Blackfoot braves came down on them and killed every one in cold blood.

Upon the arrival of the police in 1874, there was some attempt made to find and punish the perpetrators of this deed, but no success was met with, and the Indians were never brought to book.

Early in the spring of 1871, H.A. ("Fred ") Kanouse, then a young man in his early twenties, came into "Hudson's Bay territory," as Alberta was called among the American traders, to establish a trading post. The party consisted of three other white men and an Indian squaw, all well armed and with a good supply of trading stock, including whisky. They proceeded north beyond Whoop-Up for a hundred miles or more and built a strong "fort" of logs on the Elbow River, three or four miles up from the Bow. This would be within the present city limits of Calgary. Bloods, under a chief named White Eagle, came to trade, and business was good for a time, though White Eagle early displayed indications of surliness. These symptoms grew, the chief and his braves showed more signs of surliness and arrogance, an Indian characteristic which often shows prominently when the natives feel they can carry it off successfully, young Indian bucks in the war-trail season being often inclined to bully when they can. Later a white man of the Kanouse party had a slight altercation with a brave, which resulted in the white man bending a heavy six-shooter across the skull of the red man. The wrath of White Eagle boiled out, and he harangued his men with fiery vigor, demanding the blood of the whites, and urging the desirability of it commercially, in addition to honor and glory. They could get, he said, the entire stock of arms, whisky and other articles for nothing at all. Crossing the river at the head of his men, he led them toward the house on the flat, and Kanouse went to meet him, accompanied by one of his companions. Without waste of passing moments, White Eagle shot and killed Kanouse's companion, and the next instant dropped dead from a bullet from Kanouse's gun. A volley followed, and the white man's arm was shattered near the shoulder by three of the huge, smashing balls from a smooth-bore musket.

From behind barricaded doors, the traders stood off the savage charge of the Indians for three days. The house was divided into two compartments by heavy logs running clear across the building, one half of the structure being windowless and without doors. Under these logs one of the white men burrowed and hid himself in a hole he dug in the dark compartment, leaving Kanouse and one other white man to stand off the natives. The squaw did splendid service in keeping the rifles loaded, while the white coward, whose name was Fisher, stayed hidden away during the three days. So successfully did the defenders shoot that the Bloods after this period of fighting asked for a truce in which to bury their dead. This being granted, there was a lull, and Kanouse dickered with a Blackfoot brave to ride to Highwood River and notify a party of traders who had located there, the Blackfoot undertaking to do this upon receipt of a six-shooter and other valuables. The High River traders came at once, the Bloods withdrew, and peace reigned again. Then the relieving party took axes and cut through the partition to Fisher, whom they found in the last stages of fear, and who could hardly be persuaded he had been saved.

This year, too, a man named Dick Berry, quick-tempered, fearless and reckless, one of the American traders' colony, had a quarrel with a Blood named Old Woman's Child, son-in-law to a warrior named Prairie Chicken. Berry and a company of traders came to the Elbow in the following year (1872) and commenced construction of a fort near where Kanouse had built, but the Bloods again appeared in numbers and drove them away before a log had been laid. Moving westward, the traders went about twelve miles up the river and built without molestation. Among the party of quarrelsome Bloods was Old Woman's Child, a grand figure of a warrior, standing over six feet tall and being beautifully muscled and proportioned. Following his Indian nature, he lay in ambush for Berry, shot him dead, and after stabbing him repeatedly, took his scalp and vanished. Some time later a party of the tribe were at Fred Kanouse's new fort on the Old Man's River, among them being Old Woman's Child. He was apparently a bullying, blustering desperado, for the other members of his tribe feared him heartily and longed for his death, though they feared to take upon themselves the work of destruction. But one night, when he slept, they screwed up sufficient courage—probably of the Dutch sort—to hack and stab him to death, reporting happily to Kanouse the next day and presenting him with the knife which Old Woman's Child had used on Berry.

The bravery of the squaw who stood through the attack on the Elbow River fort in 1871 brings up the treatment accorded to women by the Indians. Squaws were the workers, the braves were the hunters; the women's duties were packing, skinning, preparing clothing, cooking, carrying the meat of slain game to camp, the erection of tepees, and in fact all camp drudgery. Women never were asked for advice, never were allowed to stand on a par with men; but the Indians recognized bravery among themselves, as is instanced by their treatment of the wife of Old Sun, a sub-chief among the Blackfeet. Old Sun and a hunting party of his tribesmen were away south on the trail of the buffalo herds when a band of American Indians (Gros Ventres) swooped down on their tepees, killing some bucks and taking some squaws prisoners. One of the raiders seized Old Sun's squaw and swung her up behind him on his horse, where she rode quietly until an opportunity presented itself, and then she drew the scalping knife from the girdle of her captor and stabbed him to death. Throwing the dead body from the horse, she returned to her people, and they held the deed in such honor that they gave her the right to sit in the councils of the chiefs. This is the only instance of advanced female suffrage among the Indians. Old Sun's wife still lives on the Blackfoot reserve, near Gleichen, Alberta.

The year 1872 saw the Government at Ottawa about ready to act on the many complaints received from the West. The Hudson's Bay Company and missionaries continued to send in detailed narrations of outlawry. The newest report was that the American traders on the Belly River were trading whisky for horses, and that consequently the herds of the southern Indians were becoming depleted. Resulting from this state of affairs, there was further trouble feared from the Indians, for without horses they could not hunt, and they would necessarily have to get horses or starve. Thefts and shooting might result. Reports further said that peace had reigned between Blackfeet and Crees since the decisive engagement in the fall of 1870.

Early in 1872 the Dominion Government despatched Colonel P. Ross-Robertson to investigate Western conditions. Travelling chiefly by Hudson's Bay routes, he reached Edmonton, and there gathered as much data as he could, finally sending in a long report that suggested immediate and strict enforcement of law in the West. Although he did not go very thoroughly into the Blackfoot country, he reports rather in detail on their number and manners, while of the American traders he takes his descriptions chiefly from Hudson's Bay officials, if one is to read between the lines.

"The Blackfeet," said Colonel Ross-Robertson, "numbered about 2,350 men, many old ones and some mere boys." He did not believe the tribe could put more than ten or eleven hundred capable fighting men in the field. The Blackfoot war parties usually consisted of fifty or sixty men, and when on raiding expeditions, these parties performed remarkable marches. He found, too, that murder and assassination formed a large and honorable part of the war code of all Western prairie Indians.

"The prairie Indians," continues the report, "are bold and skilful horsemen, but not very skilful with firearms. Beyond the Province of Manitoba westward there is no kind of government at all at present (1872), and no security for life and property beyond what people can do for themselves. Of late years no attempt has been made to assert the supremacy of the law, and most serious crimes have been allowed to go unpunished. Hardly a year has passed without several murders and other crimes of a most serious nature having been committed with impunity. During the present year, about three weeks before my arrival in Edmonton, a man by the name of Charles Gaudin, a French-speaking half-breed, brutally murdered his wife at no great distance from the gates of the Hudson's Bay fort. I was informed that the criminal might have been arrested, but there was no power to act. This same man had previously and most wantonly and cruelly mutilated an old Indian woman by severing the sinews of her arm so as to incapacitate her for work.

"At Edmonton there is a notorious murderer, a Cree Indian, called Ta-Ha-Kooch, who has committed several murders, and who should have been apprehended long ago. Many instances can be adduced of a similar nature, and as a result there is widespread apprehension. White men are living on the sufferance of Indians and dare not introduce cattle or break ground to any extent for fear of Indian spoliation.

"When at Edmonton and the Rocky Mountain House I was informed that a party of American smugglers and traders had established a trading-post at the junction of the Bow and Belly Rivers, about thirty miles due east of the Porcupine Hills and sixty miles on the Dominion side of the boundary. This trading-post they have named Fort Hamilton, after the mercantile firm of Hamilton, Healy & Company of Fort Benton, Montana, from whom it is said they obtain their supplies. It is believed that they number about twenty men, under command of John Healy, a notorious character.

"Here it appears they have for some time carried on an extensive trade with Blackfeet Indians, supplying them with rifles, revolvers, goods of various kinds, whisky and other ardent spirits in direct opposition of the laws both of the United States and the Dominion, and without paying duty on goods introduced into the latter country. It is stated upon good authority that in 1871 eighty-eight of the Blackfeet Indians were murdered in drunken brawls amongst themselves induced by whisky and other spirits introduced among them by those traders.

"At Fort Edmonton during the past summer whisky was openly sold to the Blackfeet and other Indians trading at the post by some smugglers from the United States, who derived large profits thereby, and on being remonstrated with by Hudson's Bay officials they coolly replied that they knew they were breaking the laws of both countries, but as there was nothing to prevent them doing it they would do just as they pleased.

"It is indispensable for the peace of the country and the welfare of the Indians that this smuggling and illicit trade in spirits and firearms be no longer permitted. The establishment of a Customs house on the Belly River near the Porcupine Hills, with a military guard of about a hundred and fifty men, would be all that is required to effect this object. Not only would the establishment of a military post here put a stop to this traffic, but it would also before long be the means of stopping the horse stealing expeditions carried on by hostile Indians from south of the line into Dominion territory, which is the real cause of all the danger in that part of the country, and the source of constant war among the Indian tribes. Indeed, it may now be said with truth that to put a stop to horse stealing and the sale of spirits to the Indians is to put a stop altogether to Indian wars in the North-West."

Colonel Ross-Robertson undoubtedly lays the correct cause of blame for disturbances among the Blackfeet to whisky, but he does not go far enough to lay the blame of the Cree and half-breed murders in the north to rum and other liquor, which was in all probability obtained through the Hudson's Bay Company. From the north to the south, and from east to west of the new Canada, the blame for most Indian troubles can be justly laid to the door of the distillers and the retailers and wholesalers of the product.

In speaking of horse stealing, he touches one of the most ancient and honorable Indian occupations. A skilful horse thief was possibly more highly honored among the prairie tribes than a successful assassin. It took brains, skill, and the coldest kind of nerve to approach the camps of the enemy alone or in very small bands and run their horses off from under their very noses. For a hundred years at least this practice had thrived, long before whisky had been introduced to the Indians of the plains. The Minnetarees (Sioux), the Blackfeet, the Piegans, Sarcees, Crees, Crows, Gros Ventres, Nez Percés, Kootenays, and in fact every tribe, stole from the others and took great credit for ability in this line. Colonel Ross-Robertson placed a harder duty on the Mounted Police when he urged the abolition of this industry than he did when he suggested the extinction of the whisky trade.

Colonel Ross-Robertson made a general summary of the numbers of Indians on the plains between the boundary and the sub-arctic circle, and the Red River and the eastern slope of the Rockies. He set the total at not over fifteen thousand, probably meaning braves, otherwise his estimate of the Blackfoot strength would mean that this tribe alone composed more than half of the native population in the vast territory. He declared that there were not more than four thousand Indians in this great country capable of bearing arms, and that while Bloods and Blackfeet had some breech-loading rifles, they were generally poorly armed and badly mounted. Few Indian villages, he added, had more than one hundred and twenty-five or one hundred and thirty lodges.

During this same year, a tragedy occurred in the Cypress Hills in southern Saskatchewan that led directly to the Dominion Government deciding to take prompt steps to enforce the laws in the Western country. Added to Colonel Ross-Robertson's report, this affair brought out very forcibly the necessity of quick and prompt action. A party of whisky traders were camped in these hills and a band of Indians stole their horses. Smarting with indignation, these lawless men went across into the States, collected a gang of as hard characters as they could, and came back to make reprisals. The white party was composed of wolfers, trappers, outlaws, traders, and general scum of humanity. They bore down into the Indian country and surprised a camp of red men, murdering in cold blood some thirty odd men, women and children. News of this affair soon reached Ottawa, and the Government decided at once that the time had come to put a force of mounted men in the Western field.

On the 20th day of May, 1873, the Bill establishing the North West Mounted Police force was concurred in, and plans for completion of the organization were at once set in motion. By September these plans had been completed, and in October one hundred and fifty officers and men were taken to Fort Garry and lodged in the stone fort. This force consisting of A, B, and C troops, NWMP Lieutenant-Colonel George A. French was appointed commissioner of the force, and Colonel J.F. Macleod, CMG, was placed in charge of the Winnipeg forces. In April 1874, D, E and F troops were recruited at Toronto, and Colonel Macleod was appointed assistant commissioner under French.

Two hundred and one men, two hundred and forty-four horses and sixteen officers left Toronto on June 6, 1874, and, picking up two carloads of additional horses at Detroit, Michigan, joined A, B and C troops at Dufferin (Emerson), Manitoba. On July 8 the force marched out of Dufferin, bound for the unknown West.

In the meantime, affairs had proceeded along ordinary lines on the prairies. The Indians hunted buffalo and each other, stole horses and raided camps, the white men traded firearms and whisky and made as much hay as they could while the sun was shining, for they knew that with the arrival of the Mounted Police, the sun of their particular prosperity would become pretty dim indeed. The Blackfeet, always ready for war, grew so restless because the Crees and Assiniboines had been so thoroughly cowed in the last great battle that they picked a cause for argument with the Kootenays, and when those fierce mountaineers swooped down from their rocky fastnesses, there was bloodshed and glory. But the Kootenays rendered so fine an account of themselves that the Blackfeet found it necessary to fire the prairies to drive them back, and by fall the entire south country was a wide blackened expanse. The buffalo had worked south in search of feed, and hunting was poor, though with the green grass in the spring of 1874 they returned, and the outlook was as bright as ever.

Traders had established posts throughout the south, and the red men of the plains and mountains came willingly to barter. There was the main post at Whoop-Up, there was Stand-Off, Slide-Out, Lafayette French's trading store at Blackfoot Crossing, posts in High River, and Fred Kanouse's post, Fort Warren in the foothills, where he had established trading relations with the Kootenays. Coming by as speedy stages as possible was the force of Mounted Police, Fort Whoop-Up being their object. Field guns, ammunition, and two hundred and seventy-four officers and men were grimly braving the Western wilds in the interests of law and order. Fort Whoop-Up was to them exactly what it had been represented, a strongly fortified fort, manned by desperate, ruffianly sharpshooters who took more delight in killing than they did in getting drunk. The belief among the officers and men was that Whoop-Up must be captured, after a desperate struggle, before the backbone of illicit whisky dealings in the Territories could be squelched. Over the wide prairies they rode, talked and imagined. Jerry Potts, interpreter and guide, led them, and he spoke little, for one quarter of his blood was Indian, and that had gone to his tongue. Through herds of gigantic buffalo, over wild rivers, and on long, tiresome trails the Mounted Police doggedly held on, surrounded by a great strange country, new in habits and conditions. They did not know if a thousand red fiends might dash out from the next coulee and try to overwhelm them; they did not know what defile or riverbank might supply sufficient concealment for a hundred outlaw white men who could empty half the saddles in a couple of volleys. Whoop-Up was their destination, and they were bound there.

"Interpreter," called an anxious officer to Jerry Potts who was leading them up a long slope, "what shall we see when we reach the top of this hill?" The force for an entire day had been expecting to drop down on Whoop-Up any time. Jerry spat and thought, and then replied laconically: " 'Nudder hill."

Led somewhat off their true course by Colonel Ross-Robertson's report that the stronghold of the traders was at the junction of the Bow and the Belly, the force lost some little time, but finally headed right and eventually came to the hills overlooking the place. All was silent, the gates were closed and grim, the battlements threatened unknown things; in the brush they found the dried bodies of four Indians who had perished in some brawl, either with whites or people of their own tribe, and such evidence of violence strengthened the belief in a desperate resistance from the Americans.

Field guns were placed on nearby eminences that commanded the fort, and a party was detailed to ride down and demand surrender. With loaded guns ready, accoutrements clattering, horses prancing, and gay red uniforms setting a bright badge against the green hills, the troopers rode into the silence, for no sign of life had yet been seen around the place. Loudly they thundered on the closed gates, sternly demanding admittance in the name of the Queen. There came a shuffling, scraping noise, the sound of a heavy bar being raised, the gate slowly opened, and an ancient half-breed, grey of whiskers and bent of frame, peered out in deep consternation. Hordes of red coats filled his startled vision, thousands of rifles were pointed straight at his old heart, so, lowering his head, he did the first thing that came into it—ran as he never had run in his youth. Ducking under the levelled rifles he vanished, and was never heard of again.

Somewhat surprised, the police pressed into the place and were cautiously feeling their way when they were saluted by a hearty voice, saying, "Walk right in, gentlemen, I won't hurt you," and they saw an unarmed ex-American soldier who walked on crutches because his legs were gone. He was the only garrison, now that the aged half-breed had decamped. Whoop-Up was captured.

The traders who usually hung out at this point had all gone on a trading tour. The International Boundary Commission, which had been working on the boundary, readjusting the lines, had about completed its work, and the traders had hastened away to be present at the disbanding of the commission, knowing well that there would be a fine market for Western wares among the home-going members.

After capturing Whoop-Up, the Mounted Police proceeded onward until they reached an island in the Old Man's River a very short distance below the present town of Macleod. Here they built barracks and stables and prepared to spend the winter, naming the new post Fort Macleod, in honor of the officer commanding the troops. The establishing of a firm respect for the laws of the land being the paramount aim of the force, and hearing of illicit whisky dealings some fifty miles distant, Colonel Macleod sent Inspector Crozier with a detail of men to arrest the culprits. They went, and succeeded in making the first arrests and securing the first convictions for whisky trading in the history of the Province. It appears that an Indian who enjoyed the title of Three Bulls had gone thirsty and desirous to these traders, and in the excess of his wishes had given two perfectly good horses for two gallons of whisky. After despatching the liquor, he felt that he had been badly beaten in the deal, so, having heard that the Queen's forces were present for the express purpose of protecting the rights of the red men, he posted to the new fort and made his complaint.

The inspector found the traders located at Pine Coulee, fifty miles north of Macleod, at a spot that was later owned by W.R. Hull, near High River. Four white men and a Negro named William Bond were arrested, while their belongings, consisting of two wagons with alcohol in them, sixteen horses, five Henry rifles, five revolvers and one hundred and sixteen buffalo robes, were seized and taken to Macleod. Bond, who acted as interpreter, was, with the two principal white men, fined two hundred dollars, and the other two men, who proved to be only hired hands, were assessed fifty dollars each. The buffalo robes were confiscated by the Police. On the day following the convictions, a gentleman named Weatherwax, known very widely in Fort Benton by the name of "Wavey," paid all the fines excepting that of Bond, who was detained on a charge of supplying liquor to Three Bulls.

Next came hurried information that the mountain Kootenays were making fierce war on Fred Kanouse's Kootenay place, Fort Warren. The Indians had grown

Seated are (left to right) Three Bulls (Blackfoot), Sitting on Eagle Tail Feathers (Piegan), Crowfoot (Blackfoot) and Red Crow (Blood), *circa* 1884. All but Three Bulls participated in the making of Treaty No. 7 in 1877.

surly toward Kanouse and the men with him, expressing their displeasure more and more strongly until finally a chief came in and, pointing a gun at Kanouse's partner, indicated his intention of killing him. Again Kanouse's ready gun flashed, and the threatener dropped, while from the outside arose the war cry.

The Mounted Police detachment hastened to the rescue, and arrived in time to count a few dead Kootenays, Kanouse and his companions having put the tribesmen to flight after a sharp encounter. Entering the place, the Mounted Police officer in charge of the relieving force met the defenders and received their explanations of the affair. Standing on a sort of counter was an open keg of gunpowder, and one of the narrators, growing excited, waved his gun about and finally, reaching the dramatic climax when Kanouse shot the chief, pulled the trigger, transmitted fire to the open keg, and blew out the side of the fort.

CHAPTER V

1692–1876

UNDOUBTEDLY, the first members of the cattle family to enjoy the succulent grasses of the Western plains were the buffalo who roamed in great herds over the prairies. Tens of thousands of Indians lived sumptuously off these hosts, and thousands of wolves grew fat and lazy because, since the buffalo were such easy prey, it was not necessary to work hard to live. From the far north to the Rio Grande and from the Red and the Mississippi rivers to the Rocky Mountains, the buffalo ranged in countless numbers.

Henry Kellsey was the first white man to see them in Western Canada when he came west in 1692, and the Indians were then killing them wholesale. La Verendrye and his sons, La Corne, Saint-Pierre, Hendry, Alexander Henry, found buffalo roaming to the Red River in the eighteenth century, and they too speak of the wholesale slaughter. Indians killed for tidbits only, leaving tons and tons of prime meat rotting on the prairies after every hunt, more even than the wolves, which followed in droves, could devour. Thousands of buffalo, even up as late as 1875 and a year or so later, were annually slaughtered by the red men for the unborn calves, which were considered a rare dainty. In 1771 Samuel Hearne found great herds of prairie buffalo on the southern shores of Great Slave Lake; in 1776 the Assiniboines of Manitoba were slaughtering them in great numbers after entrapping them in immense pounds; and, in 1811, Alexander Henry killed wood buffalo in eastern British Columbia. A century and a quarter ago, the herds were spread over a vast territory, at least two thousand miles long and a thousand miles wide; today the only wild animals of this race are a few scores of wood buffalo living in the wilderness of muskegs around the headwaters of the Hay River, in the practically unexplored northland beyond the Peace River. Here and there, in Government parks, the United States and the Dominion Governments have established herds of prairie bison in order to preserve them, and these few are all that remain of the original millions.

Piles of buffalo skulls at a railway siding in Saskatoon *circa* 1890 are a grim reminder of the long-gone herds, which disappeared quickly after 1874, slain in the hundreds of thousands.

Buffalo, which had for many years formed the staple supply for the Indians of the West, were at one time looked upon as a possible means of substituting cattle and sheep. During the early days of the Red River settlement, before the coalition of the North-West and Hudson's Bay companies in 1821, an energetic promoter named Pritchard planned an original industry—none other than that of buffalo wool. Feeling that the sheep industry would never thrive in the Red River settlement on account of the fierce wolves, Pritchard formed a company, the express object of which was to provide a substitute for wool, which could be collected from the wild buffalo of the plains and used as a substitute for clothing both for colonists and export. Ten thousand dollars were subscribed by the settlers and all Red River went to work, neglecting fields and other business to secure hides and wool. Skinners, sorters, wool dressers, experts of all sorts were imported from England and set to work; the ten thousand dollars were expended, twenty thousand more went, and a few samples of cloth had been made and sent to England. These samples, which cost the company ten dollars a yard to manufacture at Red River, brought only a dollar in England, and the great industry of buffalo wool was a thing of the past.

In the fifteenth century, the Spaniards in Mexico first introduced range cattle into America. They brought the hardy Andalusian stock and turned them loose to rustle, and from Mexico the industry crept naturally into Texas, where it is said that the first brands were used. The Texans had been considerably confused in efforts to keep track of their fast increasing herds, and were at a loss until a gentleman from Ireland by the name of Maverick hit upon the scheme of marking them with hot irons, each owner to have a separate and distinctive mark. The stockmen thought this was a splendid idea. Then Maverick, who said he abhorred cruelty, suggested that if all the cattlemen used the brand, it would not be at all necessary for him to pain his cattle with the red-hot irons, for after branding was over, they could tell at every roundup whether Maverick had any stock in the collected herds. This too appeared perfectly reasonable to the ranchers—each other man having a brand, of course the unbranded stuff must belong to Maverick. They rather respected him too for his tenderness of heart. Heartily the cattle blatted in the corrals and on the open when the white irons were pressed sizzling to the shrinking skin, vigorously the men worked, placidly Maverick herded his unmarred stock and kept away from the cruelty. Another roundup came, and the cutting out of branded and unbranded stock followed. The increase in the size of Maverick's bunch was the most miraculous in the world, and if the ranchmen of Texas had been at all biblically inclined, they would have suspected that the Maverick herds had been herded by a Jacob. Not being well read with regard to Jacob, they jumped to the conclusion that Maverick had "put one over on them," so, gamely admitting the count for the time being, they insisted that in future he must brand too, even though it broke his tender heart. From that time to the present, an unbranded animal has been known throughout the range, from north, south, east and west, as a "maverick."

It is interesting to know that the first cattle ever placed on the prairies in Manitoba, Saskatchewan or Alberta were an English bull and two lone cows that the North-West Company took into Manitoba and sold to the Red River settlers long before the amalgamation of the companies. Shortly after these animals were introduced, some energetic Americans drove three hundred head of domestic cattle into the settlement and sold them there, the settlers buying eagerly and paying as high as one hundred and fifty dollars for a cow, and ninety dollars each for trained oxen. These animals were bred, and considerable increase resulted, though it proved unsatisfactory, the progeny deteriorating in size, due—according to Alexander Boss—to want of care and cold climate and also perhaps to inbreeding. Within fifteen years, these animals had bred down almost one-third. In the first drove, the finest oxen weighed about a thousand pounds, fifteen years later the largest in the settlement weighed eight hundred and fifty pounds, the general run being about seven hundred pounds. Splendid bulls had been imported from

England and the United States, but the poorer farmers kept little, runty grade bulls, the pastures were unfenced, and the result was the same, though more pronounced, as the experience of the great ranches of Alberta fifty years later.

The second drove of American stock was taken into the settlement in 1825, but prices had dropped considerably, as the settlers had spent all the money they had received from the "Buffalo Wool Company" and must needs be more careful. Also, they now had cattle of their own. Prices paid this time were tremendously less than the first herd, the drovers realizing only thirty dollars for cows and sixty dollars for a yoke of oxen.

Sheep ranching was also first attempted in that new land before 1830 came into life. Six thousand dollars and two expert buyers were sent into the States to secure the herds, and when there, the buyers argued and quarrelled; one wanting to get the flock in Missouri and the other in Kentucky, some four hundred and fifty miles further. The "Kentucky" man won by stubbornness, and they went down to that state and purchased about fifteen hundred head. En route back, they stopped to shear, and then burned the wool because no one would pay them the price they wanted. They trailed the poor sheep for fifteen hundred miles, slitting the throats of those that dropped out and leaving them lying on the trail, and when Red River Settlement was reached, there remained but two hundred and fifty-one head, many of which died after arrival.

Next after this fiasco, the settlement took up the ranching of cattle, the turning out of herds to rustle their own living. They had heard that stock in New South Wales fed out of doors all year round and throve on it, and they thought it would be just as feasible in Manitoba. A new company called the "Tallow Company" was formed, with the view of securing herds, turning them upon the open range, and killing the fattest every year for tallow. It was a very simple method of growing wealthy quickly, and the settlers all bought shares. All knew cattle could live through winter very easily, because—did not the buffalo do it?

On a bright April day, these pioneer range animals were placed in the open, branded "TT" (Tallow Trade), and left to stray where they willed in care of two herdsmen. In May eighteen inches of snow fell, cold weather followed, and twenty-six head died. Summer brought them fat into fall, and the snows fell so deeply in winter that the settlers were convinced they must feed a little bit anyway, so they constructed corrals and crowded their range stuff in there, feeding them one-third the amount of hay they gave stabled stock. Thirty-two head froze to death, fifty-three were killed by wolves, the ragged remnant came into the next spring without ears or tails, and many had their horns or hoofs frozen off, while nearly all cows were minus teats.

That next summer (1833), the ranging stock was shoved further out on the prairies into better pastures, and in place of open corrals there were sheds constructed for them. One hundred and eleven head were lost the first year, but this second year the loss was reduced materially, though heavy enough, the wolves getting twenty head and the winter sixteen. Discouraged by the many and regular losses, the company broke up, the herd was auctioned off, and the money received was nearly seven hundred dollars less than the original cost of the stock, not counting wages paid to herders and haymakers.

Previous to the passing of the buffalo, there was little or no use of putting domestic stock on the Alberta range, because the wild animals held it in such tremendous numbers, and the Indians were unreliable as well. Buffalo bulls could and would kill range bulls, cows were swallowed up and carried away by the drifting herds of buffalo, and fences were absolutely no hindrance to the massive animals.

After 1874 the buffalo disappeared fast, slain by white men and Indians. Great kills were made; hundreds of thousands of hides were sold annually. The Montana ranges, with the exception of the river valleys, were fairly free of them, and big herds of range cattle had taken the ranges they had left; hunters followed the Missouri country and shot thousands, Indians slew ten times more. During one winter season in the early seventies, W.W. ("Bill") Tuttle (now of Macleod, Alberta, and former mayor of Fernie, B.C., during the big fire there), was hunting on the Missouri with one companion. When spring broke, they took their pack into Fort Benton, Montana, and sold 4,700 prime hides to I.G. Baker & Company. Even then, they had taken in only the very best skins, leaving hundreds of old bulls with poor hides lying rotting on the plains where they had shot them. In the spring of 1874, the I.G. Baker company shipped from Benton alone two hundred and fifty thousand prime buffalo hides.

No cattle were in the range country proper, that territory lying south of Calgary, until after the advent of the police in 1874. The first domestic stock taken anywhere near the range land was a band of fifty head of horses and cattle that David and John McDougall brought south from Edmonton to Morley, at the present Stony Indian reserve in the foothills west of Calgary. It was on November 20, 1871, that they landed with their herd, going the following year to Montana and purchasing another hundred head of steers and breeding stock from a man named Ford. This bunch they drove northward, losing a portion en route, but regaining them later on. It appears that the cattle escaped and started homeward, encountering Fred Kanouse on the trail, and Kanouse, being of the usual helpful Western breed, and knowing they belonged to the McDougalls of Morley, turned and drove them northward again until he caught up to the main herd, where he turned them over to their proper owners.

The new West was absolutely new. There were no towns or railroads, and freight was carried with long bull teams. The arteries of traffic were simply worn trails, one north, one south, one east and one west, with branch paths leading off to scattered posts, Police detachments, and settlers. In 1875 the Mounted Police force located a post at Bow River, which is now in the heart of Calgary. This was a thousand miles west of Winnipeg, two hundred and sixty trail miles from Fort Benton, the nearest point of supply, and two hundred miles from Edmonton. There were neither bridges nor ferries. Supplies for the posts in Alberta were generally hauled overland from Benton, being brought to that point by the steamers on the Missouri, though sometimes freight was sent overland from Winnipeg, but that long, dangerous journey by trail tended to make it a rather unpopular route.

Macleod was the centre of Police activity and settlement. There the head-quarters for the force were located, there the policies of the governing body were outlined. Colonel Macleod, the commander, conducted the maintenance of law and order in a stern, fair and strict manner, tempered with a tact, a judgment and a fearlessness that marked him as perhaps the best administrator of pioneer law that has ever been seen in Western Canada. With his little force of red coats, he maintained a dignity and displayed a strength that won the respect of the lawless traders and kept the thousands of warriors, used to war and blood, in subjugation.

The same year that the post was built at Bow River, he sent a detachment into the hills west of Macleod to establish a post there. Arriving at the same creek where George Houk and the other prospectors camped in 1864, they found the lost nippers, rusty and old. Little things counted in the early days, any small thing out of ordinary set its distinctive mark, so "Pincher Creek" was at once applied to the post and the stream, and since then it has ever been known by that name.

Settlers commenced to come into the country around Macleod as soon as the location of the force gave assurance of enforcement of law and order, Joe McFarland and Henry Olsen arriving from Montana with a band of dairy cattle and locating a few miles down the river from old Fort Macleod on what is now known as the Pioneer Ranch. They found a ready market at the Police fort for butter and other products, butter itself being indeed "golden." Seventy-five cents per pound was the ruling price, and they could not supply the demand. Olsen and McFarland parted company shortly after this, Olsen going into the country later included in the Piegan reserve, and squatting upon a homestead in that unsurveyed district.

Thomas Lee, an old trader, prospector and trapper, saw the new order of things that was coming into force, and he, too, decided to take advantage. He secured some few head of milch cows and located at Brockett, near Pincher Creek. Following these men, probably in the same year (1875), a Montana family named Armstrong, man, wife, and a foreman or hired man named Morgan, drove in and located on the Old Man's River just above Macleod. This herd was always known as Mrs. Armstrong's, and she too conducted a dairy business.

There were still enough buffalo to keep the Indians well supplied, and as buffalo meat appealed more to the natives than the domestic beef, they showed no indications of depleting the settlers' herds.

In July 1875, Major-General E. Selby Smith passed through the country on a tour of inspection of conditions and Mounted Police posts. He expressed great satisfaction at the manner of administration under Colonel Macleod. George W. Maunsell, a brother of E.H. Maunsell, then a member of the 1873 Mounted Police force, came to the Macleod district, where his brother was, and settled, though this was not his first experience in the country, as he had been a member of the Boundary Commission of 1873.

In the fall of 1874, a man named Shaw drove five hundred head of beef stock across the mountains into Alberta, heading for Edmonton, where he expected to find a good market. There had been cattle on the valley ranges of British Columbia for many years, even then. This drover came down through the Kootenay Pass, and turning north through the foothills, arrived at Morley, where the McDougalls were located. Winter was setting in, but Shaw intended pressing northward until persuaded out of that idea by Reverend John McDougall, who urged him to winter on the Bow River at Morley. Shaw did this, and remained over a year.

The mercantile firm of I.G. Baker & Company had by this time opened a branch at Macleod, and controlled the market of supplies. They fed the settlers and the Police, did a banking business of a sort, and handled all Police contracts for supplies. They killed the beef for the Police, and later for the Indians, and they held the beef contract to supply the Bow River post. There was then no market for beef excepting the Police posts, and in the Calgary district there was no band of cattle suitable for beef, excepting those which were run by Shaw at Morley. Naturally he was given the subcontract to supply the Calgary post, which contract he held until he went north to Edmonton in 1876.

A remarkable case of a lost man came to light at Fort Walsh, the Police post in the Cypress Hills. One day in 1875 a man, tattered, weather-beaten, dressed in rags and skins, with long matted hair and beard, appeared there, a wandering refugee from the Hudson's Bay country of the north. This person, a Manxman, had been kidnapped, probably when drunk, and had been shipped to York Factory, bound out to work for some years for the Hudson's Bay Company. Escaping from there, he had wandered for months, living on rabbits, gophers, mice, anything he could catch. Doggedly moving westward and south to escape the country of the people to whom he was tied, he came finally to Fort Walsh, where he first learned that he was on the North American continent. Ignorant and frightened, this poor fellow, James Williams, as he gave his name, had not known where his ship was bound, where he had landed, or whom the natives were. He was in a wild, new land, he was trying to escape from it, and he had walked and struggled on with the one indomitable aim until he stumbled into the Mounted Police stockade.

The suitability of Alberta as a grazing country, with her wide sweeps of rich grasses, well-watered plains, fine winters, and unlimited range, appealed to many of the early Mounted Police constables and officers, though they were convinced that it was not a feasible project as long as the buffalo roamed untrammelled over the country. Long discussions on the subject took place in barracks and in camp, and the idea of ranching came to men who are today very wealthy cattle owners. At this time there was said to be a pioneer rancher who had a band of four or five hundred head of range stock at the forks of the Red Deer and the Saskatchewan, just east of the present eastern boundary of Alberta. The Mounted Policemen knew of this, and it, too, acted as a damper to the carrying out of ranching plans.

"What is the use of going into the business?" they would ask each other. "There is no market now; there is good range, it is true, but what is the use of range without market? And there certainly is no more need of cattle here, because there are more cattle in that one bunch on the forks of the Red Deer than there are people in the whole territory."

Indications of life in the livestock business appeared in the spring of 1876, when Jim Christy and George Emerson brought in separate herds of animals, driving them from the United States. Christy came in with a band of horses, the first ever brought into the Province for trade and sale. He found a ready market among the few white settlers, the Indians, and the Mounted Police, the Police paying an average of about a hundred dollars per head. Business was picking up in the post at Macleod also, for that same spring the I.G. Baker people shipped from there a great wealth of prime buffalo and wolf hides—thirty or forty thousand buffalo and as many wolf skins.

George Emerson, who drove in a small herd of domestic cows about the time Christy drove his horses over, was a former Hudson's Bay employee, having come into northern Alberta and Saskatchewan under that company in 1869, working

George Emerson of Brooks and High River first came west as an employee of the HBC. In this photo, *circa* 1915, the pioneer rancher stands on his range with his brother Kelly to his right.

for a salary of twenty dollars a month. With the arrival of the Police he came south, and started almost at once into the business of dickering in cattle. He sold a few of his stock as he drove through the south, but the majority he kept for himself, locating near Calgary, and commencing to operate a small dairy, selling his butter and milk to the Mounted Police at Bow River post.

There was little trouble, the Indians being quiet and peaceful, well satisfied with what they had experienced of white men's rule. Negotiations were on for general treaties with the different tribes wherein they would receive food, clothing and reservations, the supposedly lawless American traders were quietly conducting business, and only small and unimportant breaches of law and order were experienced.

Some changes were made in the Mounted Police force. Colonel Macleod, assistant-commissioner of the force under Lieutenant-General George A. French, was appointed to the position of commissioner upon General French retiring to return to active service in the British army. General French was afterwards the Sir George A. French who performed distinguished service in various parts of the Empire, being particularly identified with the organization and development of the defensive forces in Australia.

In the ranks of the force, the ranching problem was still being discussed, and one constable, W.F. Parker, one of the original '73 men, took the plunge and purchased a few head from J.J. Healy at Whoop-Up, Colonel Irvine of the NWMP also securing a small number.

A few new settlers were now coming into the country. In 1875 A. Sibbald, of Barrie, Ontario, was brought out to teach the Indian school at Morley; in 1876 a Frenchman named Beaupry came with his half-breed wife and settled on the Ghost River, near the ranch later established by A.P. Patrick, west of Cochrane; and James Votier arrived, accompanied by Sandy McDonald, a Manitoba half-breed, who was a good type of Westerner. A family of L'Hirondelles arrived also about this date and settled near Calgary. Votier was a man who had spent many years in the West before the Police came, trading and hunting, but now that law was established he settled down near Fish Creek to the staid business of fanning. John Glenn, or Glynn, who had been in the West since 1854, had also started farming south of Calgary the year before, and Sam Livingstone had squatted on a choice piece on the Elbow River. Livingstone, too, was an old-timer, a wanderer and a pioneer, and he held the honor of bringing the first pigs into Alberta, having owned a drove when he was at Victoria, in northeastern Alberta, in the latter sixties.

In the fall of 1876, the drover Shaw decided to leave the Morley district. Selling some of his cattle to Reverend John McDougall, he took the rest north, together with a few head of horses that he had bought that year from traders on High River. When he left, he was accompanied by a man named Gillis, a brother-in-law to one of the McDougalls.

CHAPTER VI

1877–1879

*Indians + received them land
to the Canadian Government*

A MOST important period in Western development was the year 1877, when the Western Indians signed the treaty deeding their lands to the Canadian Government. Previous to that, the whole vast country was technically Indian property, and this they gave away for promises and assurances of support when hard times should come.

In the years before this date, there had been treaties and agreements signed with all the Indians east of Alberta, these documents numbering six in all; and the paper signed by the chiefs of the Western prairie tribes naturally has since been designated as Treaty No. 7. Lieutenant-Governor Laird and Lieutenant-Colonel Macleod were selected by the Government to negotiate this treaty, and early in the fall of the year all arrangements had been made for the assembling of the red tribes at Macleod. When the dusky clans were all present, which was on September 13, the tribes moved to Blackfoot Crossing and camped there on September 17. Accompanying them were missionaries, traders, Government officials, and a strong force of Mounted Police. The Government officials told the Indians what the Queen proposed for her children, a heated powwow lasting four days resulting. Some chiefs made extravagant demands, but men of judgment, such as Bull's Head of the Sarcees, Crop-Eared-Wolf of the Piegans, and most particularly Crowfoot of the Blackfeet, were inclined to reason, and they overbore the enthusiasts. Crowfoot was a very strong Indian character, a man who in another walk of life would doubtless have been a great statesman. He was a cool-headed, tactful, wise ruler of an unruly people, and too much credit cannot be given him for the stern hand he held over his hot-blooded young men then and in the days of the half-breed rebellion of 1885. On September 22, 1877, the treaty was signed by the representatives of the Queen and the prairie tribes, Crowfoot waiting until the last, and then saying, as he affixed his signature, "I have been last to sign; I shall be last to break."

At the very moment of the signing of this momentous document, the Indians of the United States, sixty miles away, were engaged in bloody war with the whites.

It was decided that tribes who were not agriculturally inclined should be given cattle, and those wishing to till the soil should receive the necessary implements. Reverend John McDougall was present, acting as interpreter for the Stonies, which tribe he and his father had been among for many years. At this treaty-giving, the tribes received $52,954 to divide among the 5,392 Indians represented. Doubtless the hovering traders succeeded in securing considerable of this first treaty money.

Governor Laird, well pleased with the result of his work, returned east, leaving the administration of affairs in the hands of Colonel Macleod and the Police. Settlers were coming in steadily, and the Governor predicted comfortable homesteads and great herds of cattle in Alberta before many years would pass.

Just to give the tribes a taste of the munificence of the Government, a few head of beef were slaughtered and given them as an indication of future supply. Indian-like, the red men took everything offered them, carried it to their villages, and threw the meat on the ground for their dogs to eat, for there still were many buffalo and hence no need of white men's beef.

In the meantime the Ottawa Government was commencing to take a more lively interest in the youngest child of the great Dominion. Regulations, carefully deliberated, were passed. In order to protect the fast-disappearing buffalo herds, they passed an ordinance making a closed season in all ensuing years, the closed season being between November 15 and August 14. They ruled it to be unlawful to destroy the animals in pits, to stampede them over cliffs or cutbanks, or to wantonly slay in any way; and they said Indians could kill any time until February 14, 1878. The penalty for violating this regulation was a fine not exceeding one hundred dollars or three months' imprisonment.

On March 22 they passed a prairie fire ordinance, applying to the months of December, January, February, March and April, during which time those guilty of setting out prairie fires would be liable to a fine not exceeding one hundred dollars or six months in jail. An ordinance prohibiting gambling was also passed. One can easily understand how difficult the administration of any one of these regulations would be.

When Reverend John McDougall returned to Morley, he took with him a band of horses that he had purchased at Blackfoot Crossing from a Jew named Ursinger, the second band of horses taken into the Morley district from the open prairies.

This year, too, saw the introduction of the first range cattle into southern Alberta. Fred Kanouse, the trader, went to Montana, and purchasing twenty-one head of cows and one bull from his father and brother, placed them on the range at Macleod. After this, a man named John Miller came from Montana with a bunch of cattle of the "20" brand, which he too put out to rustle for themselves and multiply to his own benefit. Five head of sheep were introduced to the Pioneer Ranch by Joe McFarland, who with Olsen moved a herd of their cattle down into the Medicine Hat country.

The first true-born Canadian, other than Mounted Police constables and officers, to arrive in the Macleod district to settle came in during that summer in the person of W.J. Hyde, now agent on the Blood reserve near the town of Macleod. Mr. Hyde commenced farming and ranching almost at once, later going into partnership with Dug Alison, an ex-Mounted Police constable.

Jim Christy, arriving with another band of Montana horses, disposed of them in a very short time, the demand being even stronger than it was when he brought over the first band in 1876. George Emerson and "Tim" (Dominic) Lynch went to Montana, and buying about two hundred head of "HW" cattle, drove them across the line to dispose of to the settlers. Most of this herd was purchased in the Macleod district, many of the Police officers and men buying small bunches. J.B. Smith purchased some; Parker, Harper and Dunn, another batch; J.D. Murray and Bell and Patterson, a few. (Mr. Patterson is Robert Patterson, now MPP for the Macleod riding.) The prices paid were low, the scale being nineteen dollars for a cow with a calf, twelve dollars for yearling heifers, and thirteen dollars for yearling steers. Christy sold his horses at prices ranging between forty and sixty dollars, and seldom reaching the hundred mark.

The next year saw the beginning of troubles between Indians and ranchers, which resulted in the majority of the stockmen in the Macleod district pulling out temporarily and going into Montana. There was some trafficking in cattle; George Emerson brought in another herd, and the I.G. Baker people brought over considerable beef stock.

One of the traders who built Whoop-Up and later sold out to I.G. Baker came to hard times this year also. Though a crafty trader and a splendid man among men, he was not so wise when it came to understanding womankind. A lady, who, if we only knew, probably had blonde hair, was the direct cause of his downfall. After selling out to the Bakers, he came into the Calgary district, bringing with him in 1878 a hundred head of splendid cattle, half-breed Devons and Shorthorns, and a good bull worth in itself some three hundred dollars. Devoted as he was to the lady in question, the trader neglected business and spent his money royally, until he had nothing left but the hundred head, which he then sold to David McDougall for thirteen hundred dollars. This money enabled the pair to get away to the States, where they finally separated when his money was gone. Such is the story among the old-timers.

Lynch and Emerson, finding importing for settlers a lucrative occupation, brought in more stock, and sold fifty head of yearling heifers to Captains Shurtliff and Winder of the Mounted Police at Macleod, and A.P. Patrick, a Government surveyor, at eleven dollars a head. These animals of Winder's and Shurtliff's were the original "Slippery Moon" brand, which designating mark they adopted. Just why it was called a "slippery" moon is not explained.

That fall, A.P. Patrick, surveying in the country at the forks where the Red Deer empties into the Saskatchewan, was troubled by a band of mischievous Cree Indians under Big Bear, a non-treaty chieftain. Big Bear refused to allow Patrick to go ahead with his work, so the surveyor sent a complaint to Fort Walsh and sat down to await results. The messenger made good time going, the Police came posting, and Big Bear, as soon as he saw the imposing redcoats, protested valiantly that he was not meaning anything serious at all. He promised to be good, and kept his word in this instance, Patrick having no more trouble from that quarter while he was in that country.

Great prairie fires having driven the buffalo south to their last great slaughter on the Missouri, the Alberta tribes of the south were starving, and the Government, not having made sufficient preparation for such a strain on the public purse, was unable to supply the crying needs of their new wards. Many of the Blackfoot, Blood, and Piegan young men followed the herds across the line, and, defending themselves against desultory attacks from Sioux, Crow, Gros Ventres and other American Indians, succeeded in slaying a fair number of animals, though of course they displayed the usual Indian prodigality and did not put much up for a rainy day. Those of the red men who remained adopted the habit of taking ranchers' stock as legitimate prey. Through the summer and early fall, these annoyances continued, the Police being unable to handle the Government's adopted children, owing to the small number of the force and the vast country covered by the Indians. Also it was thought by some of those in authority that the white ranchers had displayed poor judgment in starting in business before the Indians had been taught to act with more decorum, and that they half-deserved their loss for braving the savages. Settlers had come in, indeed, but Indians were not under control and the country was not really open. There were twelve farms in the Macleod district alone, enumerated by A.P. Patrick as follows:

"Three miles below Macleod a settler has taken up a farm and is doing well. Five miles above Macleod, on the Old Man's River, are three or four farms. Thirteen miles up (within the Piegan reserve) are three more. Twenty miles up, near the mouth of Pincher Creek, are two more farms. Seven miles up Pincher Creek is the Mounted Police farm."

The cattle that were now due as treaty payment to the Indians were delayed long in delivery, and did not arrive until late in the fall. The Government had started out with an idealistic system of giving the red men breeding herds that they could take care of themselves, and soon become self-supporting ranchers. In order to make this the immediate success it was hoped in Ottawa, it would have been necessary to shatter Indian traditions and habits that had a basic foundation of three hundred years' practice. Then, too, the tribes were starving. Apparently Colonel Macleod and the Mounted Police authorities in the West, being more in touch with the truth of things, decided that it would not be judicious to turn these breeding stock over to the Indians at once. Using as an argument the undeniably true statement that the Blackfeet, being naturally wanderers, had not yet learned to stay in one place, and would necessarily not be able to herd and watch their cattle, the white powers decided to take care of the stock themselves through that winter. Mr. Conrad, of the I.G. Baker company, philanthropically offered to herd the animals until the following summer for the increase. Colonel Macleod refused this assistance at once, and then Mr. Conrad said he would take good care of the herd through winter if he were given the calves then in the band, about four hundred in number. This, too, was refused, and Colonel Macleod employed a man named Scott, with three assistants, to take the animals and herd them till the following summer, paying the men a salary of two hundred and fifty dollars a month between them. Scott took them to a splendid pasturage at the foot of the Porcupine Hills, cut hay, put up sheds for weak cows and calves, and brought them through in good shape, though a number of the young calves perished in the early spring.

The year 1879 was one of the most unsatisfactory, from the settlers' point of view, that had yet been known. The buffalo were gone, never to return; the natives were literally starving to death. Gophers, mice, snakes, any living thing proved food to them, and it was not an uncommon thing to see some great warrior lying patiently beside a gopher hole hoping for the chance to snare the little animal so he could have food. Everywhere throughout the prairie of the south, the red men suffered, and the ranchers who lost cattle at their hands were large-minded enough not to lay all the blame on the poor natives. The year before, the southern Indians had driven the buffalo across the line, and although a few small bands succeeded in breaking north again into the country east of the Cypress Hills, and some to the South Saskatchewan, the main herd of the animals was penned in by a cordon consisting of nearly all of the Indians of the northwest states. This great herd remained cooped in within the country south of the Milk River, about the Little Rockies and the Bear Paws, extending across the Missouri River to the Judith Basin. Deplorable indeed was the condition of the Canadian

Indians, and it is really remarkable that this formerly lawless race did not arise in the anger of their suffering and wipe out every white man in the land.

On May 31, E. Dewdney, member of Parliament for Yale-Cariboo, was appointed to the post of Indian Commissioner by the solons at Ottawa, Mr. Dewdney being from British Columbia, and the Government authorities believing he would know best how to administer Western prairie affairs. Accompanied by three experienced farmers named Patterson, Wright and Taylor, who were to have charge of instruction farms for the Indians of Alberta, he left the East early in summer, arriving at Macleod on July 10, passing en route in the eastern part of the province some few bands of Indians hopelessly hunting for the great buffalo herds of the past.

In the meantime, the ranchers had experienced a considerable discouragement, what with Indian killings and bad storms. Seventy head of cattle perished in one coulee alone during the March storms, and the Indians slew scores of animals in the scattered herds. Travelling from fort to fort were great bands of red men begging food to sustain life, and there was at one time a great multitude, variously estimated at from two to five thousand, camped around Macleod, stolidly waiting for the white men to give them food to live.

The spring season was bad, stormy and very wet. The rivers flooded higher than they had in years, the Old Man's River changing its course from the north to the south side of the island on which Fort Macleod stood, and nearly flooding the island itself. A sawmill that had been erected on the island was only saved by most desperate efforts of Police and townspeople. The rotting bodies of the cattle in the coulees contaminated the waters of the Old Man's River, and a serious epidemic of typhoid fever resulted, causing the death of several white people and a number of Indians and half-breeds. Mr. Clarke, of the I.G. Baker store at Macleod, died of this disease, and was followed shortly by a man named Walsh. Considerable numbers of the white men passed successfully through the illness.

In July, E.H. Maunsell and his brother George purchased one hundred and three head of cattle and took them to their newly founded ranch. Arising at five o'clock, the two embryo ranchmen rode away to get their animals, and being green at handling stock, took a long while "cutting out" their property and then driving them across the open country to the home corrals. It was after eight in the evening when they finally closed the corral-bars behind the last of their herd, and then, hungry and tired, having eaten nothing since five o'clock that morning, they hastened to the house to prepare a meal. A broken door and a jumbled interior gave undoubted evidence that starving Indians had paid the place a visit during their absence, and it was with a heavy and hopeless feeling of unassuaged hunger that Ned Maunsell looked for food. He went to the flour sack, and lo, the red men had left enough for a couple of meals; he looked at the bacon-nail, and there dangled a small piece sufficient also for a meal or two. There was a little tea left, and a little sugar, and no men were more grateful to thieves than those two young ranchers were toward those lawless Indians who had held the power to take everything, but had thoughtfully left enough to last until a new supply could be procured.

Every day came reports of cattle killings, though there were so many Indians and so few whites that it was impossible to catch the natives red-handed, which was necessary in order to secure a conviction. In addition to the inability of the whites to produce evidence, there was the ever-present fear that the tribes would rise and wipe out every soul but their own people. The Police, though maintaining their usual fearless attitude, were fully aware of this not-too-remote possibility, and Dewdney, when he arrived, concurred in their belief. In fact, it was possibly under his orders that Colonel Macleod dismissed one case that could have formed the basis of an example to cattle-killing reds. A white man riding across the prairies saw a cluster of Indians around an object on the ground, and, suspecting a killing, he rode close. The natives were so engrossed that he rode to the very fringe of the crowd and, peering over their shoulders, saw a slaughtered beef with brand and everything plainly before him. Riding away, he told what he knew, complaint was laid before the Police, and the Indians were tried in the presence of Indian Commissioner Dewdney, though Colonel Macleod sat on the bench. The Indians were dismissed, and the stockmen swore among themselves. At that time, some thousands of the natives were camped around Fort Macleod, hoping and trusting that Dewdney would relieve their pressing needs. He tried, in a very economic manner, and satisfied them to a certain extent, though not entirely to the satisfaction of the white settlers, who still declare that if he had been more generous with supplies, they would not have suffered the loss they did.

The first roundup ever held in southern Alberta took place in the Macleod district in early August of this year, with W.F. Parker holding the position of captain. It was a poor enough roundup outfit, consisting of sixteen very indifferently mounted men and one wagon, and the result of the undertaking was very discouraging. No man had more than one mount, some were riding mares with colts running at their sides; even provisions were at a very low ebb. To help out the larder they sometimes collected the eggs of wild ducks from nearby sloughs, boiling them hard so as not to be worried by the contents. Fred Kanouse shocked many of the newcomers in the party by solemnly declaring he preferred his eggs with the birds in them, and offering to exchange good eggs for bird-filled ones whenever an opportunity arose. The sixteen men who composed the roundup party were W.F. Parker, Captain Winder, Captain Shurtliff, Jack Miller,

E.H. Maunsell, Thomas Lee, H.A. Kanouse, Robert Patterson, J.B. Smith, Joe McFarland, H. Olsen, Morgan, Sam Brouard, Bell, Alison, and W.J. Hyde. They rounded up a total of five or six hundred head and learned the disconcerting truth that there were many less cattle then alive than there had been the fall before.

Of the one hundred and three head purchased by Maunsell in July, scarcely thirty days before, there remained but fifty-nine; practically every hoof owned by Parker had disappeared; Patterson, Miller, Morgan, Kanouse, Lee, nearly every man had a sad tale of loss, assuredly, in their minds, directly pointing to Indian depredations. Still, they did not blame the starving Indians, but complained of a negligent Government and Government agents.

The stockmen decided something was necessary at once, so appointing a deputation of four, consisting of E.H. Maunsell, J. McFarland, H. Olsen, and Morgan (the latter working in the interests of Mrs. Armstrong as foreman for her), they sent them to Dewdney to present their case. They said they had come in and started in business, depending on the Government and the Mounted Police to maintain order and protect property; they said they wanted to stay in the country, and intended to do so if they could be assured of protection. Pointing to the crying needs of the Indians, and referring to the well-known fact that Dewdney intended buying cattle to supply the needy red men, the ranchers offered to sell their own animals at cost, and wait until conditions were a bit more favorable. Colonel Macleod had been approached by the ranchers before this, and he had told them they must stand their losses, as the country was not really open for settlement.

"But if we corral our stock every night, can we shoot any Indians we find killing them?" asked the ranchers, somewhat eagerly.

"If you do you'll probably hang," retorted the Mounted Police Commissioner.

Dewdney followed a similar line, refusing to make good the ranchers' losses, refusing to purchase the cattle remaining, and giving further his belief that the times were too early for ranching anyway. In high dudgeon the stockmen went away, determined to leave the country until they could be assured of protection and what they considered a fair deal. Fred Kanouse took his little band to the Kootenay Lakes (now Waterton Lakes), Maunsell, Mrs. Armstrong and Morgan, Jack Miller, Patterson and Bell, J.B. Smith, collected the remnant of their herds and drove them across to the Marias River in Montana, paying duty to get into the States. Joe McFarland, who was running milk cows and a straight mixed-farming business, decided to brave it out, and he remained. W.F. Parker stayed also, but he had no more cattle, and consequently there was no necessity of going.

After disposing of the ranchers to his satisfaction, Mr. Dewdney prepared a report on the subject of Indian killing that is rather interesting at this late day. He infers that the ranchers possibly evaded stating whole truths, and that the word of the starving Indians was more to be depended upon than that of the struggling ranchers. Taken in conjunction with Captain Winder's report of that same year —penned perhaps at the same time—this report of Mr. Dewdney's contains considerable of interest. He says:

"While at Macleod several settlers who reside in the vicinity called on me and stated that numbers of their cattle had been killed by Indians. They said they were then gathering up what were left, and proposed to drive them over the line, preferring to risk them near a settlement on the American side than keep them on our own. I asked them if they were sure that the Indians had killed them. They all stated most positively that they had, but were unable to bring a single proof. They stated that they had applied for protection to the Police, but were unable to get any.

"One man stated he had lost one hundred, another fifty, another thirty, making an aggregate of between two hundred and three hundred.

"They said, 'If you will come and look at the cattle we propose driving out we will show you they have bullet-holes in them where they have been shot.'

"I thought it very singular that these complaints had not been made to the Police, Colonel Macleod having told me it was the first he had heard of. it, that no formal complaint had been made to him.

"I said I thought it would be better to get the Indians together and tell them what the settlers had stated, and impress upon their chiefs the advisability of finding out and bringing to justice any Indians who had been killing.

"Colonel Macleod stated that in that case we had better stop payment.

"I agreed with him and sent for the chiefs. I had heard that even the night before a Mrs. Armstrong, who kept a dairy, had one of her cows shot through the head, and it was in her own corral at the time. The chiefs assured me they knew nothing of it, and that they would take every means of giving up any Indians they found killing cattle. In order that there should be no mistake I sent for the settlers and made them tell their stories before the Indians.

"My belief is that some few cattle have been killed by Indians, but do not credit reports that they have been killed by hundreds. The fact is that many settlers have gone into the stock business on so small a scale that it did not pay to keep a herder to look after them. Small bands of cows were allowed to roam all over the country, and the owners disbanded upon hearing of their whereabouts from those who happen to own large herds of cattle, and who are constantly on the ranges, and if, at the round-up, which takes place twice a year, the cows with their expected increase did not turn up the Indians were blamed for killing what were wanting.

"It was well-known that white men in this neighborhood had been in the habit of going out on the open prairie to shoot cattle, and butchered them on the spot, to supply the Mounted Police and local demands. This may account for bullet-holes being found in some of the cattle that were driven south of the line to winter, and it is thought that mistakes might have been made in the ownership of the cattle and the wrong steers killed. At any rate, it is most extraordinary that if so many cattle had been killed by Indians there was so little evidence of it.

"This we do know, that all the cattle in our southern district came from Montana, and some of them from only a few miles south of the boundary line. We have proof of many having strayed back to the old ranges, been resold, and again driven into our country. In many instances where cattle have been recognized, owners on our side of the line have been notified that the cattle were there."

Continuing, Mr. Dewdney states that, a day or two before he left the Cypress Hills on his way out in the fall, one of the sergeants of the Mounted Police, who had accompanied Captain Young, the customs officer, heard of a band of cattle being driven from Fort Macleod to Fort Assiniboine. The sergeant knew this stock, some of it being his own. The report was that a white man was selling this beef for three and four cents per pound, claiming it to be elk meat.

. The Indian Commissioner also encloses the following extract from a Fort Benton paper:

"Advices from Fort Macleod state that several head of cattle supposed to be killed by Indians were found by the North Piegans on the north fork of Milk River. The Indians drove them to Stand-Off, where A.B. Hamilton met them with a letter to Colonel Macleod."

Dewdney winds up his report with a brief account of a Calgary cattle episode, which he uses to illustrate how often the Indians are held up to blame when they are blameless. He says:

"Messrs. Lynch and Emerson, who run cattle in the neighborhood of Calgary, complained to Inspector Denny that thirty head had been killed by Indians. These ranchers wanted the Government to pay this loss. A few days later a half-breed saw some unknown brand on a band of cattle, which he at once drove to Calgary. They were Emerson's."

On the other hand, Superintendent (Captain) Winder at Macleod reports in his annual returns: "The Indians are suffering and the Mounted Police supplies are short. Complaints of cattle killing by Indians came in almost daily from March to October. Yet we could find no clue of guilty parties, though I investigated the cases personally."

CHAPTER VII

1879–1880

CROWFOOT, with his sick and destitute tribe, made no complaint of suffering when he met Dewdney, though he spoke of the Indian herds and the buffalo.

"Make a hole in the barrier of Sioux that is preventing my people from getting to the buffalo," he said, "so my young men can get through." He asked also that the Government would send him his cattle so that he could give them to Lafayette French to take care of for him. French was a trader who had lived for years among the Blackfeet, and was held in high esteem by Crowfoot, owing to a favor the white trader had conferred on him in the early days.

"They tell me I have cattle," said the old Indian chieftain, "but I have never seen them." Reverend Father Scollen, who was present when this meeting between the Indian Commissioner and Crowfoot took place, said the Indians had heard many reports of their cattle, among them being one that Captain Winder was branding the calves with his own brand, and another that the Armstrongs had taken a heifer. Upon investigation, Dewdney found that Winder had no individual brand of his own, and he persuaded the Indians to leave the stock with Scott, who was still herding them on Pincher Creek. The Commissioner was afraid to trust the herd with French, he says, as the latter was an American with absolutely no ties.

French was an independent trader and consequently was not popular with the I.G. Baker Company. Though he doubtless took advantage of the Indians, he probably did not do any worse than the others would. French's hold with the tribe dated from the time some years before when he saved Crowfoot's life on the High River. French and other traders, trading at High River, were surrounded by a large number of trafficking red men, and among them was Crowfoot, who came to French's cabin to rest. One of the traders was unpopular with Crowfoot, and consequently lost considerable Blackfoot trade, much to his annoyance. Thinking to remove the cause of his ban, this man filled a brave full of firewater and told him to go and shoot Crowfoot in French's cabin. The savage followed directions as far as going to the door of the shack, drawing a bead on Crowfoot's head, and pressing the trigger. But, as he pressed, Lafayette French came around the corner, saw the state of affairs at a glance, and swung the rifle muzzle upward as it exploded, the ball crashing harmlessly into the mud roof. Crowfoot never forgot this act, and continued to repay while he lived.

French, or, as he was afterwards named in the Blackfoot tongue, "The Lifted Arm," built a shack at Blackfoot Crossing and proceeded to take advantage of his popularity. Other traders, feeling jealous of him, are said to have entered unofficially into a conspiracy with certain of the Mounted Police to discredit him with the whites and savages alike. Two women were to go to French's store, enter, and then commence screaming, whereupon the Mounted Police were to rush the place and arrest French on charge of a serious crime. Crowfoot heard of it, and, telling French, planned a counterplot. Two Catholic priests, Father Larue and Father Scollen, were very popular among the Blackfeet, and it was arranged with one of these men to hide himself in the store and come out at the psychological moment. The day came, and with it the squaws arrived and commenced to scream as per programme, the thunder of horses' hoofs indicated the arrival of police, the thunder of heavy fists on the door and the roar of a voice demanding admittance in the name of the Queen indicated that the stage setting was reaching the climax. French opened the door as ordered, and the policeman who stormed in was confronted not only by French and two cowed squaws, but also by the Roman Catholic priest in his churchly garments. The plot died a sudden death and was not resurrected.

Dewdney found that French, taking advantage of the needs of the Indians, had secured most of their horses for a few cups of flour each. French, in explaining this business transaction to the curious Indian Commissioner, told him that he had been forced to do it through fear. Father Scollen had told him if he did not sell, the Indians would take, so he had naturally sold. Dewdney told him that he must move from the Crossing and the reserve, and offered to purchase his buildings and land to utilize it as a nucleus for an Indian Government farm. French sold out and moved back toward High River.

During the time he was in the West arranging Indian matters, Mr. Dewdney went through a tremendous amount of business. He gave treaty, bought squatters off homesteads that they held within reserves, supplied small amounts of food to the tribes and tried to straighten out the tangle all at once. He gave treaty at Blackfoot Crossing, calling all the tribes there that he could, because he thought they could be better handled at that point. Also he wanted the Sarcees to go there to take treaty because beef was seven cents a pound at Calgary and it could be purchased more reasonably from the south. Troubles also arose because the red men were growing wiser and, like whites, were willing to take more than was coming to them. Bull's Head demanded treaty for four hundred lodges, when he had only a hundred and twenty-eight by actual count; braves, standing in line and drawing treaty money and rations, successfully made the trip past the paying quarters two and three times, taking it half naked, then with a blanket on, then with a buffalo robe, using different variations of these manners of disguise.

John Glynn had settled on a piece of land at the junction of the Bow River and Sheep Creek, where he had built two cabins, a roothouse, and had broken four and a half acres, which were planted to barley and oats. This plot was partly fenced, with enough rails cut to complete the job, and he had a small stack of hay which he had cut that summer. All these things, and a milch cow with calf in addition, he sold to the Government, through Dewdney, for the munificent sum of three hundred and sixty dollars. Dewdney located Bryce Wright on this farm.

He chose another farm on Pincher Creek, four miles from the Police farm, and placed Taylor there, while on the farm at Blackfoot Crossing he put a man named Patterson.

To assuage the first pangs of hunger among the thousands of Indians who greeted him at Macleod, he gave first a hundred sacks of flour, and followed up with more later on. Two hundred and nineteen head of cattle were bought from the I.G. Baker people to feed to the Indians, and small bands were dispatched to the different tribes before Dewdney left for the East. A man named Norrish was sent with thirty-nine head of cattle "and supplies" to Blackfoot Crossing to relieve the Blackfeet, while C.W. Kettles went to the Piegan reserves with a similar supply.

The Stonies complained that the spades and ploughs given them in treaty were no good. The Government was supposed to give them sufficient tools to farm, and of course the contractors of these supplies figured anything was good enough for Indians or Government. The ploughs would not cut the sod, the spades buckled and broke when a man's weight was thrown on the handle. Dewdney promised to rectify this, and apparently did, for no further complaint is noted.

Dewdney cut the salary of Mr. Scott, who had charge of the Indian herd, from two hundred and fifty dollars a month to two hundred. Two settlers, King and Olsen, who had settled within the limits of the Piegan reserve before the reservation had been chosen, were bought off and forced to move. These men were good types of pioneer settlers, and impressed the Commissioner as such, for he gives details of their start and progress up to the time he bought them off. Olsen had come in a comparatively poor man and was then worth about five thousand dollars; King had arrived about 1876 with a spade and half a sack of flour. When he was removed from the reserve, he had three hundred and twenty acres fenced, twenty-five acres under crop, a good house, and five hundred logs in the woods for more buildings.

A man named McKenzie, brother-in-law to the McDougalls of Morley, showed a pronounced Western spirit of persistence when Dewdney refused to pay him for work done on the Government farms, in the line of breaking and planting crops. The contract price reached a considerable sum, and Dewdney thought it too much. McKenzie attached himself to Dewdney's party and declared he would stay there until they reached Ottawa if he was not paid, threatening further to stay in Ottawa until satisfied. Dewdney is said to have paid no attention until he started east in the fall, then, convinced he could not shake off the tenacious McKenzie, he arranged with the I.G. Baker Company to pay the claim.

At the close of the year there were quite a number of scattered settlers from Calgary south. Fourteen were in the country west of Calgary, most of them starting in stock-raising; a dozen farms were strewn around the wide Macleod district, mostly along the rivers, as in fact nearly all settlers located in early days. There were three or four settlers on the Kootenay bottoms, two at the mouth of the Kootenay, one a few miles up the Belly, and one at the point where the Kootenay leaves the mountains. Sam Livingstone had greatly improved his farm on the Elbow near Calgary and was raising fine crops, averaging fifty bushels per acre for oats. The crop on the Police farm at Pincher Creek, though not planted until June 7, yielded a crop of twenty-three hundred bushels of good oats off a hundred acres, even though, in its harvesting, the primitive method of scythes and homemade rakes was adopted. Taylor, in charge of the Pincher Creek Government farm, had a hundred tons of hay in stack, and eighty acres broken when winter came. The Mounted Police herding post was removed from the Piegan reserve to Willow Creek.

There was little criminal work done by the Police, their greatest efforts being directed to keeping the Indians in hand. Sam Livingstone charged G.C. King with opening and detaining a letter. King placed a countercharge of assault, and both eventually withdrew the complaints; a gentleman named Frank Delatras who sold whisky to Indians at treaty payment was fined fifty dollars at Macleod; a half-breed named St. Germain, who stole a horse, was rounded up by Corporal Heney and Constable Maxwell and given sentence of one year by Colonel Macleod. Only one conviction of an Indian for cattle killing is recorded. Nez Percés came unexpectedly upon a Stony Indian named Little Man, who was calmly killing a calf that belonged to Morgan, or rather to the Armstrongs. Little Man's family was starving, as of course most Indian families were then, and he had adopted this method of relieving the strain. Seeing that he had been discovered, the culprit repaired at once to Morgan and offered him his horse in payment of the calf, but Morgan insisted on a trial and conviction. The case came before Colonel Macleod, who fined the Stony twenty dollars.

Charles Laxton, for importing liquor, and John Hughes and Frederick Pace, for having intoxicating liquor in their possession, were fined two hundred and fifty dollars each, or six months in jail. They paid their fines.

All fall there had been prairie fires raging in various districts, but the most thrilling one was that which was started by Fred Kanouse at Macleod. An Indian came into the post one day in the fall with a nice pony that he wanted to dispose of for very little cash. Though it showed indications of mange, it had some good points, and Kanouse purchased it, later treating its mangy spots with generous applications of kerosene, which is very bad for the mange parasite. In Macleod in those days there were often people who would take in their own possession any animal that had no distinguishing brand, and so Kanouse, in order to protect himself, took his new purchase down to the edge of the town where the black-smith shop was, and proceeded to put his brand on. As he pressed the white-hot iron to the horse's side, there was a flash—and then a sheet of flame went dashing across the prairie. The forgotten mange ointment had asserted itself. Before the horse died he had ignited a half-mile of dry grass, and a great fire that necessitated the services of all the men at hand threatened the town for some time, but was eventually extinguished before any real damage was done.

One tragedy took place in the prairies that has never been solved, though the whole strength of the Mounted Police and the settlers was devoted to finding and punishing the perpetrator or perpetrators. Constable Graburn, stationed at Fort Walsh, was shot down in cold blood while riding down the brow of one of the deep coulees in that district, and though every effort was put forth and the Police were morally certain the crime had been committed by certain of the Bloods, yet sufficient evidence could not be secured to convict any of them. A young Blood brave named Star Child was suspected, and, after a year had passed, was arrested and tried, but acquitted. Old settlers who were in the country at the time still declare that Star Child was guilty.

A second story of this tragedy came to light not long ago, which, though unauthenticated, might possibly be the true solution, and Star Child might have been wrongly suspected. A man who had business in the vicinity of Fort Walsh found it expedient to winter his horses at that point, so, purchasing some small haystacks from D.W. Marsh, now of Calgary but at that time a storekeeper at Walsh, he put the horses out under the care of a man who had an Indian wife. One night the owner of the stock went out to the camp and was eating supper when a Blackfoot brave, hungry and exhausted, came in and joined them at the meal upon invitation. After supper, when darkness had fallen, this Indian went out and came crawling back in a few minutes saying that Sioux were about to steal the horses in the corral. He told the owner of the animals to follow him with his rifle, and then faded into the darkness, crawling out under the rear of the tent. The white man was stooping down to creep after him when the squaw plucked his coat sleeve and whispered to him not to let the Indian have his rifle.

They crawled into the darkness, and, approaching the corral as silently as possible, looked sharply for skulking horse thieves.

"There they are. See?" said the brave, pointing into the night. The corral, built on an elevation, was dimly visible against the sky, and the horses were quiet.

"No," replied the other.

"I see," declared the red man impatiently, "Give me your rifle."

Then the warning of the squaw came clearly, and the white man arose.

"You don't see anything," he declared, stalking back toward camp. The Indian followed, stayed through the night, and went out alone when morning came. The squaw told the white man the remainder of the story. She said that the Blackfoot went up a coulee until he met Graburn coming down, riding along the rim. The policeman stopped, chatted, dismounted, and sat down to smoke with the native; the latter waited until he saw his chance, and then, leaping toward the constable's horse, drew the carbine from the holster on the saddle and shot Graburn down in his tracks. This Indian's daughter had died at Blackfoot Crossing while under treatment of a white man, and the father had started out with a vow of sending a white man to the Beyond in her company.

"I saved your life last night," said the squaw when she had finished, "and you will not tell this. If you did you might die too."

Whether this story is the correct version of the death of Constable Graburn is not known positively, and never will be. Graburn died and his murderer was never found, but to the present day, the great dark coulee where the tragedy occurred has stood a monument to the officer's memory, for it bears the name of Graburn's Coulee.

Here and there throughout the wide country, the shacks of squatters and small ranchers were growing when the spring of 1880 came in. Log shacks, mud roofs, the most simple of primitive furniture, satisfied the pioneers, who were willing and able to bear the rough life in order to get their start in the new land. Benton and Winnipeg were as yet the nearest supply points, and great trains of slow-moving bulls and oxen creaked and groaned on the long trails as they dragged the food supplies into the little hamlets.

To all to whom these Presents shall come Sir John Walrond Walrond of Brook Street in the County of Middlesex Baronet Sends Greeting Whereas the said Sir John Walrond Walrond is the Lessee named in a certain Indenture of Lease dated the twenty sixth day of June one thousand eight hundred and eighty four and granted by Her Majesty Queen Victoria represented by the Deputy of the Minister of the Interior of the Dominion of Canada to him the said Sir John Walrond Walrond of all and singular the following lands and premises namely Townships number 8 and 9 in ranges 29 and 30 West of the Fourth Principal Meridian That part East of the North Fork of the Old Man River of Township 8 Range 1 West of the Fifth Principal Meridian in the District of Alberta North West Territories containing an area of One hundred thousand acres more or less upon the terms and conditions therein mentioned And whereas the said Lease was taken by the said Sir John Walrond Walrond on behalf of and as Trustee for the Walrond Cattle Ranch Limited and he has agreed at the request of the said Company to execute such declaration of trust as is hereinafter contained Now these Presents witness that in consideration of the premises it is hereby declared that Sir John Walrond Walrond shall hold the said Lease as Trustee for and on behalf of the Walrond Cattle Ranche Limited a Company incorporated under the Laws of Great Britain and Ireland And it is hereby further declared that the said Sir John Walrond Walrond will when called upon to do so by them execute any Assignment Transfer or Sublease of the said Lease which he may be required to do to the said Walrond Cattle Ranche Limited or to their nominee or as they may direct As witness the hand and seal of the said Sir John Walrond Walrond this _____ day of _____ one thousand eight hundred and eighty five _____

Signed sealed and delivered by the
above named Sir John Walrond Walrond
in the presence of

John Penny
Butler Bradfield
Uffculme

J. W. Walrond

The leasing system was instituted by the government in 1880. The lease shown here, to the Walrond Ranch, is for 100,000 acres.

The Government, realizing that the time for the encouragement of settlement and development had arrived, commenced their lease system of huge tracts for the purposes of ranching, renting these lands at one cent per acre per year, and giving twenty-one-year leases that were not liable to cancellation. Senator Cochrane, Dr. McEachren, the Dominion Government's chief veterinary surgeon, the Allan family of Montreal (founders of the Allan Steamship Line), and John R. Craig and associates secured leases and set about stocking them with great herds. The Cochrane lease was taken up just west of Calgary; the Oxley, or John R. Craig lease, in the hills west of Mosquito Creek; the Walrond Ranch, which was the lease taken by Dr. McEachren, was in the southern end of the Porcupine Hills; the Allan lease, better known as the North-West Cattle Company or the Bar U, was west of the present town of High River, on the river of that name.

In some cases a year's delay followed before the holders of the leases could succeed in interesting sufficient capital to start their undertakings, this being the case of John R. Craig. Sir John Walrond and associates supplied the financial sinews for the Walrond Ranch, while Dr. McEachren acted as managing director of both this concern and the great Cochrane herds. Major James Walker, a former Mounted Police inspector who had come into Saskatchewan with the first detachments of that force, and had retired from the service after doing splendid work in administration among the Crees and half-breeds and settlers in the central portion of the sister Province, was offered and accepted the position of local manager of the Cochrane Ranch. After leaving the force, he went to Benton and came into Calgary district from that point, it being the best way of entering the Province.

Calgary, which was becoming quite a centre, was kept in supplies by the freight teams of I.G. Baker & Company, which hauled all merchandise north from Benton and Macleod. The freight trains were imposing bodies of great wagons and toiling bulls. Eight yoke made a team, and four teams made a train, a foreman being in charge of each brigade and each train being supplied with cook and mess-wagon. One incident will illustrate how fast these trains moved. When Major Walker reached Macleod on his way into Calgary, he asked there for his mail and was told it had left two weeks previous, by bull-train. Calgary is a hundred and two miles north of Macleod, and Walker, posting northward on his horse, expected to find the mail awaiting him when he arrived at his destination. But surprise met him at High River, forty miles out from Calgary. Here he overtook the bull-train, placidly crawling along, and the foreman told him that with good weather he expected to make Calgary in ten days.

Major Walker, feeling that bull-trains were a little too slow to be successful mail carriers in those rushing days, set a project in motion that resulted more satisfactorily. Securing the services of an Englishman named "Lord" Hugill, he headed a subscription list with twenty-five dollars and soon had enough money promised to assure his lordship sixty dollars a trip to Macleod, twice a month.

Eggs were worth twenty-five cents each at this time, and one industrious housewife held a monopoly. Major Walker thought this was rather steep, and so he went to the woman to negotiate the purchase of some hens. It was an important matter to the lady, and two weeks of deliberations were required before she would set a price, which was a fair exchange of one hen and one rooster for a good cow in milk. This deal thereupon fell through.

Groceries, brought toilsomely from Montana, were never abundant, and some of the storekeepers had formed a habit of cornering the market in the fall, knowing that by spring, with no new freight arriving through the winter, they could fix and get any price they demanded. Sugar and bacon would soar to fifty cents per pound, and flour to twenty-five dollars a sack. This state of affairs, lasting for nearly two years after 1880, was finally broken in a peculiar manner by the actions of a Blackfoot brave, who upon receiving his treaty money went to town at once to spend it, and was refused so coldly by the law-abiding citizens that his thirst turned to sombre broodings for revenge. It seems somewhat strange that there was no whisky that this red brother could obtain, but it is probable there was a paucity then, and that the white men were keeping it for their own family uses. At any rate the Blackfoot went dry, and after wandering about for some time in deep dejection he purchased every ounce of sugar from the stores in town, paying twenty-five cents per pound and clinging tenaciously to it until the following spring, when he received fifty cents cash for every bit of it. On the heels of this tragedy the railroad came in, and there was no shortage any more.

But to resume the ranches. The Government, in order to encourage development, not only gave the great leases at the rental of a cent an acre, but also removed the duty on all cattle brought in, so as to assist the stocking of the ranges. No big herds came that year, the organizers of the companies being busy with organization and with preparing for the arrival of their stock, which was nearly all to be bought in Montana and adjacent states. Great herds of cattle ranged there, stockmen having been operating for more than a decade in spite of buffalo. The Circle cattle belonging to W.G. Conrad & Company of Great Falls and Fort Benton, an affiliation of the I.G. Baker Company, were allowing big herds to drift across the boundary, and a few small ranchers brought small bands in. John Quirk, an Irishman who became a well-known character, came in with his wife, driving a herd of cattle before him and locating on High River; George Emerson and Lynch made their third successful trip as drovers and brought in about a thousand head, a large portion of which they disposed of at once. The stockmen who had pulled out of the Macleod district were coming back, Jack Miller's "20" brand being sold to D.J. Grier of Macleod, and later being purchased by Maunsell & Browning.

The rivers, high and wild in spring, claimed their usual victims in cattle, and the summer rains, swelling them again to torrents, caused the loss of at least one man's life. On July 24 Constable O.S. Hooley, driver for Colonel Macleod, attempted to cross the high waters of the Belly River at Robert Patterson's ranch and was drowned.

The Indians kept fair order, breaking out in isolated misdemeanors but not indicating any wish to make any real trouble. They had learned very rapidly that the boundary line between Canada and the States was a most advantageous arrangement, knowing already that if they broke any Canadian laws they were more safe when south of the line, and that if they strayed into the States and rustled a few horses or shot a few hostile natives, they were perfectly safe from pursuing American cavalry and officials as soon as they reached the north side of the boundary. In their own tongue they called it the Medicine Line, and were very well pleased with the condition of affairs. Mr. Patterson, the Ontario farmer whom Dewdney had placed in charge of the Indian farm at Blackfoot Crossing the year before, was the cause of considerable pleasure to the young Indian braves, who soon discovered that he was most susceptible to the bullying penchant so prominent in young red men when they feel they can enjoy the pleasure with impunity. Patterson stood it for a short time and then hastened to Macleod, where he tendered his resignation, frankly stating that he was afraid of the Indians.

There was very little drunkenness among either Indians or whites, and Inspector Jarvis at Macleod seized and destroyed sixty gallons of whisky during the year.

Lieutenant-Colonel Macleod, Commissioner of the Mounted Police, was made stipendiary magistrate, with residence at Macleod, and Lieutenant-Colonel A G. Irvine was appointed Commissioner. On December 5, the barracks on the island in the Old Man's River were destroyed by fire, and it was then decided to remove the post to the higher bank of the river a short distance up the stream, where the town of Macleod now stands.

Preparations on the big leases continued unabated when spring broke in 1881. Frederick Stimson, manager of the Bar U for the Allans, arrived, looked over the range, and decided where to build his ranch buildings; Major Walker, for the Cochrane people, went to Montana to buy cattle, and Stimson arranged to get his through Lynch. A.P. Patrick, the Dominion Government surveyor, who had been in the West on Government work since 1878, went east and purchased two hundred and eight head of good stock in the neighborhood of Galt and Guelph, Ontario. This band was the first of the Mount Royal Ranch herds which ranged on the Ghost River west of Calgary in the edge of the hills. Patrick shipped the stock to Winnipeg on the Canadian Pacific Railway, which had reached that point, and then drove them overland, assisted by Sandy McDonald and Jack Ellis. The trip lasted from April 10 to the early fall.

Captain Winder, NWMP, who had come into Alberta with the first Police force, resigned from that body to take up ranching. His wife and family had arrived in 1876, and he had decided to make his home in the new land. Going to Montreal, he formed the Winder Ranch Company, secured a lease and returned, then purchasing twelve hundred head of mixed cattle from a man named Mitchell who was ranching on Trout Creek. Seventy-five head of Oregon mares were also purchased to form a breeding herd.

The Indians were beginning to feel the results of peace and to long for excitement, the result being that the Mounted Police had considerable worry from them, which, added to the annoyances of white stock thieves, gave the force a busy season. Macleod district was overrun with horse thieves, smugglers, illicit liquor traffic and cattle killings, white men, Indians and half-breeds all doing their part. Jingling Bells and Marrow Bones, two well-known Bloods, stole a large band of horses at Morley and took them as far south as Macleod, where they were caught while passing the town. They offered to put up a fight, but were overpowered before they could act. A number of horses belonging to the Mounted Police force, which had been kept at the Police farm near Pincher Creek, were run off and taken to Montana, where the authorities, acting with the Canadian authorities, succeeded in returning the stock, though the thieves escaped. *Mounted police*

As a result of these troubles, the Mounted Police sent recommendations to *urged* the Ottawa Government urging the passing of extradition laws that would enable *Ottawa* them to secure and punish cattle thieves and horse thieves who escaped over the *gov't to* boundary with their booty. *pass extradition laws to secure cattle properties*

"Star Child," the Indian suspected of murdering Constable Graburn in 1879, *and* was arrested on the Blood reserve by Corporal Patterson and two constables, accom- *will encourage* panied by a guide and the interpreter Jerry Potts. The Indian was tried at Macleod, *cattle* but evidence proved insufficient, and he was acquitted, the jury disagreeing. *business*

The Blackfeet almost broke out into open war over a beef head, which resulted in the arrest of an Indian named Bull Elk. A white man named Daly who was killing beef on the reserve saw Bull Elk walking away with the head in question, and made him bring it back. Bull Elk was sulky, and not many moments had passed until he took a shot at Daly, narrowly missing him. Mounted Police officers were sent out by Superintendent Crozier and hurried to the scene, where they proceeded to arrest the red man, but the Blackfeet clustered around in threatening hundreds, pressing close with scowling faces and weapons in their hands. The policemen took charge of Bull Elk, the Blackfoot young men immediately took him away, and the Police, seeing that a fight must inevitably follow if they insisted on making the arrest then, and knowing they would be able to arrest the man in the near future, allowed him to remain in the hands of his friends. A day or two later, Bull Elk came to Macleod voluntarily, acting on the sane advice of his chiefs. He was tried and given a mild sentence, and the episode passed off.

The buffalo, that great food supply for the Indians, were now gone, the prairies of Canada having seen only a solitary bull or a lone cow during the preceding year or two. Range cattle formed a fair substitute, and the young men of the tribes carried off some successful killings that went unpunished simply because it was impossible to locate the guilty parties. The Police and whites now had a good grip on the administration of laws in the country, and the Indians would never again be able to do any bluffing, try though they might. The times were no longer like those of the starving days of '79, and the Indians realized it. Consequently, when they did break laws and go on raids, they took pains to do it when there was every chance of concealment.

Across the boundary, the slaughter of the penned-up herds of buffalo still continued, and the white and red hunters made huge kills. One hundred thousand buffalo robes were sold out of the Yellowstone valley from the winter hunt of 1880–81, but never again would there be such a shipment, for the herds of millions had now dwindled to a few thousands, and the day of the buffalo was gone.

Sitting Bull, the great Sioux chieftain who came to Canada as a refugee shortly after the Custer massacre, decided to go back to his own country in this year, and, taking his tribe with him, crossed the boundary and gave himself up to the United States forces at Fort Beaufort, Montana.

The Marquis of Lorne, the first Governor-General to visit Alberta, came through the country during the summer under escort of the Mounted Police. He made an extensive and comprehensive journey, visiting all important points. Superintendent Wm. H. Herchmer, NWMP, was the Mounted Police officer in charge of the tour, and John Longmore acted as guide, assisted by Poundmaker, the Cree chief who later made a reputation in the rebellion. The Governor-General spent two months in the Territories, leaving the end of steel at Portage la Prairie on August 8 and returning in October. September 12, 13 and 14 were spent at Calgary, and the party reached Macleod on September 19, where Superintendent Herchmer handed over charge of the escort to Superintendent Crozier, who conducted the return.

CHAPTER VIII

1881–1883

THE Cochrane Ranch Company bought about six or seven thousand head of cattle in Montana, under the arrangement that the herds were to be delivered at the boundary to I.G. Baker Company's men, as that company had contracted to deliver them at the ranch west of Calgary. The stock cost sixteen dollars a head in Montana, delivered at the boundary, and the Baker people charged two dollars and fifty cents per head for delivery. Howell Harris was in charge of the herd to the line, and Frank Strong, foreman for the I.G. Baker Company, was in charge of the drive from the boundary north, with thirty cowboys under him and three hundred horses. In those days the wages of cowboys ranged from forty to sixty dollars a month, while foremen drew not less than a hundred and not more than a hundred and twenty. Strong divided the herd into two divisions, in order to expedite travel and make as rapid a trip as possible. The steers were placed in the first batch, and George Houk was put in charge of this "dry" herd, as it was called. Cows and calves followed under a foreman named Jack Allen, now of Macleod, Alberta, and Strong took charge over these men. The stock was shoved along at a merciless rate, the steers in the dry herd averaging fifteen to eighteen miles every day, while the cows often did fourteen. This drive has remained the criterion for hard driving, as no such great numbers of cattle have ever since been moved so rapidly by trail. The poor animals were "tin-canned" and "slickered" from morning until night, kept on the move from daylight until dark, and were usually so weary when darkness came that they preferred resting to eating. Then, too, they were herded so closely that there was little chance of a square meal even if they did have the energy to look for it.

Straining along behind the "dry" herd came the "drag"—the calves and the cows—who performed marches as remarkable as the steers, according to their relative strength. A number of wagons tailed along behind this herd to pick up straggling calves, but these were not sufficient to hold all that lagged. The little calves, too young and weak and hungry to stay with the grown stock, dropped out, were piled in the wagons, traded, or left to perish, hundreds dying in this latter way. The cowboys in charge of the drag gave and traded scores of calves for a pound of butter, a drink of milk, a cup of tea, or more unworthy exchanges. Whisky traders met with along the line of the drive accepted the calves as legal tender and reaped considerable profit thereby, the loss to the Cochrane Ranch running up into large figures.

The worn and weary stock was jammed across the Bow River near where the present Mewata Park is in Calgary, and were turned over to Major Walker and his men, who were waiting there. The counting took place near the present CPR depot, and then Walker sent the animals west to get located on the new range. Several of the cowboys who participated in the drive remained to work for the Cochrane Ranch. Winter came on before the stock could recuperate from the gruelling days on the trail, and hundreds died before they had located shelter and water, and when spring broke there were many heaps of bodies in the coulees. One thing that this drive impressed upon the cowmen who accompanied the herd was the relative powers of endurance of the different breeds. The black Polled Angus beasts stood up best of any, the Herefords rated second, and the Shorthorns suffered most of all. When winter came down on the range, there were six thousand head of fine cattle on the range, but by the middle of the next summer this had been reduced materially by weather, misfortune and, sad to relate, white men, half-breeds and Indians.

The Montana range stock that was purchased to place on the great Alberta ranches was as goodly a lot of grade animals as has since been seen on the ranges in the Province. They were big, fine animals, well finished, well developed, and carrying weight easily. From various reasons that will be explained later, the range stock of Alberta deteriorated for some years, and has not even yet recovered its lost ground entirely, though there are individual herds that would average better than the general run of those thousands from the Montana ranges in the early eighties. To keep up and improve on this quality, it was deemed necessary to import some good bulls, so the Cochrane people sent to England, and, with the Allans of the Bar U, purchased one hundred and thirty-six Hereford and Polled Angus bulls, which they shipped at once, en route for the West. Twenty-one of these bulls belonged to the Bar U, the remainder being Cochrane stock.

Owing to the fact that the cattle purchased in Montana had been sent so hurriedly through to Alberta, it was impossible to brand them rightly. Major Walker and his superiors were satisfied when a "hair brand" had been put on each animal, as they intended to brand properly when they arrived. A "hair brand" is one that is simply scraped in the hair with a knife, or with acid, and disappears when the new coat comes. Finding it impossible to do the branding in the fall following their arrival, it was decided to make the big branding in the spring. Here it was that dishonest men obtained many good animals, for a brand was the only indication of ownership, the cattle wandered wide in the strange land, and there was no sign of the hair mark when spring trimmed off the old coat. Another thing also caused considerable loss, an occurrence that the settlers were not really ashamed of.

In the spring of 1882, Major Walker set out with his men to round up the scattered stock and brand every hoof on the range, following the explicit directions he had received from Dr. McEachren. In this he was assisted by the settlers in the district, who had little to do for themselves, and therefore turned out to help for the pastime. Some of these men owned cattle, too, but not in nearly such great numbers as the Cochrane Ranch did. One rancher who had a fair-sized band owned a cow and calf that he had made family pets of, and they stayed around his place all winter. A Cochrane foreman known as "Tex" had called at this ranch house and had often seen the animals, so when the owner saw this pair swept into the Cochrane corrals and roped for branding, he raised considerable objection, carrying his plaint to Major Walker in person, and insisting on receiving his property unmarred by the big "C." "Tex" was called into the conference also, and he upheld the contention of the small rancher, but Major Walker had received his orders, and he intended to carry them through with military precision. "Brand everything unbranded on the range," had been the order, and he would carry it out.

The settlers, considerably worried by this state of affairs, quit helping the Cochrane hands, and before night told Major Walker that their own private interests demanded their immediate and personal attention at home. Then, mounting their horses, they scattered to their several ranges. Many a secret coulee, many a distant ravine and sidehill, still held Cochrane cattle, and these settlers knew the country much better than the Cochrane men did. Straight to these isolated bands they urged their horses, and, in self-defence, as they declare, began to reap a harvest of calves and unbranded cows and steers that should have rightly received the big "C." Having started, they kept on gleaning, until a very respectable number of the big company's animals were diverted to individual ranches, without the exchange of a dollar. The settlers say that if they had not done this, they might have been wiped out financially, for all their stock, branded and unbranded, was on the Cochrane range.

The cattle business, once started, went with a swing. When the Cochrane herds first came, the valley of High River was dotted with the herds of Emerson,

Lynch, Quirk and French, while the valleys of the Belly, the Old Man's and the Bow had scattered bands. A year later there were many thousands more in the Porcupine Hills, and herds continued to grow in the other districts, big companies locating every season for some years.

Major Walker, who had been more or less hampered throughout his administration of Cochrane affairs by Dr. McEachren, his superior manager, went again to Montana when summer came to buy more cattle. He arranged to purchase four or five thousand head from Poindexter & Orr, large ranchers in that state, and, when about ready to close the deal, was called three hundred miles to Benton in response to a wire from Dr. McEachren, who informed him in the telegram that arrangements had been completed between Dr. McEachren and I.G. Baker Company to have the latter firm handle the purchasing of the stock. This company had been considering the advisability of stocking an Alberta ranch for themselves, and Dr. McEachren thought the purchase of more cattle for the Cochrane Ranch could be done more profitably if they were purchased in conjunction with the I.G. Baker Company. Arriving at Fort Benton, Major Walker learned they had decided not to buy, so he hastened back to Poindexter & Orr's, only to learn that cattle had risen and that the herd he wanted would cost twenty-five thousand dollars more than it would have cost if he had closed the deal when he started for Benton. This caused Major Walker to send in his resignation, to take effect as soon as a successor could be appointed.

The fall of 1882 and the winter of 1882–83 was a most disastrous one to the Cochrane Ranch Company. Other ranchers in the district around Calgary suffered very little, only the great herd experiencing extreme loss. The Poindexter & Orr herds were driven hard and reached Fish Creek in a weary, exhausted condition, arriving at the creek in September, just as a bitter snowstorm swept down from the north directly in their faces, burying the trails, drifting in great banks, and blocking the weak animals as though by a mighty barrier. Poindexter, a cowman of long experience, suggested that he leave the animals there for a month, saying he would leave his cowboys that length of time in order to herd the stock. Major Walker did not think this necessary, as his orders were to get the stock on the range as soon as possible, so Poindexter decided to go ahead.

"Your contract is to deliver them at the Big Hill," said Major Walker, referring to the ranch headquarters, which were known by that name until the town of Cochrane replaced it. Poindexter, declaring he would surely deliver them, sent men ahead, picked up some scores of lusty native steers, and jammed them south through the snowdrifts, turning them back again at Fish Creek and throwing the weak, trail-worn herds into the path they had made. Driving fiercely, they forced the herds to the Big Hill, and then Poindexter, riding to the head, said to Major Walker: "Here they are. I have carried out my contract and delivered at the Big Hill. Count 'em now, because half of them will be dead tomorrow!" By this time there was no chance of feed on the Cochrane range, the snowstorm lasting until October 15. Following this came a thaw, which simply softened the snow, and then the bitter weather came down, forming a crust on the snowbanks that was strong enough to hold up a team. The cattle, with their hoofs worn to the quick, moved only with difficulty, finding the sidehills almost impossible.

Frank White, a railroad man with but little experience among stock, was appointed to fill the position that Major Walker had vacated. W.D. Kerfoot, a Virginian with experience on Montana ranges, came out that winter and acted as practical adviser to White. Though by the time that White took charge, the company directors in the East knew the condition of the range, yet White received telegraphic orders to hold the cattle there. There was fine grazing no further distant than Blackfoot Crossing and the Little Bow, and the starving thousands of cattle drifted persistently in those directions, seeming to know that their salvation was to reach those points. The work of keeping the stock on the Cochrane range was a gigantic one, and White found it necessary to establish camps of cowboys at Calgary, at the mouth of Fish Creek, and at the mouth of High River, their only duties being to turn back the Cochrane stock. The cowboys worked all winter long, night and day, the numbers of suffering animals dwindling visibly. Every morning long strings of bawling cows and steers could be seen by the people of Calgary walking downstream along the tops of the riverbanks; every evening the people of Calgary saw the Cochrane cowboys shoving the reluctant beasts back on the iron-bound range.

When the spring of 1883 came, the snow still stayed, not being entirely gone until June, and then it was seen how terrible the loss had been. Dead bodies were heaped in every coulee, thousands of head having perished. Some of the long ravines were so filled with carcasses that a man could go from the top to the bottom, throughout its entire length, and never have to step off a dead body. Indians made very good wages for some time, skinning the animals for twenty-five cents each. Out of twelve thousand head of stock that had been purchased and placed on the Cochrane range, there remained now but a scant four thousand, counting natural increase.

In the meantime, other ranches had established and had experienced but little of the losses the Cochrane people had suffered. F.S. Stimson, manager for the Bar U, sent Tom Lynch of High River down to Lost River, Idaho, to buy herds to stock the range. Then he sent the twenty-one head of thoroughbred bulls being held at Benton on to the ranch in charge of Herbert Miller. Jim Minesinger, Miller, Phil Weinard, and one other man were sent to work at the ranch, putting up a house, and everything was prepared for the arrival of the herds, though there was little grub, and Minesinger called wrath down upon his indifferent Western head when he killed one of the thoroughbred yearlings for food.

Lynch bought three thousand head at Lost River, and, leaving there in May, arrived at the ranch west of High River early in September. Shortly after their arrival, the great September snowstorm that met the Poindexter & Orr herds at Fish Creek swept down through the Porcupine Hills and scattered the Bar U herds. The snowstorm lasted eight days. After the storm, the Bar U men went out to sweep in the drifted stock, finding some had gone as far south as Macleod. Bringing them back into the hills, they turned them out to rustle, leaving only two cowboys, Deever and Moodie, to keep them from drifting too far south. These two cowboys rode the Old Man's River all winter, turning back all Bar U stock that came down that far.

While at Lost River, Tom Lynch hired a Negro as a cowhand. The black man was a great, powerful fellow, and, though he had neither horse nor saddle, claimed to know something of cattle. Being short of hands, Lynch took him on and placed him on night-herd in company with a cowboy named Preston, the latter taking the Negro for a green hand and naturally letting him do all the herding. Next day they gave him an old, battered saddle and a very quiet horse. Looking at his outfit he remarked:

"Ah say, boss, ef yuh'll jest gimme uh little bettah saddle en uh little wuss hoss, Ah think mebbe Ah kin ride him."

This was sufficient for the cowboys to commence hazing their new brother-in-arms. They speedily rustled a good saddle, and just as rapidly produced the very worst horse that was in the whole outfit, and there were some bad ones, too. The black man, John Ware, later known as "Nigger John," and one of the most popular men in the cattle country, and a champion rough-rider, swung up to the top of the wicked mustang and rode him from ears to tail. The cowboys, recognizing a master hand, took John among themselves, and before they had gone very far toward Alberta, he had shown he knew cattle so well that he was placed in charge of one of the herds.

Shortly after he had been promoted, he had a slight argument with a couple of Montana cattlemen, which showed that the big Negro had the right kind of courage. Sometimes when large herds were being driven through a district, it was possible for the local stockmen to garner a few head by putting up the bluff that they were cutting their own stock out of the strange herd. Two Montana

John Ware with his boarhound, Bismark, in 1891. By this time, Ware had a ranch of his own with 400 head of good stock.

men tried this, going straight to John and telling him he was taking off some of their stock. They made their complaint in the morning, just as John's men had started the cattle on the trail, the aim of the Montana men being to have John tell them to cut out their stock and go, instead of stopping the herd and sorting out the mythical animals himself. Should he stop the herd, the fat was in the fire, anyway. Upon hearing the tale of woe that was expounded by the strangers, John was sorry, but said he could not stop his herd. If the men would follow until he bedded down for the night, he would gladly help them pick out their own property. This they refused to do, and, expressing great indignation, proceeded to drag out their "artillery" to hold the herd by force. But before they could bring their guns to bear, old John was among them, using the heavy end of his quirt so effectually that both the disturbers fell out of their saddles, and were left lying senseless by the trail as the herd moved slowly on. The two never showed up again to claim "their stock."

On March 25 John R. Craig succeeded in interesting sufficient English capital to start the Oxley Ranch, and came West to get it in running order. Mr. Craig, who remained in charge of this ranch for three years, was acknowledged to be one of the best judges of cattle who ever came into the ranch business. He took no cattle to his range during this year, though he purchased one hundred and sixteen head of British Columbia horses, paying seventy dollars each.

Another ranch that came in about this time and located near Calgary, on the Elbow, was the Chipman Ranch, which was later bought by R.G. Robinson, who holds it yet.

Two Alberta tragedies marred the year, one a drowning and the other a double murder, both taking place in Montana. McKenzie, the man who had the dispute with Mr. Dewdney in 1879, went to Montana to invest the money he received from the Government in a band of cattle. He was on his way back, driving his new herd, when he came to Hell Gate River, which he found he must swim. The water was thick with the washings from hydraulic mining that was being carried on upstream, and the mud and silt gathered so thickly on the swimming man's clothing that it dragged him down, and he drowned before help could be given him.

The other occurrence was the murder of Mrs. Armstrong and her foreman, Morgan. These were the parties who were ranching near Macleod in 1879 and who left when the disgruntled ranchers pulled out to the Marias. When the Canadians came back to Alberta, Mrs. Armstrong remained in Montana, living with two young girls whom she had adopted. One day these girls appeared at the settlement near where the Armstrongs lived and told how both Morgan and the woman had been murdered the night before, the girls escaping before they were seen. The town was full of men, and a roundup of every one in the place was conducted by the authorities, who placed the girls where they could see every one, and then marched every one past that point. Upon seeing a man who had recently been employed at the Armstrong ranch as a hired man, the girls declared he was the guilty party, so without loss of time he was summarily strung up.

In certain districts in the ranch country there was considerable loss from anthrax or blackleg, numbers of young stock perishing from this disease. Some small herds were reduced fully fifty percent.

Alexander Begg, who had been located near the mouth of High River for some time, and who was doing some Government work of a kind, reported to

Ottawa that twelve hundred settlers came into the territory during the year, the majority settling north of Calgary, although a goodly number located in the south and the Bow River Valley. Among those who came to the Macleod district was a young Englishman named Walter Huckvale, who became one of the most prominent ranchmen in the Province.

Malcolm McInnes arrived in Alberta about this time from British Columbia, where he had been engaged in ranching. He settled on Fish Creek next to John Glynn's farm and put a band of horses on the range there, later purchasing a herd of cattle.

Before the close of 1883, the ranch herds of Alberta numbered fully twenty-five thousand head between the Bow River and the boundary. Settlement received a great impetus, and large numbers of people arrived. Progress in the towns was marked, and by the close of the year there were three newspapers being printed within the boundaries of the country that five years before had been the unsurveyed kingdom of the Indian tribes. The *Edmonton Bulletin* was established some time before this year, the *Macleod Gazette* had been started in 1881, and the *Calgary Herald* commenced publication in 1883.

Railroad construction on the Canadian Pacific main line was pushed at a remarkable rate. In January the track-laying gangs were twelve miles east of Maple Creek, and when winter stopped the work on November 28, the steel had been spiked down as far west as the heart of the Rockies, the official figures showing steel to be one and one-eighth miles from the summit when work ceased. Passenger and freight trains crawled into Medicine Hat, Calgary, and as far west as construction had reached, homesteaders and merchants unloaded their wares at their various destinations, and the prairie country was open at last. From the east there was direct railway communication, though Edmonton must yet be reached by a long trail haul, and the ranch country south was supplied with no transportation excepting bull-trains and horses. Benton as a source of supply was practically wiped out, though some stuff was hauled from that point for a year or so longer. But the railroad had made Calgary the distributing and shipping point, had broadened the markets of the ranchers from local demand to British Columbia and the East. The industry was on the high road to success, and the pioneer ranchers began to plan for big results very shortly.

Major Walker, upon leaving the Cochrane Ranch Company, purchased the company's sawmill and timber limits and started operations in that business, supplying much of the timbering for the CPR bridges. Frank White resigned the control of the ranch shortly after the spring and early summer had shown the disastrous results of following orders from the East, and the ranch herds were under charge of W.D. Kerfoot and a Mexican half-breed named Ca Sous, the latter a splendid stock-hand. Discouraged by the results of the two winters in the Calgary district which, by the way, seemed to strike only the Cochrane herds, the directors of the ranch, far from giving up, located and leased a new range near the Waterton Lakes, southwest of Macleod. They secured an immense tract under lease, and a home ranch with two hundred acres of timber, and four hundred acres of hay land, two hundred acres of which was naturally irrigated. Early in the summer they commenced sending the stock south, Ca Sous, with seven cowboys, Fish, Raikes, Anders, Smith, Ables, Lawrence, and Barkley, taking the cows down first and remaining to put up hay, putting up five hundred tons before fall. By the time all the stock had been moved, it was learned that four thousand and a few score more were all that remained, including increase, of the twelve thousand head purchased during the preceding two years. Jim Dunlap, foreman after Ca Sous left, accompanied the drives of the two bands, and W.F. Cochrane, later on the manager of the ranch, went down in the capacity of bookkeeper. Mr. Kerfoot remained on the Cochrane headquarters west of Calgary.

The Walrond Ranch, known better as the "Waldron" or the "WR," in recognition of their brand, was formed by Dr. McEachren and financed by British capital, Sir John Walrond being the president. Their range, on the north fork of the Old Man's River between the Porcupine Hills and the Livingstone range, was stocked this year with a large herd, James Patterson, now Dominion Livestock Inspector at Winnipeg, being employed as foreman under the local manager, G.W. Fields. Patterson was an experienced cowhand who had worked on the Montana ranges. In addition to beef, they intended forming a large herd of grade Clydesdales. George Lane, now owner of immense ranch interests in Alberta and breeder of the finest Percheron stock in America, came over from Montana this year and looked over the country as far as Kipp, with an idea of locating in Canada if it pleased him. Returning to Montana for the winter, he came up again the next spring, F.S. Stimson of the Bar U having secured his services as foreman. The Allans and Stimson seem to have started their big ranch with a better judgment than some of the other ranches of the time were begun. They sought the best stockmen they could get, and George Lane's employment by them was the result of a request made to the Montana cattle association for the best stockmen they had. Jim Patterson was employed by the Waldron people following a similar request by the officials of that concern.

W.R. Hull, now of Calgary, made his first trip into Alberta, he and his brother bringing twelve hundred horses through the Crow's Nest Pass and selling them to the Mounted Police and the North-West Cattle Company (the Bar U). The

Hulls had been ranching in British Columbia for about ten years; at least, they had located there in 1873 and been interested in livestock since that time. They started their Kamloops ranch in 1880. *Law and order being established against the stealing of cattle*

The Pincher Creek settlers experienced their first troubles with Indian cattle thieves that fall, or rather early winter. A cowboy riding in the dry forks of the Kootenay found some Indians butchering a beef, and hastened to Pincher Creek, where the settlers, under John Herron, an ex-Mounted Policeman, went on the trail of the red men, rounded them up, and brought them in, fourteen in all. They were Stonies who had drifted south while hunting. The entire party was taken to Macleod and tried, some of them receiving two-year sentences.

Ranches continued to spring up in the Calgary district and around Macleod. Squatters and lessees were quarrelling, the holders of the leases objecting very much to those homesteaders who settled on the leases, claiming that the settlers were there simply to take advantage of free use of thoroughbred bulls and to lift a few calves whenever chance offered. Sproule and Walsh had a ranch near Calgary; A.P. Patrick and a man named Baynes had gone into partnership in the Ghost River country; the Military Colonization Company, under the management of General Strang, had leased and stocked a large ranch east of-Calgary in the valley of the Bow; and J.D. Geddes had also come in with a goodly band of livestock.

The Cochrane Ranch people, wishing to utilize their lease and property west of Calgary, were making arrangements to have the land stocked with sheep, which they intended placing under the care of W.D. Kerfoot. Just to show that they were keeping up with the times, the settlers in the Sheep Creek country had the name of their post office changed to "Okotoks," which is the Indian word for "Rocky."

Joe Healy, "Clinker" Scott, and "Old Man" McLaughlin contrived to start a mining excitement just west of Banff, at a point known as Silver City. Healy, who always carried many splendid specimens of ore, which he had secured in the mineral belt in Montana, held claims at Silver City and was not at all particular, it is said, in making sure that the specimens he showed interested parties were from Montana or Castle Mountain, which was at Silver City. Many old-time settlers still hold, uncherished and unsaleable, those shares they purchased during that mining fever that raged in Calgary and the small towns of the West during the next year or two. Silver City sprang that year from nothing at all to a town of two hundred and thirty-seven men, seven married women, five girls and two boys. There were one hundred and seventy-five houses in the place. Today all that remains is a Canadian Pacific Railway siding, a flag station, and one shack in which old Joe Smith, ex-mayor and early pioneer of Silver City, still lives and hopes, between rheumatic twinges, for the roaring days of mining-camp life and the demonstration of the riches of the mountain that he loves.

The townspeople of Calgary were presented with an exciting and unprecedented sight during the summer. Whisky business was booming, "under wraps," of course, for the product was not legally recognizable and all drinking must be done silently and behind closed doors. The Mounted Police, enforcers of the law, were ever on the watch, and woe to the unwary whisky merchant who was caught selling his merchandise, or who was even discovered with liquor in his possession without a permit. Permits were given, "for family use," five gallons per year being the amount allowed to each householder. The Calgary Mounted Police force, try though they might, found great difficulty in securing evidence against any of the men whom they were morally certain were breaking the liquor laws. The best-laid plans of both officers and men seemed to go always awry, and nothing resulted from various deeply and darkly laid schemes. Finally, and very reluctantly, the men of the force became convinced that some member or members of their own body divulged the raiding plans before they were carried out. This conviction led to suspicion of two men, and simmered down at last to Constable Taylor of E Troop. One night a band of sturdy redcoats swept down on Taylor when he was placidly sitting in a building in the town, tied him firmly, and then mirthlessly and grimly ducked him in the cold, swift waters of the Bow until the dampness reached his very soul and chilled it as cold as his feet. Large numbers of civilians stood on the bank and watched this affair with deep interest, the whisky people declaring to nearby friends, who were not policemen, that it was a shame to abuse a man thus. This soaking of the suspected "leak" appeared to be the correct remedy, for no more information of police doings and intentions seeped to the outside world.

The Sarcee Indians were placed on their reserve eight miles from Calgary and maintained pretty fair order among themselves, with the exception of one instance when a very sturdy brave named Crow Collar broke into the ration house and took supplies. Mounted Police went to the reserve and demanded the surrender of Crow Collar, who was sulkily refusing to be arrested. Bull's Head, chief of the tribe, supported his tribesman's position, and the police gave up temporarily, to prevent bloodshed. Two days later Bull's Head and Crow Collar, having thought better of their hasty decision, came voluntarily to the barracks and gave themselves up, spending two days in the cells as a slight indication of what might come to them if such actions were persisted in.

In the spring of 1883, John R. Craig placed the first herds on the Oxley Ranch when he brought 2,500 head of picked range stock from Montana. The Stewart

Ranch at Pincher Creek, of which Jim Christie was a partner, was breeding fine horses, many of whom made very satisfactory Police mounts; the new barracks and post built at Macleod were completed, and the Hudson's Bay Company opened at that point with E.F. Gigot, now of the Hudson's Bay Company at Nelson, in charge.

During the summer and autumn, the Macleod district was infested with horse and cattle thieves who made some successful raids, though often prompt pursuit by the police led the culprits to desert their stolen property on the trails. Several times the members of the force took the liberty of escorting men with bad reputations back to the boundary line and telling them to stay at home, else prison would result. American Indians made their annual raids into Canada and were quite successful in securing a number of horses, while the Canadian Indians were not entirely disappointed at the results of their own forays south of the Medicine Line. Three parties of Crees stole a number of horses from the I.G. Baker Company's Montana ranch, but the NWMP got after them so promptly that they failed dismally in retaining the stock. Twelve hours after the theft was reported, Sergeant Paterson of the force had secured all the animals but three, and had arrested eleven Crees, who were each eventually sentenced to two years' hard labor. One war party of Crees raided widely in the States and returned flushed with spoils, but were forced to disgorge, chiefs Piapot and Little Pine turning "state's evidence" and giving the police a leverage with which to force the return of the stolen property. The Bloods gained a number of fine horses by travelling through the Indian country of the south, and were able to keep possession of most of them.

In addition to watching the red men, the police did some good work among the whites, if the records count for anything. J.D. Murray was fined one hundred dollars at Macleod for buying potatoes from an Indian; Alexander Doyle was charged with embezzling oats, and jumped his bail of four hundred dollars as soon as he was turned out. John McDonald was given four years for horse stealing, and F. Watson and Charlie Cameron were given six months each. A half-breed named Patrick received two years at Macleod on the charge of killing cattle. Although the Police had been unable to fix the murder of Constable Graburn on Star Child, they succeeded in catching him stealing horses this year, and he received four years, his companion Man-With-Knife being let down with six months. A number of Indians who were suspected of misdemeanors were let off and dismissed on account of insufficient evidence. Gamblers had increased throughout the ranch country, and a considerable number of convictions took place in Macleod and at Calgary. At the latter point, during the year the Police arrested and fined the following gamblers: Gus L'Hirondelle, Wm. Fisk, Wm. Houston, W.G. Smith, Joe L'Hirondelle, Wm. Calder, Gus Erniston, E. McGrath, J. Raspriem. Henry Wheeler and Hiram Bosenthal were each fined for selling liquor, and Jake Fortier and J.G. McKlintock were fined for setting out a prairie fire. Charles Anderson and Sam Brady also paid two hundred dollars each for selling liquor.

The history of the Macleod district would never be complete without a mention of Dave Cochrane, a man who settled down to the life of a farmer and rancher as soon as he left the Mounted Police force. Cochrane was no relation to those of the big ranch, but he was a man who succeeded in keeping his name before the public, whether belonging to a wealthy family or not. Dave had traits—idiosyncrasies, one might charitably call them—which tended to make him at least an interesting character. He first came before the notice of the public when he squatted on the Piegan reserve and refused to move until the arbitration board, which Indian Commissioner Dewdney had appointed to settle the matter, had awarded him thirty-two hundred dollars and a free homestead. Dave then cast about for a homestead, found one, sold it for a few hundred dollars, and moved up into the Waldron lease, where he squatted. We will leave him there for the time being, but will hear of him again.

CHAPTER IX

1884–1885

By the beginning of 1884, the success of the cattle industry in the Fort Macleod district, and generally along the base of the Rockies as far north as Calgary, was assured. Forty-one companies and individuals, holding under lease from the Government a total of 1,782,690 acres, were then engaged in the business, and many thousand head of cattle, horses and sheep were on these leases. In addition to the above-mentioned acreage under lease, on which cattle had been placed, there were eight hundred and seventy-five thousand acres leased on which no stock had been placed. These latter leases must be closed out within three years if no cattle were placed on them in the interim.

The North-West Cattle Association, composed of the ranchers of the south, had been formed with headquarters at Macleod, and, in accordance with strong representations made by this body to the Government at Ottawa, restricted districts were laid out for sheep. The Government also passed new lease regulations, which were briefly as follows: No lease could exceed one hundred thousand acres of land, and could not exceed twenty-one years. In case a lease was unsurveyed when taken up, the lessee must pay for the surveying. Rental rates were ten dollars for every thousand acres per year, and the lessee must before three years have one head of horse or beef stock for every ten acres of the lease. After placing the prescribed number of animals on the lease, the lessee had the right to purchase a home ranch within the limits of the lease for two dollars per acre. If two people applied for the same lease, it was to go by tender.

No sheep were allowed to range from the boundary line on the south to the summit of the Rockies on the west, from High River on the north and the north fork of the Bow, east along the north fork of the Bow to the eastern boundary of the Province.

A.M. Burgess, Deputy Minister of the Interior, after a tour of the West, states that mixed farmers would in his opinion help the big ranches to a marked degree. He instances the winters when the Cochrane stock died in thousands around

Calgary, stating that if there had been a hundred small farmers in the vicinity of the Cochrane range at that time, there would not have been nearly so great a loss, because the ranch company could have purchased small quantities of hay from each of these hundred, and could have brought most of the stock through in good shape. Mr. Burgess also expresses an opinion, which, though fully believed in at that time among most of the big ranchers, has since proven to be a wrong policy. He says: "It would not be profitable for stockmen, in view of the infrequency of severe winters, to make specific provisions against them every year, but in order to escape occasional disaster they must either do this or encourage the settlement of agriculturists in their districts."

(The mistake in Mr. Burgess' remarks is made in that portion which declares that it is not profitable for the ranchers to prepare hay every year; his statement about the desirability of farmers in the districts is good, and is a condition of affairs which the large ranchers are today endeavoring to encourage. Many a worthless stack of straw could be turned into revenue if the farmer would take a hundred or two hundred head of range stock from some nearby rancher and feed them through the winter months.)

Continuing, Mr. Burgess says: "Antagonism between ranchers and settlers is purely theoretical, such antagonism being forced by small, independent speculators forcing themselves upon the leases of the ranchers and entering into competition with them, and then demanding restitution for being forced to move, or refusing to pay Government rent."

The following companies and individuals had cattle on leases by the end of 1884:

NAME	NO. OF ACRES
Mount Head Ranch Company	44,000
NW Cattle Co.	59,000
Ryan & Whitney	3,000
Alex. Begg	1,440
W. Mitchell	42,000
F.W. DeWinton	15,000
Stewart Ranch Co.	23,000
G.R. Davies	10,000
Rocky Mountain Cattle Co.	73,000
Anglo-Can. Ranch Co.	64,000
Jones, Inderwick & McCaul	100,000
O.S. Main	22,000
Military Colonization Co.	92,000

G.F. Wachter	7,000
Eastern Townships Ranch Co.	33,000
F.S. Stimson	55,000
Moore & Martin	33,000
C. Martin	66,000
Halifax Ranch Co.	100,000
Wm. Steed	10,000
Cochrane Ranch Co.	100,000
Cochrane Ranch Co.	34,000
J.M. Browning	55,000
E.A. Baynes	12,000
Alex. Staveley Hill	80,000
Winder Ranch Co.	50,000
Bell Bros.	5,000
Ives & Clarke	5,000
Brunskill & Geddes	13,000
Bell & Patterson	6,000
M. Gallagher	6,500
E. Maunsell	6,500
Sir John Walrond	100,000
Oxley Ranch Co.	100,000
Viscount Boyle	5,000
Walrond Ranch Co.	100,000
W.S. Lee	25,000
Muirhead Ranch Co.	22,000
Garnett Brothers	20,000
F.W. Godsall	20,000
D.E. Akers	5,000
W.F.N. Scobie	12,000
T.B.H. Cochrane	55,000
G.R. Davies	47,000
J. McFarlane	13,000
A.L. Staunton	8,000
Alberta Ranch Co.	27,750

The ranchers' markets were the Indians, the Mounted Police, and the CPR construction camps. The principal ranches selling were the Stewart Ranch Company, the Bar U and the Oxley. Several carloads of fat beef were shipped to

Winnipeg as an experiment, and the result was fairly satisfactory, despite the fact that the railway was not in very good shape for handling them speedily.

In March, A.P. Patrick of Ghost River killed a record two-year-old steer that had rustled all winter and been killed right off the range. This animal dressed six hundred pounds of prime beef.

High waters in the rivers in June and July caused the usual troubles, on account of lack of bridges. Cattle were swept away and drowned, men risked their lives in the hazardous fords, and settlers on low bottoms were flooded out. A Blackfoot Indian who had come into Calgary to loaf and absorb as much liquor as possible was lounging at the rear of the store on the banks of the Elbow when Roget Rocelle fell in the stream and was being swept away to the hungry waters of the nearby Bow, but the Indian threw off his blanket and plunged in. Stemming the strong current with powerful strokes, he seized the exhausted Rocelle and dragged him ashore, where he resumed his blanket and his loafing position. White men who saw the Indian's action decided something was coming to him, and he was greatly astonished later in the day when he received a present of ten dollars and a wagon-load of provisions and blankets.

A.C. Sparrow, J.D. Lauder, and the Mount Head Ranch Company are mentioned as starting ranching in this year, though Lauder had been in the country for some years, coming with the Mounted Police before 1880. These men all ranched in the High River and Calgary districts. At Macleod, Walter Huckvale branched out into ranching, purchasing about two hundred head from O.S. Main and ranging them on the Belly River a short distance below Kipp. W.R. Hull, whose venture in horses during the preceding year had been very satisfactory, came to the Province to stay, locating a ranch some two years later. The Military Colonization Company on the Bow River east of Calgary was, under the management of General Strang, making the ranch a sort of educational point for young English gentlemen who came from England with the idea of learning to ranch. The newspapers of that day mention that "Mr. Pruen, late of the South African Police, Mr. Goldfinch and Mr. Newbolt, the latter a son of General Newbolt, are working on the Military Colonization Company's ranch." That fall the ranch pasturage was entirely destroyed by a prairie fire that swept over the entire lease and made it necessary to remove the seven hundred cattle and three hundred horses, which the company had, to unburned districts. No explanation is given for the setting out of these fires.

The first thoroughbred Percherons that came into the Province were brought in during this year by the Chipman Ranch people, who bred splendid stock, though horses continued to be very low in price. The Cochrane Ranch Company, operating under the name of the British American Horse Ranch Company, continued to hopefully develop their holdings west of Calgary. Senator Cochrane and his associates sent A.E. Cross to Cochrane in the spring to act as bookkeeper and veterinary surgeon under W.D. Kerfoot, the manager of the new ranch company.

Kerfoot, with eight thousand head of fat-grade sheep from Montana, crossed the Bow River at Calgary in a blinding September snowstorm and turned them out on the range. Two hundred Shropshire Down rams, pedigreed and imported, were taken to the ranch, and then the second venture was watched carefully and with interest, both by those directly interested and the settlers in general. Alexander Begg, at the mouth of the High River, had been successfully breeding and ranging sheep in small numbers, but this was the first great drove to be turned loose in the district, and the outcome was problematical.

In the fall of 1883, D.E. Gilman and Company sent two hundred and fifty head of horses from the Hay Stack Valley, Grant County, Oregon, to the Province of Alberta via the Crow's Nest Pass, in charge of a young horse wrangler named Frank Ricks, the drive lasting from May to September. Ricks was a top rider, one of the best rough-riders in the entire West, and probably the equal of Johnny Franklin of Macleod in his prime. At any rate, both Ricks and Franklin are two of the few bronco-riders who continued to ride for many years and never seemed to be any the worse for it. Ricks had ridden bad horses from California to Oregon before he came to Alberta, his bucking-horse honors beginning to come to him in 1878 when he rode at the state fair at Sacramento, California, winning first prize, a hundred-and-fifty-dollar saddle and bridle, and a hundred dollars in gold. Ricks remained in Alberta, securing work on the Mount Royal Ranch and later going into ranching for himself, growing in time to be one of the prosperous horse ranchers in the rich country west of Calgary.

The Cochrane district seems to have been ever able to furnish excitement. The lambs and ewes had not finished investigating their new range when an outlaw horse was brought in that was as bad as the worst animal that had ever come into the Province, and that is saying a good deal. He was a fine, active, long-barrelled dark chestnut, a wild horse and an outlaw of some ten years' experience. Not only did he buck in a thousand different and original twists, but he was a man-killer—a savage, untamed brute. He had been backed a few times, had broken up a few men, and had been ridden twice before, once by a man named Fish, who dismounted as soon as the animal quit after the first spasm. These two failures to dislodge riders simply had added to the bad horse's meanness. Up to this time "Nigger John" Ware held the palm for rough-riding, though he had never been at the ranch to ride this horse. The Cochrane manager, W.D. Kerfoot, kept the brute in an open corral after he had discovered his characteristics, and one of the wonders of the men employed around the place was why such an animal should have ever been bought.

One September day Frank Ricks came along on foot and asked for the loan of a horse. "Take the chestnut," offered Kerfoot, and Ricks promptly roped and saddled. Then he rode as the Cochrane hands present had never seen before. Ricks was a superb horseman, riding on weight and balance, not by main strength like John Ware and other hard horsemen. He rode that ugly chestnut until it could scarcely stand, he cut it from tail stump to ears with his spurs, he temporarily beat the spirit out of it with his heavy quirt. When the session had been completed to Ricks' satisfaction, he left a wreck in the place of the thousand pounds of fighting horseflesh he had mounted. Perhaps this was a gentle hint to Kerfoot not to try those jokes on every footman who strolled along. A.E. Cross of Calgary was at that time the veterinary surgeon and bookkeeper at the ranch, and his duty for some time after this was to treat the battered outlaw, whose spirit, even after the terrible man-handling he had received, was not broken. In fact, he seemed more bitter than ever towards humanity. To attend his hurts, Cross would step into the corral, retreating then before the biting, snapping, striking whirlwind into the squeeze-gate, and, when followed there, shutting that great clamp on the maddened horse and then putting soothing lotions and bandages upon the wounds that Ricks had left.

In the Macleod district the cattle continued to come in, chiefly from Montana. The Mounted Police who were stationed at the boundary to clear incoming herds reported that the Oxley Ranch brought two thousand and eighty head of particularly fine cattle. John R. Craig, in his book *Ranching with Lords and Commons*, says they brought thirty-one hundred head across, quite a discrepancy, unless he counted calves, and the Mounted Police officer tallied only the grown stuff. I.G. Baker Company sent sixty-nine head through, O.S. Main three hundred head, and the Military Colonization Company three hundred and thirty head. George W. Houk located on a ranch on the Belly River, fifteen miles from Lethbridge.

John R. Craig, rounding up his stock in the hills, nearly lost a band of thirty head by a rather clever ruse of some horse rustlers. A boy came one night to their camp and asked for work, which was given him with the horse wranglers. A day or two later a couple of strangers, who were ostensibly passing through the country, camped near the roundup outfit, apparently knowing none of Craig's men. The next morning the lad and the strangers were gone, the only reason to regret their departure being the absence of thirty head of horses. Ranchers and the Mounted

Police searched south toward the boundary, figuring that the thieves would strike straight for the safety of that line, but no success met their efforts. Craig thought then that they might have cached themselves in the hills and, acting on this, he caused a search to be made through the foothills. Sure enough, they found them, horses and men, and the thieves were arrested and tried, the two stranger men getting off on the grounds that the boy told them the horses were his, and the lad receiving a sentence of six months at hard labor.

A number of immigration delegates from the British Isles were taken through Alberta during the summer as guests of the Government. These gentlemen were very favorably impressed with everything they saw, and predicted a wonderful future for the Province. But one of them, evidently a tender-hearted Irishman, for his name was Peter O'Leary, reported himself as being very greatly shocked at Macleod by the way stockmen handled horses. He urged that something be done to prevent cowboys using whips, quirts, ropes and spurs.

Nothing has been done as yet.

In the spring, when John Herron of Pincher Creek was engaged in a widely advertised wrestling match with a man who was many pounds his superior in weight and considerably ahead of him in height, the settlers and ranchers first met George Lane, now a cattle king of the Province. The entire settlement had turned out to witness the match, and considerable money had been put up, the bigger man being favorite. The gladiators were stripped and ready to step into the ring when a tall, lank young fellow sprang onto the platform and waved a roll of money which he longed to place on the "little fellow" (Herron). This was George Lane, whose judgment has since proven almost proverbial.

That fall a second record-breaking steer was marketed in Calgary. This was a three-year-old from the range of the Mount Head Ranch near High River. The steer, known by his admirers as "Captain Higgins," dressed sixteen hundred and eight pounds, off the grass. Two two-year-olds from the same range dressed over a thousand pounds each. The biggest single sale of beef on the Calgary market was two hundred and fifty thousand pounds, which Captain Stewart sold to Angus Sparrow and to Samples & Company, beef contractors for CPR construction.

Two thousand people came to Alberta this year, many of them taking up mixed farming. The Government and the Canadian Pacific Railway Company were doing all in their power to encourage this class of immigration, knowing well that eventually it would form the backbone of the Province. The Government farm on Sheep Creek, which was in charge of John Lineham, raised the biggest crop in the Calgary district, three thousand seven hundred and seventy-one bushels of oats off a hundred and forty acres, and one hundred and sixty bushels of wheat off seven acres.

Land was broken on the CPR experimental farm on October 22, 1883, and the crop in 1884 on this new breaking was an average of twenty-eight and a half bushels of wheat and fifty-six and a half bushels of oats per acre.

Four murders marred the peace of the south, only one being committed by a white man. A war party of Indians picked up some horses at the forks of the Red Deer and were taking them away when a half-breed galloped after them, endeavoring to regain the animals. Killing him with a rifle bullet, they escaped, though the Mounted Police found the horses in the possession of a war party of Bloods, who were stopped as they neared Fort Macleod. These Indians declared vigorously that they knew nothing of any killing, though they added that a war party of South Piegans from Montana were in the neighborhood of the forks when they were there. Shortly after this, the Mounted Police convinced themselves that this was true, as they received word from Montana that a Piegan named "Big Mouth" had done the killing. No action was taken to bring him to justice, probably because it would have been impossible to convict him.

Early in the year a brutal murder took place at Calgary when a Negro named Jesse Williams shot and killed a man named Adams. Williams was traced by his hobnailed boots, and confessed shortly after he was placed in prison. He was hanged at Calgary on March 29.

William Reid, alias Buckskin Shorty, was stabbed and killed near Calgary in a drunken brawl with a man named John McManus, who was tried and sentenced to six months imprisonment for manslaughter.

Montana Indians, raiding in the south end of the Province, shot and killed a settler named Pullock, escaping scot-free, the authorities being unable to fix the blame on any particular tribe or party.

There were a number of minor offences, chiefly on the part of whisky dealers, gamblers, etc. B.S. Jones gave liquor to Indians near the reserve just out of Macleod, and then finished up by shooting and wounding one. Mr. Jones was fined three hundred dollars for supplying the firewater, and then received a twelve-month sentence for being so light with his trigger finger. Liquor violators who fell in the toils at Macleod this year were handled strictly, because the demand for whisky had encouraged the taking of big chances. There was big money in it, a five-gallon keg easily bringing sixty dollars wholesale at Stand-Off, where a thriving trade continued to flourish in spite of the most violent efforts by the Police. The softening influences of settlement and law had swept over Whoop-Up and Slide-Out, had put French out of business, had taught Kanouse and other traders that times had changed, but Stand-Off was the hardy perennial, and was worse; it was a perennial thorn to the Police. Six men, some of them still living in the Macleod district, paid two hundred dollars each for the privilege of watching the Mounted Police constables destroy the liquor they had striven so vigorously and secretly to bring to the thirsty of the district.

Law was asserting itself, and cases multiplied. The first case in the history of the south of the Province in which an employer was sued for wages took place at Calgary, where Tom McHugh was the defendant. In Montana the ranchers were also getting busy, seeing quite plainly that the time of order had arrived. The Montana Stock Association, having been annoyed for some little time by an especially well-organized and grasping band of rustlers, succeeded in rounding up the whole band at Rocky Point, and there, after a very brief trial indeed, unofficially hanged every man.

Railroad development in the south was commenced, the Crow's Nest Pass Railway starting construction between Medicine Hat and Lethbridge, opening up the way to the great beds of coal that were at Lethbridge, and also opening up to the settlers a splendid ranch and farming country.

The winter of 1884–5 was the best, from a stockman's point of view, that had ever been experienced. Throughout the entire ranching country, the cattle and even sheep rustled on the frozen grasses, all cattle going through without a spear of hay being given them, and sheep rustling until nearly February. At no time was there more than six inches of snow on the ground, during the year there was no disease among the stock, and the spring loss of calves did not average two percent of the crop. By the middle of February the snow had vanished, and the streams and springs in the south had re-opened. The sheep on the British American Ranch did splendidly, little feed being given them, while the majority of ranchers, though they had put up some hay, never fed a forkful. The weather up until the time the thaws came in February was a steady, invigorating cold, staying in the vicinity of zero and a degree or two below, and some remarkable drives of beef cattle were made during the cold months.

In January two hundred head of cattle from the Stewart and the Halifax ranches left Pincher Creek for Calgary, having been sold to Angus Sparrow. Driven carefully, as beef cattle must be, they found their own feed along the trail as they came, making the trip of one hundred and ninety miles by trail in twenty days, and arriving in good killing shape. The price paid for the three-year-olds in this bunch was sixty-five dollars per head, and the average weight of the dressed carcasses was nine hundred pounds.

In February a herd of one hundred and fifty beef cattle from O.S. Main's range at the mouth of the Little Bow were trailed into Calgary, via High River, making the ninety miles between the latter town and the ranch in seven days and the remaining forty miles in four. These are astonishing examples of the conditions of the weather and the abounding nutriment of the prairie grass, which cures like hay and retains its strength after the frost has come in the fall and stopped growth.

The Stony Indians, west of Calgary, suffered severely during the winter owing to the Government giving them more credit than was really coming to them. These Indians, an intelligent body who took to white men's ways as soon as they learned how, had, under the leadership of Rev. John McDougall, progressed considerably as agriculturists. They worked hard on small plots of broken ground, planted crops, and did their best to make a success of the new order of living. Such promising reports of their advancement were sent to Ottawa that the Government in 1884 decided they must have progressed enough to be self-supporting, and consequently it would not be necessary to pay treaty. The result of this action was that the winter of 1885 saw droves of starving Stonies wandering through the hills, from below Pincher Creek to the main line of the CPR, begging food from the ranchers, picking the flesh from the bones of dead horses and cattle, even eating coyotes when they could get them. Sometimes parties numbering as high as a score would be fed at one time at some of the larger ranches.

Difficulties between squatters and ranchers grew more frequent and bitter, the squatters, though doubtless in the right in many instances, suffering in contrast with the ranchers because men frequently squatted on leased lands for the simple and only reason of being bought off by the lessees. Naturally, whether for blackmail or for farming purposes, all settlers picked the finest locations, springs and river flats. Between Fort Kipp and Slide-Out on the Belly River, there were seventeen settlers and twenty-five miles of continuous fence shutting the range cattle away from the water and the shelter of the valley. Wm. Pearce, Superintendent of Mines for the Government, investigated these conditions, and suggested that the differences in this particular instance could be settled if the squatters would leave their fences open in the fall so that the range stock could get down to the sheltered bottoms. But the settlers did not approve of this generally, claiming the range herds would take their own few head away. Hon. Thomas White, Minister of Interior in this year, put new regulations in effect regarding ranches and leases. These regulations were tending toward encouragement of homesteaders and small farmers, and were to the effect that homesteaders could take up homesteads on any leases granted after that day, and could hold possession. He further ordered that no more ranching leases be issued for as long a period as twenty-one years (the original term), and further that the department was to cancel the old twenty-one year leases whenever opportunity offered. Prior to this, the Government leases secured by Senator Cochrane, Dr. McEachren and others included the clause that homesteaders could not settle on these leases without permission of the lessees, though this clause was seldom enforced, the lessees usually buying the settlers off whenever they located within their boundaries and proved undesirable, either on account of personality or because they had selected a particularly choice plot of land.

For some years, ever since the buffalo had left, the timber wolves in the foothills proved to be a great nuisance to the ranching business. One full-grown wolf, in good health and strength, was considered to be usually responsible for a thousand dollars loss each year to the ranchers in the particular district in which he operated. They were cunning, powerful animals whose forte was to strike terror into young stock and then drag it down, usually by hamstringing it. Colts were the easiest and fell frequent victims, owing to the habit of the mare to run away from the danger, taking her colt with her. This open running was what the wolves wanted, and they soon pulled the little animal down. Cattle were more difficult, for when the presence of wolves became known, the old cows would bunch up with their calves and present a very dangerous front of flashing horns and ready feet. In such a case the wolves would worry the band until some young and unsophisticated yearling would break loose and dart across the prairies in a panic; then the chase, and finally a cheerfully bloody slaughter of the poor brute.

A bitch wolf with a litter of cubs that were just budding into maturity was the greatest scourge, for she killed more frequently in order to teach her cubs how to do it when they would be cast upon their own resources. Thousands of dollars worth of stock was killed every year by these beasts, and the ranchers were at a loss to know how to dispose of or shut off this channel of depreciation. This year they petitioned the North-West Council to pass legislation granting a liberal bounty for wolf scalps, but the council turned them down, and they were forced to stub along as in the past, growing desperate every now and then, and paying out of their own pockets for the destruction of some particularly heavy killer.

The unfortunate Cochrane people found the "hoodoo" west of Calgary had not entirely passed them up yet, for with spring the herds of sheep west of Cochrane suffered a tremendous loss through fire and flood. A prairie fire, starting near the mouth of Beaupry Creek, swept up that valley and destroyed a thousand head, though this was not the extent of the loss, for on April 3 another large batch died. A large corral had been erected near the head ranch at Cochrane, and a Montana sheepherder named Mike Maloney was in charge of a herd that he corralled every night in this enclosure. The night of April 2 was very stormy, a snowstorm coming after dark and drifting heavily before a high wind. The corral walls formed a sort of break and swirl, dragging the driving snow to a resting place in the inside. By morning the drifts had reached the top of the corral and the sheep had stepped over the fence, wandering before the hurricane until they came to a deep slough or small lake three miles away, where they perished in hundreds by drowning, three hundred dead bodies being dragged out after the struggling flock was finally driven away. Another heavy loss was experienced later when hundreds and hundreds of ewes died that spring while lambing.

The greatest roundup that ever took place in the history of Alberta was held early in the summer, the men meeting at Macleod on May 25, the roundup consisting of one hundred men, fifteen mess-wagons and five hundred saddle horses. Jim Dunlap, foreman of the Cochrane Ranch at the Kootenay Lakes, was elected captain, and they swept the country, starting from Rocky Coulee and going to the mouth of the Little Bow, thence to Mosquito Creek, and then to Eighteen-Mile Lake, throwing the cattle on Willow Creek at Cut Bank. There the roundup split into two, one under Dunlap going up Pincher Creek, and the other under George Lane combing Willow Creek. John Quirk of High River caused a tremendous sensation among the gathered ranchers and cowboys when he and his hired man arrived attired in a most economic class of nether-clothing, which they had built out of seamless sacks.

The roundup collected great herds, aggregating about sixty thousand head, and after it was over, it was decided that the cattle business had reached such proportions that general roundups were too cumbersome. The men had covered all the country from the Little Bow to High River, the country between the Belly and the Old Man's rivers, and between the Belly and the Kootenay. It was too large to be done satisfactorily with one general roundup, and it was decided to round up in districts in the future, with each ranch sending representatives to each district roundup. A few cattle having drifted over the boundary, it was found necessary to send representatives to the Montana roundups also, though the number of Canadian cattle found at these up to this time did not make it profitable to drive them back, and they were sold to the Montana ranchers on the ground.

W.R. Stewart, a well-known character in the Macleod district who in later years ranched west of the town of Claresholm, was a vigorous youth in those days, and, feeling the spirit of adventure calling resistlessly, arranged with a man named Jake Crownover to travel into the Peace River country. They went, and all was smooth sailing until they arrived at the Yellowhead Pass, where the charms of a little squaw, with the bloom of red blood appearing through the brown of her cheeks, proved too much for Crownover, and he decided to settle down right there and then. After an energetic quarrel, Stewart and he parted, Stewart starting back home by the British Columbia route. He lost his pack-horses in the quick-sands of one of the northern B.C. rivers, built a raft of logs and embarked on the North Thompson, drifting along quite speedily until the raft and his provisions were all destroyed in Hell's Gate Rapids, Stewart escaping by a miracle, but losing all his belongings in the river. Without food and unarmed, he commenced his long journey south, eating raw fool hens until he found an Indian cache and enjoyed a hearty meal again. Shortly after finding this cache, he came across a band of horses, and, catching one, rode down the riverbanks until he arrived at Kamloops, bearded, dirty, and most disreputable-looking, his clothing hanging on him by mere shreds. In the meantime, the Indians had discovered the loss of their horse, and, following Stewart to the town, had him arrested for horse steal-ing shortly after his arrival. Things promised to go hard with the young fellow, for the authorities were inclined to look upon his story as a fabrication, but J.D. McGregor happened along just then, recognized Stewart, squared up his difficul-ties, and enabled him to get back to the south.

The financial backers of the Oxley Ranch were Alexander Staveley Hill and Lord Lathom, Lord High Chamberlain to Queen Victoria. During the summer, these gentlemen came to look over the new venture in which they had placed so large an amount of money, and John R. Craig, their manager, naturally treated them as considerately as possible. Staveley Hill, having been in the West before, knew something of conditions and people, but Lord Lathom was new, coming right from the dignity of the royal court. One day the two visitors, with Mr. Craig and E.H. Maunsell, went to Macleod, and of course remained in town to lunch. "Kamoose" Taylor, a man who had been through a varied experience from preacher and whisky trader to "squaw man," was proprietor of the eating-house at Macleod, and was greatly impressed with the honor of feeding such distinguished guests. Feeling that his big-thumbed waiters would not be able to "do the honors" as satisfactorily as he could, he took upon himself the duty of serving the party. Throwing a dirty towel over his arm, he approached, saying to Mr. Hill:

"Soup?"

Mr. Hill would take soup, and so it seemed would Mr. Craig and Mr. Maunsell. Taylor then came to Lord Lathom, flourished the towel gracefully, and again queried:

"Soup?"

Being British to the core, Lord Lathom wanted to know what was going to be the result before he committed himself, so, turning in his naturally dignified way, he asked:

"What kind of soup?"

Taylor, essentially Western, and perhaps a bit flurried, too, replied with strong emphasis: "Damn *good* soup, yer lordship!" while Maunsell and Craig choked, Mr. Hill smiled behind his beard, and Lord Lathom looked at the boniface in deepest astonishment.

Slavery had been dead in the United States for nearly twenty years, but Alberta holds a record of having come near to this obsolete condition. In September a mulatto named Harrison was ordered to leave Calgary, which order he followed so strictly that he did not stop until Lethbridge was reached. At that thriving village, he succeeded in running a debt amounting to seventy dollars to a barber named Charles Bryer, who decided to get his money back. Harrison had nothing but his body to realize any capital on, so Bryer put him up at auction, knocking him down to a restaurant keeper for a hundred dollars, and the mulatto then set to work and earned his freedom.

"Buck" Smith, now of Elk, B.C., who had been whisky trader, prospector, and everything else in the early days, opened the first stopping-house at High River, giving a "house-warming" in January at which one hundred and fifty cowboys and ranchers attended, with feminine companions grading up from half-breeds and squaws. J. Barton's celebrated bronco-buster, a cowboy named "John" (probably John Ware), astonished the assemblage by displaying such pronounced knowledge of parlor tricks that he was made the official "caller-off."

John R. Craig severed his connection with the Oxley ranch this fall, shortly after he had made a purchase of some "TV" stock that were unusually poor stuff. Mr. Craig was a splendid judge of cattle, and how his judgment slipped in this purchase is not known, though it is said by old settlers that the buying of this herd was one of the paramount reasons for his quitting the big ranch. The "TV" stuff was a herd of scrubby, scrawny Tennessee cattle that had been brought into Montana for sale and that no Montana rancher would have. E. Stanley Pinhorn, an experienced Australian stockman, was made manager in Craig's place, and he at once secured the services of Jim Patterson as foreman, to take the place of Johnson, Mr. Craig's foreman. A real stock war almost sprang up over the change in control of this ranch. John R. Craig, just before his retirement, sold a large band to Angus Sparrow of Calgary, and Sparrow did not go to get them until after Pinhorn had taken the reins of government. Pinhorn refused to let the stock go at the price that Craig and Sparrow had agreed upon, and excitement rose high, the High River stockmen naturally standing back of Sparrow, their neighbor, while the Mosquito Creek and Willow Creek cattlemen were behind Pinhorn to a man. Matters came to such a stage that Sparrow went with men to take the stock willy-nilly, and Pinhorn rode into the herd with his guns loose and threatened to shoot the first man who took a head. Things were certainly ripe for war when George Emerson, the father of all the stockmen, poured oil on the troubled waters and maintained peace. Pinhorn kept the cattle. He was a popular man among his ranch neighbors, and until his death, a few years later, divided leadership of the Willow and Mosquito creeks ranchers with a man named Charlie Sharples, then manager of the Winder Ranch, and previously one of the first Cochrane Ranch cowboys.

The Riel Rebellion, which occurred during the spring and early summer of this year, caused a ripple of excitement among the settlers of southern Alberta, though the seat of the trouble was so far removed that no actual inconveniences, so common in wartime, were experienced, excepting the recruiting of various forces, the shifting of squads of Mounted Police, and certain restlessness among the younger Indians. Poundmaker, a leading Cree chieftain, did his level best to get the Bloods, Blackfeet and Piegans to rise with the half-breeds and Crees of Saskatchewan, Manitoba, and north Alberta, but wise old Crowfoot could see a bit in advance of his time, and he refused to sanction any rising. At first he had considerable difficulty in restraining his young men, being forced at one time to send trusted runners to the settlements along the Bow, west of the reserve, telling them to be prepared if his young men arose. Crowfoot's decision was followed by the Blood and Piegan chiefs, who never really had much use for Crees and Cree half-breeds anyway.

One cause of the rebellion was over the "river lots" that the half-breeds had settled on. These lots, instead of being square, as quarter-sections are, were about ten acres wide and ran back from the rivers fully two miles, the reason for this being that the half-breeds wanted to live in communities where their houses would be close together. They claimed this right, the right of squatters before survey, and they had in many cases settled and started to cultivate when the Government surveyors came along. These men had been sent out to survey in correct sections, perfect squares, and when they came to the half-breed river lots, the surveyor in charge insisted on surveying nice, perfect squares for the squatters. The half-breeds resented this vigorously, and finally grew so desperate that they called Louis Riel back from Montana, where he had gone after his earlier troubles with the Canadian Government. Riel was the half-breeds' king, their leader, and when he said "war" was necessary, they followed him.

After the rebellion had been smothered and Riel was hanged, the Government appointed a commission to investigate these river-lot claims of the half-breeds, the result being that they retained the narrow strips they had originally demanded.

Beef prices mounted remarkably high during the rebellion, there being a big necessity for quick delivery, and the herds of the West being the only supply. Butchered animals, cut into four equal quarters, brought a price of fourteen cents per pound, whether bull or cow, stag or prime steer. Meat was meat, and the ranchers who supplied the meat were not at all particular in picking out their finest stock for the soldiers to eat. Later on, the Government cut prices to thirteen and to twelve cents, which made some ranchers quit taking such contracts and look for other markets. The railway construction gangs of 1884 were supplied from Western ranges, also, and they, too, paid fourteen cents per pound, taking the run of the herds.

The Duck Lake massacre took place on March 25, and the first thrill of possible Indian outbursts aroused the settlers of the south. Batoche and Cut Knife followed, and again the fear was felt. Before these latter engagements, a field force had been raised in Alberta, under command of General Strang, composed of settlers and Mounted Police. Major Walker was appointed his second in command, and when Strang and his troops were ready to march to Edmonton and then down into the war zone, it was decided that Major Walker should remain at Calgary in charge of the home guard, as a precaution against the rising of the Blackfeet and other tribes. F.S. Stimson of the Bar U formed a Company of rangers who, known as "Stimson's Rangers," patrolled the Porcupine Hills; the Mounted Police force was culled down to a few men on detachment, and as many as possible went against Riel and his savage allies. Inspectors Perry and R.B. Deane were ordered on active service, and the Police who were left in the south had to do almost impossible feats to patrol the wide country. Rev. John McDougall of Morley volunteered as a scout and did valuable work under Inspector Perry.

Though they did not take up the cudgels with the rebels, the young men of the Blackfoot and other southern tribes were not above taking advantage of conditions to the extent of stealing as many horses as they could and bullying whites when they had a safe chance. United States Indians, Gros Ventres or Assiniboines, shot at a Mounted Police scout thirty miles west of Medicine Hat; railway construction gangs lived in terror of the wandering bands of Indians who, travelling in large war parties, stole horses boldly and rode into the States in search of more. More than once the presence of a lone Mounted Policeman kept the railway navvies from stampeding in frenzy and almost forcing the overbearing young braves to commence killing. The police did such eminently satisfactory work in maintaining order that in the heat of the rebellion, the Southwest Stock Association expressed official and high appreciation of the work of the NWMP, and especially of that portion of the force under Major Cotton at Macleod. This

Seen here is sheep rancher Frank White's Merino Ranch in 1891. White had been the manager of the Cochrane Ranch during the disastrously hard winter of 1883.

action was taken at the regular meeting of the Association, being moved by F.C. Inderwick, seconded by John Gamett, and unanimously approved by the ranchers.

Business did not slacken up to any extent, herds continuing to move in to stock the ranges and move out to supply the ever-broadening market. The Glengarry, or "44," Ranch Company was formed by A.B. Macdonald and a good-sized lease secured west of Nanton; the Bow Park Ranch, A.T. Cable, manager, located between the mouth of Pine Creek and High River; B.M. Godsal was now ranching in the Calgary district; the Little Bow Ranch Company had located at Mosquito Creek and the Little Bow; W. Skrine and W. Iken had started in the High River district; Somerset and Picard were located up the Elbow. Frank White, former manager of the Cochrane Ranch, now resident at Ghost River, bought twenty-five hundred head of Montana sheep and put them in his Merino Ranch, west of Cochrane, near where Mitford now stands.

The stockmen north of Mosquito and Willow creeks met at Skrine's ranch on September 7 and formed the North-West Stock Association, electing Alexander Staveley Hill president, F.S. Stimson vice-president, and G. Levinge, of the Mount Head Ranch, secretary. This association was composed of the Oxley, Bar U,

Sheep Creek, Mount Head, Little Bow Cattle Company, Military Colonization Company, Winder, Skrine, Emerson, Iken and Quirk ranches.

Some difficulty was experienced by fall fires, Godsal's range being swept clean, his cattle being saved with great difficulty, and the entire Blackfoot reserve burning, with the exception of one bottom. When winter came, the Indians were in a serious condition owing to this lack of range, and were very bitter against the railway for setting the fires, which started from locomotive sparks.

In November, J.D. McGregor purchased three carloads of horses from the Little Bow Ranch and shipped them to his farm at Brandon, Manitoba, and Gilman brought 175 Oregon horses into the Province, which he sold to the Military Colonization Ranch. At the boundary, the Mounted Police inspected 1,109 head of splendid stock that W.O. Main was bringing in, 104 head for Frank Strong, 309 head for I.B. Corcoran, 241 head for J.B. Pearce, 118 head for I.G. Baker, and 100 head for E.H. Maunsell.

Considerable bother had been caused the Mounted Police by Indian horse thieves and white whisky smugglers, the Indians, especially the Bloods, doing wholesale execution among the Montana herds. Mounted Police constables patrolling the boundary kept down depredations from white American thieves, refusing entry to a considerable number of prospective raiders, but they could not handle the Indians, Piegans and Bloods making many trips across the line and almost always securing a lot of whisky and a considerable number of horses. A whisky trader who settled just across the line did a lot to damage the moral fibre of Canadian Indians until the United States authorities, acting on a request from the NWMP, removed him from his advantageous position. Several successful Indian raids were made among the herds even of the Mounted Police, and the settlers in the eastern portion of the Province were so pestered by the red rustlers that they took to sleeping in their corrals with loaded guns as company.

Iron Shirt, an Indian who was arrested for some misdemeanor, was so impressed with his one night in the restriction of a cell that he made a futile attempt to end his life, trying to hang himself with a moccasin string, but being too heavy for the job. A Montana whisky trader named Al Lester was caught at Macleod selling liquor to Bloods, and was sentenced to six months in jail. His wife, who accompanied him, seems to have considered this to be practically a divorce, for she eloped at once with a party named Shanks, and returned to the United States.

There were two murders, the murderers escaping in each case. An Indian named Window Bellows brutally killed his squaw one cold November day and succeeded in reaching the Medicine Line and safety. Bob Casey was shot and killed at Medicine Hat on July 9 by a man named Ben Hale. They had been drinking and discussing a horse race, the argument going so strongly against Hale that he resorted to his gun and shot Casey dead. Hale escaped across the Montana line, and the Mounted Police communicated at once with Sheriff Healy, the same man who held command at Whoop-Up before the police came, and who was so busy at Silver City during the mining fever. Mr. Healy informed the Canadian authorities that he knew where Hale was, and could get him in a short time—if there was a reward. Because of the lack of this inducement, Hale escaped scot-free.

The death of Captain Winder, manager and founder of the Winder Ranch, occurred during the fall, and one of the early builders of law and order in the Province thus passed. As a Police officer, he had established an enviable record, and as a ranch manager he was proving to be one of the most progressive, his horses especially, bred from good Oregon stock, grading by good breeding into choice remounts.

CHAPTER X

1886–1887

GREAT strides were being made in ranching, and there were over a hundred thousand head of horned stock on the ranges, the ranchers found their markets broadening and their stock demanding fairer prices. A new difficulty was arising, due to the great bands on the best ranges, the grass being damaged by fires and tramped out by the herds that were beginning to crowd in the choice districts. Troubles between lessees and squatters remained an open sore, the holders of the Cochrane lease west of Calgary being especially annoyed, as their limits came down to within seven or eight miles of the town. Homesteaders demanded rights, and stayed stubbornly on the lands they had taken; the lessees tried to make them move and failed. In some cases, though, the squatters, evidently willing to make ready cash, allowed themselves to be bought out. Thirteen settlers went in on Skrine's lease on High River, but Skrine experienced no difficulty, stating that he and the settlers were proving mutually helpful.

William Pearce, after a personal visit to the ranches, reported to the Government that there were 104,000 head of cattle, 11,000 horses and 24,000 sheep on the ranches of the Province south of and including the Bow Valley; J.S. Dennis reported 25,000 sheep in Calgary district alone. During 1886, 34,000 cattle, 3,500 horses and 7,000 sheep came in, of these 26,000 cattle, 2,000 horses and 6,500 sheep coming from the States, and the remainder from British Columbia. Included in these figures Mr. Pearce counts 11,500 head of horses and cattle belonging to seventy-one non-leaseholders, half of whom never homesteaded, and he adds that be did not include a large amount of stock that was owned by homesteaders.

From the time he started ranching on Mosquito Creek in this year, A.E. Cross lost through wolves every year ten to twenty-five colts out of a band of horses ranging in numbers from two hundred to two hundred and fifty. This is said to be a good average of the loss from this source throughout that district. Dr. McEachren, whose interests were there and a little south, recommended to the Government the establishing of a bounty on wolves, but again the cry of the stockmen in this particular was ignored.

A very good winter was experienced, the loss in the calf-crop throughout the districts not exceeding two and a half percent. Mange, which had appeared in several localities during the preceding year or two, was reported by Dr. McEachren, chief Dominion veterinary surgeon, to be entirely eradicated. He laid its introduction among the herds of the Province to the Blackfeet, who, he declared, brought it from Montana ten years before.

The Government enacted quarantine regulations against American cattle and ordered the Mounted Police to enforce them. The *Macleod Gazette* had no hesitation in declaring that this was a move to protect the ranches held by the big companies, now that they were well stocked. It would keep back American ranchers who would settle in the Province with big herds, and it would better enable the big ranches to handle the markets and let their cattle range wide.

Other action was also taken by the Government, including the raising of the rent on new leases to twenty dollars a thousand acres per year, and the placing of a twenty percent duty on American cattle and other livestock from that country. Prior to this date, leaseholders could bring stock in duty free so long as none were sold within three years. This action of the Government brought an immediate rise in the price of beef.

Two officers of the Imperial service, Colonels Ravenhill and Phillip, visited Alberta during the summer, looking over the prospect for army remounts, and expressed themselves as exceedingly pleased with the outlook for the future, being particularly impressed with some splendid mares of Irish hunter strain that had been placed on the Quorn Ranch range for the particular purpose of breeding remounts.

The open range was still open, with perhaps the exception of homesteaders' fences and the fencing along the Belly between Stand-Off and Fort Kipp. The Cochrane Ranch had endeavored to introduce an innovation the year before by putting up a splendid wire fence, twenty-five miles long, running from the Old Man's River to the Porcupine Hills, and being the start of a complete fence around their gigantic leasehold. One necessity of this was a marked deterioration in the grade of steers, due directly to careless breeding, which the Cochrane Ranch, as yet untouched, wished to guard against. In every instance, the large ranches had headed their herds with splendid imported bulls, and then had turned all out on the open range, where every breed mixed democratically. Not only was this true, but the range was becoming populated with a great many bulls of the scrubbiest scrub stock, homesteaders and small ranchers being satisfied with any kind of a

Members of a roundup crew on the Lucas, Eastman and Wallace Ranch in the Porcupine Hills in 1886 take a rest before heading out on the range again.

beast to head their herds, figuring that bone and weight was fully as good as blood and conformation. Being a sort of pioneer, this Cochrane fence experienced a hard though brief existence, lasting about as long as a snowbank would remain in the streets of Honolulu on a midsummer day. Whenever a cowboy had a sharp instrument and was within reaching distance of that fence, he took a crack at it, and before long it had been so seriously dismembered that its usefulness was gone forever. The fenceposts made splendid fuel in the timberless land.

The Cochrane herds, having formed a sort of habit of running up against hard luck, experienced their usual share that winter. Having passed through two seasons of bitter experience and loss in the Calgary district, they rested during the first winter they were on the Kootenay range, and now were ripe for more hard times. Escaping from the Calgary district when they did, they failed to experience the grand winter seasons that followed their exit, and now, in this winter of 1885–86, the weather gods smote them again. The north and south ranges were in fine condition, with good pasturage and good shelter, little snow and mild days, but in the district around the Waterton Lakes, the snow clouds piled and broke, heaping the ground with six-foot drifts and absolutely burying all feed from the most experienced rustler among the herds. Herded closer when the snowstorm struck the district, the cattle milled and yarded on the edge of the lakes, browsing the willows to stubs, bellowing in their hunger. A loss that would practically wipe out the entire herds was more than threatened; it was almost an assured fact.

Hearing of the straits the ranch people were in, Frank Strong of Macleod offered to save the cattle for a thousand dollars. His bid was accepted readily enough, and Strong rounded up a band of about five hundred cayuses, hired a few experienced hands, and shot the herd into the hills. The Piegan reserve was an open and splendid pasturage—as was nearly any other section in the district, with the exception of the Kootenay Lakes, where the cattle were locked in. Strong's band of ponies reached the hills and floundered ahead for two days, urged by whip and spur and vociferation, cutting a wide swath in the crusted snow, which lay three feet deep on the level and much deeper in the drifts among the hills. Reaching the lakes and the imprisoned herds, Strong turned his band of equine rescuers homeward, and they went back on the gallop, making the return trip to the Piegan flats in eight hours. Right behind them the hungry steers travelled on a high trot, stepping far and fast toward the open prairies and food, and before twelve hours had gone the thousands of starving Cochrane cattle were eagerly filling their thin stomachs with the cured upland grasses of the reserve.

The High River and Mosquito Creek roundup worked their district in June, and two wagons, the Bar U and another supplied by a pool of small ranchers, comprised the outfit. Tom Lynch drove the Bar U wagon, with "Nigger" Green as cook, and the pool wagon was driven by Sam Howe. The latter wagon was sent down in the new country in the forks of the Bow and Belly, which was an illustration of how far the cattle had strayed from their base, owing to increase in herds and overstocking of some of the choice ranges, this year being the first time that a roundup outfit had found it necessary to go into the country so far south and east. Consequently it was new to nearly every man on the roundup, and before they returned, they became very thoroughly lost.

As usual on a roundup, there were a large number of raw horses in the remuda, and these animals were ready to stampede at any time. One night, when well down between the rivers, Jimmy Johnson, the horse wrangler, tied his horse to the wagon wheel while he ate supper. The horse became frightened, tore himself free, and dashed away, stampeding the entire band. A.E. Cross was a young man then, in good condition for prolonged exercise, so he and another started out to turn the running animals. They ran ten miles before they could get around the herd, and they steered them gently back to camp, where Fred Ings proceeded to rope one, making a botch of it and starting the animals again. By that time nightfall had come, and the horseless roundup outfit retired to their blankets to await morning and a long walk. But, during the night, two half-breeds, Hunter Powell, later a Bar U foreman, and Henry Minesinger, went out and turned the herd, walking them placidly back to camp, where they were waiting for the cowboys when they awoke.

The next day they went deeper into the country toward the forks, and became absolutely lost. Sam Howe, driving the wagon, had taken a lazy man to ride with him, the tired person claiming to be ill. He was not too ill to give advice about the route Sam was taking, for he continually declared the driver was going too far to the right. Sam allowed his better judgment to be overruled to such an extent that he bore more and more to the left until, at nightfall, he found himself at the point he had left in the morning, having completed a circle. Fortunately the cowboys, bending in all the time to keep track of the wagon tracks so as to know about where to go for supper and sleep, had performed a wider circle and rode in about dark. The following day, in a drizzling rain that kept up steadily, they took the right direction and came out on the high banks just where the yellow floodwaters of the Bow join the Belly and form the South Saskatchewan. As the cowboys sat and looked upon the flood, they saw the ferry boat from Lethbridge go whirling past, the high water having broken it from its cables, evidently while it was being used, for a wagon was draped disconsolately over one side and the dead team dangled. No one had been drowned when the ferry went out, though the driver and the ferryman had narrow escapes.

The first big ranch in the Medicine Hat district was started this year by Thomas Tweed, W.P. Finlay, John Ewart and Ezra Pearson under the name of the Medicine Hat Ranching Company. Mr. Pearson had been in the West for some years, and drove the first stagecoach between Medicine Hat and Lethbridge in 1883. He was in charge of the new ranch, and he stocked it that year, bringing one hundred and fifty head of cattle from near Elora and Drayton, Ontario, in the spring, and more in the fall. With the first shipment that came in during April and was placed on the range east of Medicine Hat, Mr. Pearson had brought the Shorthorn bull, Royal Gem, the first registered animal ever taken into the district. The animal was exceedingly handsome, being snow-white and having particularly splendid points. He was sired by the great Young Ablesburn, champion of the American continent. In November a Clyde stallion was also brought from the East and placed on the ranch. The "MHR" was the first and biggest cattle ranch in the Medicine Hat district, a section of the Province that within a few years had as many head of range stock running over it as the most densely stocked ranges further west.

About the time this ranch commenced operations, W.N. Nicol, now a resident of Vancouver, came with six hundred sheep and located nine miles down the river from Medicine Hat town, he being the pioneer sheepman of the district.

In the spring, A.E. Cross left the employ of the British American Horse Ranch Company and commenced ranching for himself, starting out on a fairly generous scale. Securing about fifty thousand acres of leases on the Little Bow and the headwaters of Mosquito Creek, he homesteaded a quarter on the latter stream, four miles from the stopping-house kept by Joe Trollinger and his squaw Lucy, and stocked it with four hundred mixed cattle and a hundred calves. Trollinger about this time leased his farm to the Powder River Ranch Company, an American organization that, feeling rather crowded in the States, was looking for more room. This company, in addition to leasing Trollinger's place, secured a lease of seven townships in the district, and had six thousand head on the land before winter set in.

The Glengarry Ranch, under A.B. Macdonald, secured their leases in March, and by fall had twelve hundred head of splendid breeding stock as a nucleus for the greater herds they intended to run.

The first consignment of stock ever shipped into Alberta by rail from the West was four hundred head brought in by Hull and Trounce, and delivered at Calgary, where they were sold to McDermott and Ross, Harry Raikes, then foreman for the latter outfit, being on hand to brand them. During the season W.R. Hull and his associates brought five hundred horses and three thousand cattle into Alberta, selling two-thirds of them and utilizing the remainder to stock their first ranch, the "25."

W.C. Wells and Nelson Brown purchased the Mount Royal Ranch from A.P. Patrick and added five hundred head to the herd on that range, bringing them from British Columbia. Pettapiece & Potter and Lafferty & Brown brought four thousand sheep from Montana between them, and threw them on their ranges at the head of Nose Creek. The Oxley Ranch alone branded over two thousand calves at the fall roundup. When the year ended, the following individuals and companies had stock and held leases in Alberta, in addition to the leases enumerated in 1884:

NAME	NO. OF ACRES
Durham Ranch	33,000
Vowell & Eberts	29,000
D. McEachren	30,000
Allfrey & Brooks	10,000

Jacob Eratt 5,000

British American Ranch (3 leases) 189,000

Mount Royal Ranch Co. 12,000

Francis White, New Oxley,

Canada Ranch Co. (2 leases) 180,000

Bell Bros. 5,000

Ives and Sharp 5,000

John Hollies 29,000

Moore & Macdowell 22,000

Wm. J. Hyde 3,900

W. Bell-Irving 5,280

D. McDougall 6,000

S.E. St. Onge Chapleau 100,000

J.W. Ings 1,920

Thorpe & Bedelle 24,000

NW Land & Grazing Co. 24,500

M. Oxarart 11,000

Geo. Alexander 44,000

W.C. Skrine (2 leases) 16,900

Rev. J. McDougall 7,680

Union Ranching Co. 100,000

Hand-in-Hand Ranch Co. 100,000

J. Ick Evans 66,000

A.J. McKay 38,000

Brown Ranch Co. 33,500

E. Meek 88,000

Tom McKay 50,000

T.P. McHugh & Co. 9,700

T.H.Logan 100,000

Geo. Scheetz 100,000

E. Hausman 100,000

H.M. Taylor 100,000

P. Doyle 60,000

Potter & Pettapiece 5,100

F.W. Craig 11,000

J.J. Sullivan 23,000

D. McEachren 16,640

A.R. McDonnell 20,800

Lafferty & Martin 7,000
Ingram & Chambers 1,280
A.C. McKay 76,000
Glengarry Ranch 52,320
J.H. Conrad 89,300
S. Spencer 22,000
Herb Spencer 40,000
T.C. Powers & Bro. 24,500
Greeley & Wood 21,920
A. Adsit 1,920
A.P. Patrick 5,120
Lieut-Col. Irwin 3,000
W. Carter 100,000
C.W. Saunders 3,040
W.L. Nichol 4,000
G.W. Quick 5,972
W.G. Conrad 32,580
Curry Bros. 11,000

Murder, theft, highway robbery and whisky smuggling marred the order of the year, Calgary district being particularly unfortunate, not only in suffering from law violations, but in being unable to fix the crimes. On May 22, Sergeant Gordon of the NWMP arrested two suspicious characters, Gallagher and "Crackerbox" Jones, handcuffing them and starting for Calgary. En route they stopped for dinner, and Jones complained that he could not eat with the manacles on. Gordon kindly unlocked the irons on both Jones and Gallagher, and Jones promptly plucked the revolver from the policeman's holster, shot and disabled him, and then escaped with his companion.

On August 14, two men held up and robbed two brothers named de Rambouville near Calgary. On August 23, a man named Burns reported to the Mounted Police that the Edmonton stage had been held up and robbed by two men, one of them being masked in a British flag. The police found the flag but not the men. On August 25, "Clinker" Scott (Scott A. McKrenger) was murdered near Shagganappi Point, just west of Calgary on the Bow River. Undoubtedly these crimes were all committed by one pair of men, even the Scott murder, but the perpetrators were never brought to book.

The Oxley Ranch had a number of cattle branded with a quarter circle, and this so appealed to a gentleman who used the "Q" brand that he "worked over" some scores, carrying the quarter-circle around to a circle and then adding the tail. But he was too grasping and careless to succeed, for he rebranded too many, and did the job so crudely that the marks of the old brand were quite clearly distinguishable under the new one, and everyone realized the change as soon as the roundup took place. This gentleman, though addicted so pronouncedly to rustling, was nevertheless quite honest, after his own peculiar lights, for, upon making a quick and successful trip across the boundary line, he sent back one hundred and two dollars that he owed to Ned Maunsell.

Indians broke out in drunken brawls, causing worry to the Police. A band of yelping, drunken Sarcees attracted the attention of some policemen who were on the mission grounds at Calgary. Proceeding to investigate, they went up a ravine until the two of them, Inspector Moodie and Constable Green, were greeted with whizzing bullets and the crackle of rifles, whereupon they retired. Some Blood Indians on the Belly River caused a slight flurry when they descended upon the camp of a couple of white men, demanding grub, "and pretty quick, too." The white men refusing, the Indians opened fire, but were driven off by the return volley which the whites turned loose, no casualties being suffered on either side.

Six hundred gallons of perfectly good liquor was spilled on the ground by the Calgary police detachment alone, the forbidden drink being brought into the country through every conceivable channel, often through the collusion of railroad men. The heaviest individual sufferers were C. Lafferty and J. Young, who were caught with fifty-five gallons. Hugh McLeod, manager of the Grand Central Hotel at Calgary since its erection in 1883, fell in the toils quite frequently this year. First he was fined one hundred dollars, next two hundred dollars, then, finally, he received a fine of four hundred dollars and costs with six months attached. McLeod appealed this sentence, and the appeal was sustained. Thomas Peers was also fined two hundred dollars for selling liquor, and Howard Douglas had to pay one hundred for importing. It might be mentioned here that it never was and never had been considered a disgrace to be caught and punished for having or selling liquor in the "dry" days, everyone wanting it, and everyone not connected with the Mounted Police and the authorities assisting, when required, in protecting the lawbreakers.

John Cowdry opened his private bank at Macleod early in the year. D.J. Grier, a very old-timer, was appointed justice of the peace at that town, being one of the first batch appointed in the territory, and Sir John A. Macdonald passed through on a tour of the West.

Among the new ranchers to arrive were the Rosses, who founded the High

River Horse Ranch. This was started by the father of Major George Ross, the latter developing into one of the greatest polo players in the world.

The first agricultural fair in southern Alberta was held at Macleod on October 15 and 16.

In connection with the succeeding paragraphs, an assertion that appeared in the *Macleod Gazette* late that fall is interesting as an example of how much the most experienced old-timer could tell of the weather of the future. Says the scribe: "Old-time cowmen foretell a mild winter. A dry summer is always followed by a good winter. Even at that a stack of hay will not be amiss." Note the saving clause so essential in a newspaper article.

In March of this year, John A. Turner, now one of the most prominent breeders of Shorthorn cattle and Clyde horses in the West, settled in the Calgary district.

The old North Wind, thinking that Alberta had enjoyed too many chinooks and mild seasons, produced from the darkest corner of his varied storeroom the bitterest weather in stock, and poured it down on the herds of Alberta, keeping it up from before Christmas till the end of February, and then changing to wild floods and fierce snowstorms. A few weeks after the optimistic article previously mentioned was published in the *Macleod Gazette,* deep snows had fallen and drifted, crusting so heavily that no steer could "muzzle" through to the hidden grasses, though the wiser horses managed to make pretty good shift with their hoofs. Very little hay had been put up owing to the ranchers' belief that it was unnecessary, and also to the fact that so much good hay-land had been grazed over during summer and fall. Only a few stacks of tame grasses were in the whole country, and very little wild hay. John Herron of Pincher Creek had a few stacks of millet and other tame hay, the first grown in the district, but as a rule every rancher was woefully short on feed. Ten years had passed since Fred Kanouse turned the first range cattle loose to rustle on the plains around Macleod, and during all that ten years there had been no heavy, general loss, so the ranchmen naturally figured there was not much likelihood of one coming.

Consequently the storms and drifted, crusted snows were a terrible blow, especially among the new or "pilgrim" stock that had been brought into the ranges in vast herds, totalling thirty thousand head north of the Old Man's alone when winter started. These were the first "stocker" herds in Alberta, and they suffered tremendously, much more than the native stock, though the latter died in thousands. When the bitter weather was at its worst, there were forty thousand starving horned creatures within a radius of twenty-five miles of Macleod.

The rabbits died, the lynx left, the herds of antelopes starved in hundreds, the poor brutes wandering into the very settlements, where they were often killed in the streets. The I.G. Baker Company's cattle, the first "beef" herds in Alberta, were scattered widely through the south and suffered frightful loss. A beef herd is one made up of cattle that are being fattened for market, such as the present herds of P. Burns & Company and J.H. Wallace, who do no breeding, but simply buy young or mature stock and round them into prime shape. Sixty percent of this first beef herd was wiped out in the winter of 1886–7.

Yarding like moose, the cattle in the great country north of the Old Man's River died like flies, bitter winds, no food, and a continual extreme cold soon huddling the stock in suffering bands, unable to travel far because of the mighty drifts. Ezra Pearson, manager of the "MHR" at Medicine Hat, left that town on the early morning of February 4 to drive the twenty-three miles to the ranch, the snow two feet deep on the level, with great drifts here and there and the thermometer fifty-four degrees below zero. Driving four horses to a light sleigh, it took Pearson three long days to make the distance. This ranch had put up hay for their stock, but even they suffered considerable loss, chiefly among the last bunch of stock they had taken in that fall.

All over the ranch country, the price of hay soared to phenomenal prices, ranchers standing willing to pay forty dollars a ton for all they could get and being unable to secure a single wisp. Those fortunate enough to possess a few stacks hoarded it like gold, and stood ready to fight to save a single forkful.

Clustering in the coulees or huddling on the open, the animals suffered and died in enormous numbers. Some, breast-high in packed and crusted banks, died as they stood; some who were sheltered somewhat by bluffs or coulees starved pitifully, ravenously searching for food until the frost had reached their vitals. The bodies of great steers were found in the spring, heaps of them, with their throats and stomachs punctured and torn by sharp splinters from dried and frozen branches and chunks of wood which they had swallowed in their anguish. The coulees showed the most bodies in the spring, for naturally the animals sought their shelter, crowding close together for the warmth of each others' bodies. One would succumb, others would crowd in on that body, others would drop, and when the winter broke the bodies lay piled six and eight deep all up the bottoms of the ravines. Hundreds of animals, helplessly endeavoring to find sustenance, sucked the hair from the hides of their dead comrades, dying finally with their throats, mouths, and stomachs lined with it. Though the ranchers and the cowboys performed prodigies of endurance, riding wide and far and hard in their supreme efforts to save such of the herds as they could, they did little good, being only able to watch their stock perish and hope for a break in the weather that never came until nearly the first of March. John Quirk, of High River, in

Ezra Pearson of Medicine Hat was the manager of the pioneering Medicine Hat Ranching Company. He also drove the stage between Medicine Hat and Lethbridge in 1883.

his efforts to save his herds, drove and dragged some of the most exhausted ones to his ranch house, where he and his wife attempted to rejuvenate the exhausted creatures by pouring warm water on their frozen limbs.

The loss in Alberta was a fearful blow to the stockmen, but the cattle people of Montana, Wyoming, and Dakota suffered even more severely, many men being forced entirely out of the business, which happened but rarely in Alberta, though twenty thousand cattle died in the country north of the Old Man's River alone.

In the Calgary district, west of that town, the snow was not very deep, but the exceedingly cold weather coming about January first and lasting steadily through to the middle of February crumpled up even the fat cattle in spite of the fact that there was sufficient food. Well-conditioned beeves stood three weeks of the continual cold fairly well, but after that they gave up and died in scores. The Stonies and the Sarcees lived sumptuously on the fat beef that lay scattered all over the hills in the West, while the Blackfeet, Piegans, and Bloods found more ready meat on the prairies than they had since the buffalo went out.

Wolves, coyotes and Indians grew fat and lazy with food wasting on the ground, and in spring, the ravens had rich pickings, the birds coming in flocks of thousands and doing scavenger work throughout the whole country.

J.D. McGregor, running cattle at the mouth of the Little Bow, suffered a heavy loss; W.N. Nichol's sheep at Medicine Hat were hit hard; A.E. Cross on Mosquito Creek lost sixty percent of his herd of five hundred; W.C. Wells and Nelson Brown on the Mount Royal Ranch lost twenty-five percent; the Glengarry Ranch lost six hundred head; the Quorn Ranch, which had imported large numbers of "pilgrim" cattle the fall before, lost nearly every hoof; Skrine lost seventy percent; Tom Lynch, out of a band of two hundred and ninety "dogies" that he drove in during the preceding fall, rounded up a scant eighty head. The N Bar N cattle outfit, which had come from the States the fall before with six thousand head, which they located near Wood Mountain, pulled out in discouragement in the spring, taking the remainder of their herd, two thousand head, with them. A Montana rancher named Scobie, of Dawson County, counted 3,747 head bearing his brand at the fall roundup, and the spring count produced 159. Dr. Kennedy of Macleod suffered the most "sweeping" loss of any Alberta rancher, his herd of two calves perishing in Joe McFarland's pasture field.

In Calgary and High River districts the snow came on October 26, continuing until the middle of November, and then going before a chinook, after which the hard weather came down and stayed steadily for two months.

There were isolated cases and districts where the loss was not as heavy as on the big ranches. In the Pincher Creek country, a number of small stockmen pulled through with only ordinary loss. In the Davisburg district south of Calgary, the few stockmen had put up hay and managed to come out in good order, and on Fish Creek, Malcolm Innes, profiting by his experience in British Columbia, had prepared for hard weather and consequently suffered much less than the majority of neighboring ranchers.

The best range in the whole country had been the Blood reserve, where there was good grass and few stock. Walter Huckvale wintered his cattle there and suffered but very little loss.

The rivers were booming by the end of March, and the ranchmen with their cowboys performed prodigious feats of recklessness in saving their remaining cattle from the flooded bottoms and islands in the valleys. A thaw broke the rivers on March 11, and all the low land between Pincher Creek and Kipp was flooded deep, Morgan H. Long, keeper of a stopping-place at Macleod a hundred yards from the bank, and well above ordinary high water, being driven out, the river rising five feet in the house. Three telegraph poles on the west shore and two on the east shore of the river at Macleod were snapped off by ice masses weighing four or five tons each being hurled against the posts by the gigantic pressure of the flood. On the 10th of March, John Cowdry left Macleod in weather that was twenty below zero, bound for Lee's Creek (Cardston). That night the thaw came, a strong chinook that took every bit of snow off the ground and filled the coulees swimming-deep with water. For days it was impossible to cross any of the southern rivers, as the tremendous icefloes would grind any small boat to pieces before it could make the passage.

As near as can be judged, the average losses in the Province were about twenty-five percent; in the Calgary district, fifty to sixty percent; from High River to the Old Man's River, twenty to twenty-five percent in the Pincher Creek country, and fifty percent in Medicine Hat, the latter district suffering considerable loss in their horse herds as well.

On top of the winter losses came a drop in beef, and the big ranchers began to look toward England as a good market. Seven hundred head were exported as an experiment, the stockmen realizing forty-five dollars per head, clear, and proving that the market was an entirely feasible one.

Far from being discouraged with their losses, the cattlemen and sheepmen proceeded to stock up again, a large number of cattle being brought in, and huge quantities of hay being put up in the fall, one lesson as to the destructive abilities of a bad winter being plenty to teach the stock owners. Ira Brown of Medicine Hat imported two thousand five hundred sheep, and a number of thoroughbred studs and other animals were also brought in, John A. Turner bringing a shipment of thoroughbred Shorthorns, Clydesdale stallions and mares, the latter being the first to come into the district. The Canadian Pacific Railway, always ready to help the country, continued with their policy, which they still carry out, of importing purebred stock at half-freight rates. The High River Horse Ranch imported three hundred and ninety-two horses from the States, including the famous racehorse Grey Eagle. Several ranches stocked with mares from British Columbia, Oregon, Eastern Canada, and Ireland, the popularity of the Oregon and British Columbia stock being based on the fact that the best saddle horses were coming from those parts, the breeders there having been using thoroughbred studs with the native mares for so much longer than the breeders of Montana and Alberta that the grade was much higher. C.W. Martin of Calgary also imported some fine horses. New ranchers were T.S.C. Lee on Bow River; Blunt and Holmes on High River; A.H. Goldfinch on the Bow; the Primrose Ranch Company on the Big Lake and the Little Bow; Inderwick and Leatham between the middle and north fords of the Old Man's River; W.F. Rivors between the Elbow and Sheep Creek; the Ailsa Ranch on Pine and Sheep creeks, and B. Pruen at the mouth of High River.

The wolf problem remained unsolved, and the packs were stronger and more bold, fed up as they had been on the dead bodies on the ranges. Wolfhounds were brought in for the first time, the Cochrane Ranch securing a pack in an endeavor to handle this wolf and coyote problem.

New Government regulations concerning ranchers and homesteaders were passed, allowing a lessee to purchase a home ranch for two dollars an acre, or his whole lease for two-fifty, and enabling homesteaders to homestead or pre-empt on leases, such homestead or pre-emption rendering that portion of the lease void.

Cattle from the Montana ranges were commencing to filter across the line with the more-than-connivance of their owners, as the Montana ranges were becoming grazed off, and there were vast untouched tracts in Canada where the feed was grand. Sam and John Spencer of the Double Rowlock, of Sun River, Montana, were near the boundary, and were pressing their herds into Canada, and the Mounted Police and Customs Department were greatly annoyed. At last Dr. Allen, Inspector of Customs, seized one hundred and eighty-three head of Sam's stock, took them to Maple Creek and sold them, thereby making the Spencers much more careful for some months.

The Indians were fat and saucy, commencing the year in good shape due to the unfortunate cattle, and asserting themselves with much more assurance and boldness than they had ever done before. Police outposts had been moved further out, one going to Pekisko at the Bar U headquarters in November 1886, and the detachments at Chin, Forty-Mile and Bull's Head being moved to Kipp's, Milk

River Ridge, Writing-On-Stone and Pendant d'Oreille in 1887. Glanders broke out in the horses of the Brown Ranch on the St. Mary's, and two were destroyed by Inspector Sanders and Veterinary Surgeon Jackson.

Whisky traffic was still persisted in, and the police caught some smugglers, though they were frequently outwitted by the crafty lawbreakers. The miners who reached Lethbridge in July of that year to open up the mines went on strike, and bloodshed was feared until a sturdy force of Mounted Police put the fear in the roughs. Red Crow, chieftain of the Blood tribe, accompanied by Indian Agent Pocklington and Inspector Sanders of the NWMP, went to Fort Assiniboine, Montana, to recover horses that had been stolen by Indians from the Bloods. The trip was quite successful, the horses being regained and Red Crow entering into an iron-bound, "hope-to-die," treaty with the Assiniboines and Gros Ventres, whereby all tribes interested should forever eschew stealing each others' steeds.

Impudence and boldness characterized the actions of the young bucks, particularly among the Bloods, who with spring felt the war-raid fever pouring through their veins. The Police winked at much of this rudeness and arrogance, not because they lacked in nerve, but simply because the settlers were scattered so wide, and were in such a helpless condition, that if even a small band of the Indians should really rise, many white souls would perish before the renegades could be caught. Charlie Miller of the Kootenay Lakes was annoyed by finding that Indians made free to enter his house and take his food; complaint to the police brought investigations, which were dropped when it was learned that Miller had not locked his door; several young Bloods, on a spring raid, stole horses and killed cattle; a strong band of these young braves prepared to cross the line and wipe out the Gros Ventres, who had killed six Bloods in Montana the previous year, but the Mounted Police persuaded them not to go, the treaty trip of Red Crow and Inspector Sanders resulting from these arrangements. An Indian armed with a knife held Corporal Hayne at defiance, though the corporal retreated with the Indian's horse, and, securing reinforcements, returned to arrest the red man, but found him gone. On April 27 three horses were stolen from Watson, of Medicine Hat, and one from Rev. Mr. Todor, the latter's horse being found a few days later on the Blackfoot reserve.

On April 26, a rancher named Gow, a former Mounted Police man, came from his ranch at Graburn's Coulee and reported that Indians had killed two head of his stock. Corporal Spicer went out in pursuit of the thieves and cornered them in a coulee, whereupon they opened fire and he wisely awaited for reserves to come up. Naturally the Indians made a clean escape, though strong suspicions pointed to a couple of particularly saucy Blood braves as being the guilty parties. These suspects, Dog Rib, or The Dog, and another Indian named Big Rib, were caught a few days later by Inspector Saunders, who was going over the reserve. They were tried first on the charge of shooting at Corporal Spicer, but were dismissed, the case falling down for lack of evidence, but four days later they were given five years each for stealing the horses from Watson at Medicine Hat, the sentence to be served at Stony Mountain, Manitoba. Sheriff Duncan Campbell of Macleod took the pair of convicted braves away to the penitentiary—or rather started to do so—losing them at Dunmore, where they escaped from him and succeeded in crossing the boundary line, coming back later in the year to visit their friends and relatives, and to boast a little bit among the Bloods on the reserve. In September, Sergeant Williams and three Mounted Police constables came upon Big Rib and ten other Indians on the reserve. Williams at once grabbed the bad red brave, and commenced to take him to jail again. Big Rib put up a stiff resistance, and, aided by his ten friends, managed to get clear of the four policemen without a shot being fired or a gun drawn. Eagle Rib, who assisted in the rescue, was caught later and sentenced to three months.

The Dog and Big Rib having preached their doctrine of scorn into the ready ears of the younger Bloods, the braves looked upon the Mounted Police with contempt and grew more and more arrogant and bold. Calf Shirt, war chief of the young Bloods, went into Montana and returned with much whisky, which it was suspected he intended to use as a war medicine; Good Rider, leader of the band that raided in the Kootenay Lakes country in the spring, went out and shot a Cochrane steer. The police, seeing the chance to hand a salutary lesson to the leaders of the restless young mischief-makers, arrested both, tried and convicted them, and forced a much more respectful demeanor from the Indians in the future. Calf Shirt served a month and Good Rider a year.

Another Indian incident was a quarrel near the mouth of the High River between two white men, Tucker Peach and Thompson, and two Blackfeet, Trembling Man and Bad-Dried-Meat (Deerfoot). Tucker Peach was shot in the arm by Deerfoot, and Trembling Man was shot by Thompson, Deerfoot escaping a second time that year, having just got away from the Police after they had arrested him for larceny.

In spite of the troubles with the great mass of the young Bloods and a few young Blackfeet, the older Indians remained quiet, many of them beginning to work small farms, and a few being employed for the first time in history as Mounted Police scouts, a departure that assisted the force very much, for when once in uniform and in a position of importance, the average Indian is bitterly stern with less fortunate ones—though he might steal for himself, he is decidedly harsh with his brethren if they do so.

On October 23, James Dunlap, foreman-manager of the Cochrane Ranch and one of the best cowmen in the country, suffered the misfortune of freezing his feet while driving from Macleod to the ranch. Dunlap was taken to the hospital at Macleod, where he was given every attention and was expected to pull through up until the time he was told he would lose both feet. Then he simply gave up and died within two days. His position on the ranch was then filled by a man named Paul Lee Morrison.

During the last three days of October, the settlers and townspeople of Macleod fought night and day to protect the town from entire destruction from a prairie fire which had been started by a hunting party in the hills near Pincher Creek, burning over a tract sixty-eight miles long and fifteen wide. The shooting party themselves laid information against their cook for setting this fire out, and the man was fined fifty dollars before Captain Scobie, J.P.

CHAPTER XI

1888

A GOOD winter followed the disastrous one of 1887, and the cattle wintered well, the calf and foal crop being particularly good, though in some alkaline districts anthrax proved fatal to a considerable number of young stock. The cattlemen tried inoculation with blackleg vaccine to eradicate this disease, but found it inoperative, the vaccine not insuring safety in the least. Then they tried the old method of moving their herds whenever blackleg appeared, and this treatment proved entirely efficacious, new pastures assuring relief and safety. Leaseholders and squatters were still squabbling, though the homesteader was gradually getting in on the ranges—unwelcome, but persistent and stubborn. Thousands of acres of leased land were still unstocked, held by Eastern people simply for speculation, settlers declaring that even the rent was unpaid, though this is not known positively. Many people were forced to occupy the position of squatters because they could not secure the written permission of leaseholders to homestead, the latter nevertheless fearing to order them off because of the widespread disturbance such a move would cause. In February an Order-in-Council was passed making it unlawful for a lessee to homestead or pre-empt within another's lease, and on March 19 another Government regulation was passed changing the requirements for stocking a lease from one head per ten acres to one head on every twenty acres. Forty-four leases were cancelled during the year.

J.D. O'Neil, living near Calgary, aired a bitter grievance against Dr. McEachren. It appears that no homesteader could locate on any lease granted before 1886 without written permission of the lessee, and that in 1883 O'Neil went on the Waldron Range (though the correct word is "Walrond," the general habit is to use "Waldron"), picked a quarter, erected a two-storey house and a good barn, built corrals, and had everything ready for his family by Christmas time. Then he was told by the Waldron people that he could not stay. Nevertheless he went east for his family, returning with them and a couple of carloads of stock and implements, meeting Dr. McEachren on the train and taking up the matter of the homestead directly with him. O'Neil claimed he asked for permission to homestead where he had built, but was refused. He asked for compensation for the improvements on the land and received another refusal. He then homesteaded near Calgary.

E. Dewdney, Minister of the Interior, that year cancelled nearly six hundred thousand acres of leases in the Macleod district alone, the districts thus thrown open to homesteaders being as follows: "North of Macleod along the east side of Willow Creek to the Winder Ranch, from Winder Lease to Mosquito Creek, and all the lands east of that between the Little Bow, the Old Man's and Belly rivers, with the exception of one lease in the forks of the Big Bow and the Belly. All of the land east of a line drawn from Macleod to Belly River, with the exception of two small leases held by Bell and Patterson, and Dug Alison. All the land between Pincher Creek and the south fork as far as Lee's Creek west is under the new leases, which are subject to settlement excepting one small lease in the names of Ives and Sharpe."

The stockmen continued to improve their herds, importing purebred animals in large numbers. The only female stock on the large sheep ranches were Merino ewes of various degrees of impurity, which were being crossed with fine Shropshire rams. The cattlemen were bringing fine bulls of every breed to grade up the range animals. Ranchers did no dairying or farm work, declaring that cowboys would never would never work on foot. "Until recently," reports Superintendent Herchmer of the NWMP in his report for the year, "it was contended that cowboys would not cut hay. In the near future they will have to attend gardens and domestic animals." What a fall from the independent, spurred, belted, chapped existence of cowboyhood is here predicted!

Shorthorns and Herefords were the prime favorites with the ranchers and homesteaders, though the Polled Angus was gaining favor, and some of the most successful breeders were using Highland bulls and meeting with marked success, even though it took a Highland steer five years to properly mature, while a Shorthorn or Hereford steer was often ready at three, and never took more than four years to reach his prime. The popularity of the Highland stock was due to their rustling qualities, their tendency to range close, and the exceeding fineness of the meat and robe, buyers always paying top prices for these steers because of the admitted superiority in quality.

All ranchers now cut much hay, and nearly all had shelter for weak cows and calves, A.M. Burgess' declaration that it was impracticable for big ranchers to prepare every year for hard weather being to a certain extent proven to be wrong. The advanced among the big ranchers were yarding all calves and feeding them through the winter, the Bar U even bringing in all their weak cows. This company fed seven hundred calves and two hundred cows in sheds, and their loss in the spring was practically nil.

Some small ranchers, men with forty to a hundred head on the range, complained about this date with regard to the roundup regulations, which had been practised for some years by the stock associations. Mavericks were always found on the roundup, some often belonging to owners of small, fenced herds that had broken out and joined the range cattle. This meant a loss to the owners, and one that they could not well stand. On the roundup, the outfits always killed mavericks for their camp supply, and further, if a maverick escaped the camp butcher and ownership was not claimed and proven, the stock associations sold the animal and turned the money into their general fund. The regulation which caused most discussion at this time, though, was that one concerning representation. The small ranchers claimed that, governed by big ranch companies, the stock associations had been led to adopt an objectionable rule, a rule that worked to advantage for the big companies in so far as it enabled them to get along with less men of their own on the work. Owners of five hundred head or any fraction thereof must have a man at the roundup to protect his stock, the small owner of fifty head thus being put to as much expense of labor and time as the owner of five hundred head. Efforts were made to change this regulation, but they were without avail.

By the spring of 1888, the country south of High River, between that point and the Waterton Lakes, was populous with ranchers and cowboys, the following men being engaged in the business to a greater or less degree: A.M. Morden, Stewart Ranch, George Lavasseur, E. Wilmot, Alberta Ranch, M. Legrandeur, Morgan & Cummins, Wm. Berry & Sons, Chas. Smith, Bruneau & May, C.G. Geddes, Martin Holway, J.H.G. Bray, N.F.M. Scobie, on Pincher Creek; W.S. Lee, F.W. Godsal, D.A. Blain, Jones & Sharpe, Garnett Brothers, between the south and middle forks of the Old Man's River; I.G. Baker & Company, cattle scattered everywhere; Cypress Cattle Company, on the Belly River between the Big and Little Bow rivers; Ricardo & Williams, north of the Bow River; A.J. Whitney, H.W. Savery, at Kipp; A.W. Draper, James S. Norris, Pioneer Ranch, at Macleod; A.E. Browning, Cornish Cattle Company, Trefoil Ranch, Winder Ranch, Willow Creek Ranch, C.A. Lyndon, F.W. Craig, New Oxley Ranch Company, Victor Ranch, on Willow Creek; F.C. Inderwick, Waldron Ranch Company, Macleod Cattle Company, Whitney & Daly, L. Brooke, Maunsell Brothers, Dug Alison, Peter McLaren, on the Old Man's River, and its north and middle forks; Sheep Creek Ranch Company, Lucas, Eastman and Wallace, on Sheep Creek; Glengarry Ranch Company, R.W. Whitney, on Trout Creek; Alexander Ranch, Cross Brothers, Powder River Cattle Company on Mosquito Creek; Jerry Potts, on the Piegan reserve; Tom Lynch, John Quirk, Little Bow Ranch Company, J.J. Sullivan, W. Skrine, North-West Cattle Company, on High River; Primrose Ranch, W.G. Conrad, W. Podger, Hubert Samson, on the Little Bow; R.M. Patterson, James Bell, J.D. Murray, Walter Huckvale, Main and Dennis, R.H. Wilson, Frank Strong, Sam Brouard, on the Belly River; A.C. Sparrow, between Elbow River and Fish Creek; Brown Ranch Company, St. Mary's River below the mouth of Lee's Creek; Cochrane Ranch, Five-Mile Creek, Porcupine Hills.

The death of one of the most energetic stockmen in the province occurred in July of this year, when Frank Strong died at Victoria, whence he had gone for his health. Strong, who went to Montana from Michigan in 1871, went to Alberta in 1880 and was thereafter a leader among the ranchers, acting as manager for I.G. Baker's stock business and at the same time operating a horse ranch of his own. Strong's horses were among the best saddle animals in the country, he being a very careful and judicious breeder. Many of his horses were in the Police force, and they proved to be among the best stock there.

Exportation of beef to England took a great forward step during the fall, though a certain prejudice had arisen in that country against Canadian ranch beef on receipt of the first shipment owing to its color, which did not appeal to the fastidious beef lovers of Britain. This prejudice, which sprang up at the time of the sale of the ranch stock the year before, had disappeared to a considerable extent by the time the fall shipments of 1888 had been marketed. Five thousand head of prime Alberta steers were sold on the British markets, the average price to the ranchers on the range being between forty and fifty dollars net. Some few ranchers who took advantage of this new source of disposal of their surplus were unsuccessful in their sales owing to the fact that they did not thoroughly understand the market and failed to use proper judgment as to the best time for shipping. In July Jim Patterson and Charlie Sharples, representing the Oxley and Winder ranches, left Macleod for Medicine Hat in search of a good trail to that point over which they could drive their beef herds for shipment. They picked a route, and the Oxley and Cochrane outfits shipped a large number of fat wild steers from that point early in the fall. Both these shipments were too early for the best markets, arriving about the time heavy cargos of Argentine steers landed. In addition to this, the wildness of the Cochrane cattle was such that they had to be slaughtered at Deptford immediately upon arrival, which tended also to prejudice the sale.

The Bar U, which ranch seemed always to land "in clover," when there was any clover to be found, delayed their export beef until late in October, when they shipped the second trainload ever sent east out of Calgary for England, the first shipment being one of fifteen cars which Sir Lister Kaye had sent earlier in the month from his Mosquito Creek Ranch. Both these trainloads reached the British market at the right times, and netted good prices and profits, the Bar U shipment possibly bringing in a little more than the "76" steers from further south.

All the steers shipped to England were the absolute "tops" of the various herds they represented, and the general run of the steers on the big ranches seems to have improved considerably. An Eastern cattle buyer named McCormick came through the hill country in September on a purchasing tour and was markedly impressed with the excellence of the Bar U and Cochrane cattle, especially the latter, the superiority of which was doubtless due to the isolated range of the Cochrane herds, they being far away from the country where the scrub bulls ranged, and being headed by the finest of purebred animals. McCormick expressed himself as satisfied that range stock generally on big ranches was better bred and better beef than the usual run of Eastern grade stuff.

The Bow River Horse Ranch, under the management of G.E. Goddard and located on a portion of the old Cochrane range west of Calgary, was breeding horses for the English market, and in an effort to prepare them properly for this outlet, the ranch employed only Englishmen to handle and break the animals. They shipped thirty or forty head this year, and found a very fair sale.

The CPR, though it had not a very complete equipment for shipping stock, did its best and rushed the stock, horses and beef through as fast as possible, the Cochrane cattle averaging thirteen hundred pounds when they left the train at Montreal. These were the animals that met the poor market. One remarkably successful shipment was made very early in September, arriving at Aberdeen, Scotland, on September 20 and being sold for prices ranging as high as a hundred dollars each, the average price for the whole three hundred and eighty-six head being seventy-five dollars.

During that summer, a number of fairly large ranchers came into the Province and entered the ranch business, in some cases purchasing established ranches. C.J. Reach, ranching near Macleod, sold out to a man named Dunbar; F.C. Inderwick, on the north fork of the Old Man's River, sold ranch and cattle to A.B. Few. Among the ranching settlers to arrive were Joseph G. McIntosh, a brother of Duncan S. McIntosh, who had arrived a year or two before him and settled on Willow Creek; Barker and Donovan, who located on Lee's Creek with eleven hundred sheep and thereby caused a temporary furor among the cattlemen; Bill

Brady, John Lamar, Johnny Franklin and a hundred and twenty-five Mormons, the first lot to arrive under the leadership of Joseph Card, in whose honor the name of Lee's Creek settlement was changed to Cardston. Two American settlers, William Dawes and Donald Kuntz, came into Macleod looking for a location, expressing an intention of bringing a band of two hundred head in as soon as the next spring broke. The necessity of paying duty on this stock did not please either of these men, and they expressed themselves forcibly, declaring that a thousand men and a hundred thousand cattle would come and make Alberta their home if it was not for the "Chinese wall" of the tariff. Montana was overstocked, and the ranchmen there were eager for new range, if they could get it without too much outlay of cash.

Importations continued, particularly among horses, for the grade was poor, and there was not a ranch in the country that could not profitably weed out a goodly proportion of mares and stallions. The Waldron Ranch brought in some Clydesdales; the Quorn, English thoroughbred stallions and three hundred Irish mares; a drover named Beak, of Oregon, sold nearly four hundred saddle horses at Macleod and district; W. Black, of Macleod, brought a Devon bull and some Holstein thoroughbred cows and heifers, and in October the Waldron received a Shire stallion, a Shire mare, three Polled Angus heifers and five Herefords, thoroughbreds all, direct from England.

Local prices were fair, the Calgary market holding pretty well between six and a half to nine cents, Montreal four to five cents, Chicago five to six cents. I.G. Baker & Company paid as high as thirty-two dollars for two-year-olds and forty dollars for threes on the range, Dave Cochrane receiving this price for a small band.

Horses improved a little, the average price for a good saddle horse being one hundred and twenty-five dollars, teams grading in value between two hundred and two hundred and seventy-five dollars. R.G. Robinson, who bought the Chipman or Elbow Park Ranch this year, offered Goddard of the Bow River Horse Ranch a hundred dollars each for a band of brood mares and was refused, whereupon he bought a stock from John Lineham for forty dollars a head, the nucleus of the present high-class herd he owns. Robinson's idea was to breed a good grade workhorse, and he made a success of it from the start, his ranch coming to be one of the show points in the vicinity of Calgary.

The editor of the *Macleod Gazette* grew very bitter toward Dr. McEachren during the summer, and he and the doctor had a vigorous campaign of editorials and letters, the scribe accusing the Government's chief veterinary surgeon with partiality and nearly everything else, and Dr. McEachren replying with denials and threats of libel. Some mange was appearing among the range stock; some say now that it was from the "CY" cattle originally, and the editor took Dr. McEachren to task for saying that mange had entirely disappeared the year before. He accused the doctor of confining his inspections of cattle and other stock when on inspection trips in the West exclusively to the animals of the Waldron Ranch, the doctor being personally interested in that company. Dr. McEachren denied these insinuations and accusations, and letters followed editorials with great promptness and verbosity.

A few colts died of strangles, hoof disease killed a number of northern horses, and the flies in the foothills west of Calgary were so bad that the cattle and horses there did not put on any flesh until the frosts of fall killed the pests. Cutworms appeared in a few isolated cases and caused some damage, though A.B. Macdonald of the Glengarry Ranch soon disposed of them by turning the water from a little irrigation system he had built onto the affected fields, the worms then coming up to the surface and dying in millions.

The first mention of hailstorms and resultant damage appears in the records of this year, which is a very clear indication that farming had grown in importance, for though doubtless there had been hailstorms before this, it was never thought necessary to mention them in Mounted Police or livestock annals. Hailstorms in July destroyed forty acres of crop on the C.P.K. farm at Gleichen, forty acres for W.N. Williams, ten acres for Magnus Begg, twenty-five acres for Dan Martin, ten acres for Al Dufour, twenty acres for J.V. Lind, and fifteen acres for Grass & Walbeck. John Clark, the Crowfoot horseman and rancher, lost thirty acres of splendid oats, and the following night Gleichen received another attack, the hailstones coming with the force of bullets. The reports state that they were driven six inches in the ground, and thousands of panes of glass were broken in Gleichen. Sheep, horses and cattle in Johnson Brothers' corrals broke through the strong walls and stampeded across the prairies, and the Blackfeet on the reserve suffered from the loss of many tepees, ruined by wind and hail. Hailstorms at Kipp destroyed Long and Urche's promising crops, and quite a number of farmers in the Calgary district lost their grain in this manner. Blackleg carried several hundred calves into early graves, but the progress of this plague was stopped as soon as the ranchmen learned a change of pasture was sufficient to assure recovery for all excepting the dead ones.

The Indians, with the exception of the Sarcees and some of the young Bloods, showed reasonable improvement, cattle killing remaining their one besetting sin, outside of a liking for whisky and other people's horses. The Cypress Cattle Company complained early in summer of the loss of cattle killed on their range by wandering Indians, but the Police were unable to find the perpetrators. Again the company objected to supplying free beef to Government wards, and again the police failed, though this time they did find a carcass, skinned, disembowelled, and stripped of much meat, the animal's heart sticking jauntily on one of its own horns. Nobody suspected the poor beast of such an acrobatic method of suicide, and Indians had been in the near vicinity, so suspicion pointed to them, though that is all that resulted, for the red men were becoming cunning indeed in killing stock. Particularly in the fall was the Indian presence most regretted by the ranchers on the Little Bow and the country beyond, for then the calves were fat and sleek, toothsome and lusty, and the hides were fine for winter wear. Bloods and Blackfeet, securing permits to leave their reserves to hunt, journeyed into this country where there was no game at all, and just sort of hung around, doing nothing apparently, though the ranchers sometimes found the fresh skins of calves in their camps. Strong complaints from these suffering stockmen brought a Police patrol and considerable relief.

Calf Shirt, ex-war-chief of the young Bloods, after one month in the barracks in the preceding year, came out a different man and accepted a position as a Mounted Police scout at twenty-five dollars a month, rations found for horse and man, and horse supplied by the man. Calf Shirt knew he could keep a horse supply all right. Star Child, suspected murderer of Graburn in 1879 and innocent child of the prairies, was wanted by the Police for some trivial offence, and eluded the force until one day in Macleod, when, recognizing Superintendent Neale and wishing to demonstrate his friendliness, Star Child offered his hand in peace, and Neale hung onto it until a passing policeman relieved him of the struggling warrior.

The Dog still remained a coveted fugitive, though he entered Macleod openly one day and was promptly grabbed by a policeman, whereupon his Indian admirers flocked to his rescue, hauled him free from the white man, and bolted across the open prairie, one of the Indians' horses carrying away a heavy bullet from the officer's revolver. Deerfoot the Blackfoot was free, too, though an Indian scout named Giveen had met up with him on the open range and wounded him in an exchange of shots.

Sam Bedson, a restless young Blood, took a trip to Medicine Hat and stole a couple of horses, rode to the section house at Winnifred, broke into and looted that place, continued on to Macleod and shot a calf belonging to a man named Henderson, and then retired to the reserve with the expressed intention of remaining a hero in the eyes of his male and female compatriots. But Calf Shirt, ex-war-chief, now Mounted Police scout with a blue coat and brass buttons, was sent on Sam's trail, and Sam bowed to the law and accepted arrest after a physical argument with his former chieftain.

During the year, the police recovered one hundred and eighteen stolen horses, most of which were found on the Blood reserve, though twenty-seven head were animals stolen from the Bloods themselves and picked up among the Indians across the boundary. Fifteen horses were returned to the Crees, nine to the Stonies, twenty-one to the Gros Ventres, twenty-seven to Frank Strong, seven to O'Hara and Corcoran of Montana, one to Turner of Lethbridge, one to McNab of Slide-Out, two to Moss of Morley, one to McDougall of Macleod, one to McCullough of Pincher Creek, one to J.B. Smith of Macleod, one to R. Richards of Macleod, and two to the Halifax Ranch.

One white horse rustler had a very narrow escape, and showed a marked respect for his skin. Edward Austin, more widely known as "The Kid," lifted horse and saddle from Tom Purcel and fled across the boundary, hotly pursued by the grim old rancher. Shortly after arriving on the "safe" side of the boundary, "The Kid" dismounted for a few minutes and was left afoot by his horse, a most embarrassing position for either cowboy or horse rustler to be left in. He labored in a high-heeled way over the plains until he reached a ranch house, and was greeted with the usual Western hospitality, fed, bedded, and made welcome to the whole place. Purcel, doggedly pursuing even across the line, arrived shortly afterwards, explained his errand, and proved his case so satisfactorily to the ranchers that they immediately did what they could to square matters—gave him his saddle which "The Kid" had himself sweated under during the last few miles of his journey. Displaying a remarkable lack of wisdom, "The Kid" at once stole the saddle back again, caching it in a haystack, and so arousing the ire of the Montana stockmen who had befriended him that they asked Purcel as a personal favor to leave the man in their hands to punish as they deemed fit. Montana stockmen still had a habit of giving the "rope-cure" to rustlers, and Austin began to display more than mere glimmers of reason when he decided he would go back to Canada with Purcel and face the music of a charge of horse stealing. He was tried at Macleod and given a stiff sentence. The willingness of the Montana stockmen to handle this case to their own peculiar satisfaction will be understood when the following advertisement that appeared in Montana and Alberta papers that summer is read:

"$3,000 reward.—In March or April last about two hundred head of horses were stolen from the Teton range. The brands have probably been changed. A reward of $3,000 will be paid by this association for the capture of the horses and thieves, or $1,000 will be paid for the capture of thirty head of these horses and $250 for each thief, either dead or alive. If as many as a hundred horses and the thieves are captured the entire amount of $3,000 will be paid.—Montana Stock Growers' Association."

Whisky smugglers were doing a rushing business, and the police were in despair, though they made some splendid "catches" quite frequently, being particularly pleased with the capture of a wholesaler named George Lewis, who was wrapped in the toils at Whoop-Up when he had two hundred and fifty gallons of whisky in his wagon. The liquor laws were very unpopular with the general public—even the police, if the truth be entirely known—and indignation was often expressed by settlers, who demanded the right to get drunk if they desired. The Government, seeing how matters were drifting, threw a little sop in the widespread dissatisfaction by permitting the manufacture and sale of four percent beer, hotels with twelve or more sleeping rooms and stabling for five or more horses being given permits to sell this mild beverage, the Government charging a tax of ten cents per gallon. In addition to this, the long-suffering public was given the privilege of buying brandy from druggists if backed with the certificate of a qualified doctor, and the medical profession at once became very popular indeed. The four percent brewery started business at Lethbridge, and did remarkably well from the very day it opened its doors, Indians particularly favoring it because they certainly and undoubtedly became very happily drunk on this brew, and the police were helpless to stop it, as the Government analysis had shown it to be a non-intoxicating beverage. The solons at the seat of government had been trying to solve the problem of Montana "forty-rod" for some time, and they were convinced they had at last struck it in this legalizing of four percent brew.

In the meantime whisky, rolled into the land in every conceivable way. It came branded as red ink, it came in in bales of hay, in loads of oats, even in kerosene cans, boots, boxes and kegs. A shipment of Bibles arrived at one town, and the settlers purchased them so avidly and hastened away so rapidly to commune with themselves and the book's contents that the Mounted Police smelled a rat, the study of the Bible not being a pronounced habit of the majority of Westerners in those days. Upon purchasing one of these coveted volumes, the Police smelled more than a rat, for the "Bible" was a little metal cask, formed like a book, branded like a book and entitled "Holy Bible" in gilt letters, but filled with whisky!

Horrible concoctions were made by the whisky smugglers to make quantity take the place of quality. Raw alcohol was usually smuggled, and the trader, once safely cached in Alberta, proceeded to lengthen his stock by mixing it with water, bluestone and tobacco. Instances of men being killed by the last draught from such kegs were not at all rare. The Indians drank anything with a "bite" to it. Red ink, Painkiller and Florida water were favorite beverages, and the Sarcees—the most pronounced drunkards of all the tribes—spent two glorious days with a homemade brew, every grown person and three-fourths of the young fry being drunk for that period. They secured a case of Painkiller, some Florida water and some raw alcohol, emptied it in a huge "ten-skin" cauldron, added a dash of tobacco and a pinch of bluestone, placed a fire beneath the kettle, and then remained beside it until the bottom had been reached and the entire tribe was incapacitated.

Through all these tribulations, the drinking public stuck loyally to their pestered brethren, the whisky smugglers and dealers, as is shown by the following patriotic appeal, which appeared in a Moosomin paper of the date of 1888:

"Indignation meeting.—A meeting will be held in Orange Hall tomorrow night to protest against the late mean and despicable action taken by the Police in subpoenaing respectable and worthy citizens to give evidence as whisky-sneaks, thus interfering with the liberty of free-born subjects, and as likely to intimidate good citizens from entering hotels. Every one should attend and protest against such a resurrected tombstone, iron-heeled law, to bear which is to suffer worse than the slaves in Siberia. Arouse ye all!"

Two interesting events took place in the United States that were of considerable interest to certain of the citizens and settlers of Alberta who knew Texas and the southwestern stock country of the United States. One was the breaking-up of the notorious "Kingfisher" gang, a bold and reckless and very proficient band of stock rustlers who had kept the ranchers in their particular territory in terror for years. The leader of the gang met a quick, sudden, and generally undeplored death in a saloon in El Paso, Texas, and his followers promptly scattered, some coming directly to Canada, where they proved splendid stockhands and became respectable and honored citizens. The other interesting occurrence was the closing of the famous Texas cattle trail, a mighty road, six hundred miles long and one mile wide, which the Government had surveyed and reserved for the early stockmen to drive their great herds to market over. Railroads had finally rendered this romantic and doubtless fairly bloody trail useless, and the United States Government decided that year to close it to traffic and eventually throw it open for homesteaders.

The fall in Alberta was mild and fine, the Old Man's River roundup not commencing until October 6. Mild weather continued on into December, when snow and ice came to the country north of Sheep Creek, though the south continued warm and open nearly to Christmas. In fact, even the day after Christmas was "green," and the new-formed ice on the Old Man's River was insufficient to support the Calgary stagecoach, which broke through and nearly caused the death of Braden, the driver.

John Herron resigned the managership of the Stewart Ranch to go into ranching for himself, and Alexander McLennan succeeded him. Middleton, who had been instructor on the Piegan reserve for some time, resigned to take control of Sir John Lister Kaye's farm at Gleichen. Macleod people were somewhat hurt at the appointment of Robert Evans as quarantine inspector at that point. They looked upon it as political favoritism, pointing out that Mr. Evans had been a near-candidate for the Conservative party at Carswell, and that the Government should have appointed a Western man at least.

W.F. Parker of Macleod, who had acted as captain of the eastern Pincher Creek roundup that fall, was honored with the presentation of a pair of handsome silver-mounted spurs from the roundup members, in commemoration of the fair and able manner in which he had conducted the work. Late that fall another honor fell to Mr. Parker when Lieutenant-Governor Royal appointed him and W.F. Cochrane of the Cochrane Ranch members of the brand committee for that district.

This spring one of Alberta's prominent horsemen, G.E. Goddard, took charge of the Bow River Horse Ranch west of Calgary.

CHAPTER XII

1889

EARLY in January of 1889, the many ranchers around Calgary and High River commenced an agitation for the formation of a stock association that would be the leading one of the Province. Calgary was now the largest town in the central and southern section, and the settlers began to feel that the north district of the ranch country should take a leading part. Already there had been, or were, two associations, the first being the South-West Stock Association, formed of men in the Macleod district, and the second being the Alberta Stock Growers' Association, which had been formed at Macleod also, and into which the South-West Association had merged. Since its inception, the ASG Association had rather lagged in its work, and the ranchers, feeling that the business of ranching required more ginger in its general administration, and more interest and protection from the Government, desired a resurrection and a stronger, more active organization.

F.S. Stimson, George Lane, Barter, D. Macpherson, Ross, Podger, Alexander, Skrine, Goddard, H.B. Andrews, W.R. Hull, A.E. Cross, and Mollison held a meeting at High River for the purpose of discussing the most likely means of reviving the Stock Growers' Association, which then had headquarters at Macleod. The meeting urged that the headquarters be made Calgary, but deferred arriving at a definite decision until they had heard from the more southern ranchers.

The Government did a few things to further protect the holders of leases, passing an Order-in-Council to the effect that no person could graze stock on the public domain without the consent of the Minister of the Interior, with a penalty of seizure and forfeiture in event of violation.

The ranchers sadly needed stock detectives and hide inspectors, but these requirements were ignored by the Government. Brand ordinances had been passed late the previous year, dividing the country into districts, appointing brand recorders, with two men in each district to act as a brand committee with the recorder. Penalties for using another man's brand were made legal for the first time, the punishment being not more than one hundred dollars or forty days in jail. Another long-required regulation was passed, one that the ranchers had been crying for since ranching commenced, and one without which it is difficult to see just how they had continued business as long as they had. The new ordinance read: "The presence of a recorded brand on any animal shall be prima facie evidence of ownership." This was a law that the ranchers had been asking for since they started, eight or ten years before, but had never received until 1889. It helped matters much with regard to thefts, but it also placed the homesteader who had unbranded stock in an embarrassing position.

There was little snow during the winter, and some of the cattle did not come through as strong as they had in preceding seasons when snow lay a little deeper. The ranchers decided that this was caused by the lack of snow, the hard, well-cured prairie grasses requiring considerable moisture to be properly digested, and snow being the only practical means of supplying this need. Wolves caused general havoc, though the pack of wolfhounds on the Cochrane Ranch killed very many coyotes during fall and winter. Despite the precautions of the stockmen, these pests increased, and though there was no Government bounty, the owners of the larger ranches did not hesitate to place rewards on the heads of certain well-known and particularly destructive wolves. There was one huge white brute with a black-tipped tail who for some years had made her headquarters up a certain coulee in the Porcupine Hills, coming out on the range every spring accompanied by six or eight well-grown pups whom she at once proceeded to instruct in the art of killing colts and calves. This animal and her numerous progeny did such marked damage to the herds of her particular district that rewards aggregating seventy-five dollars were placed on her life, and even that inducement failed to bring men who could get her until Joe McIntosh and Charlie Brown went after her with ropes, running her down and roping her after a bitter chase. The old she-wolf fought to the last when she found herself caught in the cowboys' rope coils, but it was no use once they had hold of her, for the horses soon dragged and bounced her across the prairies and hills until she was dead. McIntosh and Brown figured they had done a pretty fair piece of work when they secured this seventy-five dollar prize, but they proved to be doomed to considerable disappointment, for the only rancher who paid up his share of the promised reward was Stanley Pinhorn, manager of the Oxley Ranch, who had promised and promptly contributed twenty-five dollars.

Early March storms killed a few early calves, and lump-jaw commenced to appear in the herds, but not in the alarming numbers it finally grew to. The calf crops in the Calgary, High River, Macleod, Lethbridge and Medicine Hat districts were the largest and healthiest in the memory of the Alberta ranches, though the sheep ranches had not such good results and were already beginning to reduce preparatory to withdrawing, owing to the continued low price of wool and the possibilities of heavy winter loss. Some sheep ranches, in fact, disappeared this year, while horses commenced to attract more attention, and many were brought in to stock new and established ranches. In 1882 there was only one purebred stallion in the whole Pincher Creek district, but now the number had grown to scores. This state of affairs in this one district was illustrative of the condition in all other districts as well.

The range cattle were not showing the improvement that was hoped for, because of the loose methods of breeding, though the Cochrane herds maintained the highest standard, because they always used their own bulls and very few outside ones drifted on to their range. With this one exception, there were all sorts of bulls on the prairies, a traveller passing nearly any large herd being almost sure to see Shorthorns, Galloways, Herefords, Polled Angus, sometimes a Highlander, and always a large number of runts and scrubs.

Stock troubles grew with the increase of the herds, the newest development being white men who made small killings of beef to supply the local butcher shops. This sort of thievery, petty though it might have been had only one man taken it up, was done so widely that a considerable loss was experienced. Another difficulty was a shortage of hay in the Porcupine Hills, which worried those ranchers, because they all needed hay now, every ranch that pretended to the name at all bringing calves and weak cows into the home place to feed during the winter, and some of the most advanced weaning all calves early in the fall, so as to give the cows a better chance to face the winter. It was an interesting and nerve-racking experience to spend a day or a night at some ranch house where five hundred unweaned calves had just been thrown into a corral, and five hundred affectionate mothers circled wild-eyed around the outside, bawling madly for their blatting offspring.

Alberta exports, while no heavier than those of the preceding year, had proven more successful than usual, the fact that numbers remained so small being entirely due to the limited space obtainable on shipboard. Reaching England after a bad voyage, and striking a late, bad market, the Alberta beef still realized very satisfactory prices.

A new departure on the livestock industry was the importation of Ontario dogies to fatten for market. Dogies are outland stockers, the name being applied indifferently to those from Mexico, Texas, Ontario and. Manitoba. J.D. McGregor, the pioneer stockman and breeder, was the first to engage in this line of the industry, as he brought these cattle in and placed them on the Quorn range to fatten. Before fall he had imported two thousand, mostly two-year-olds. L.W. Herchmer, speaking of these shipments, says: "The change of pasture will do wonders for this stock, but I doubt if skim-milk-raised steers will ever equal our range cattle."

Following the death of Frank Strong, the ranch owned by the latter was placed under the management of Steve Cleveland, a well-known stockman. Cleveland kept the business up to its high standard, and also discovered the greatest "bronc" rider who ever came into the Province, John Ware and Frank Ricks not excepted, though the latter never had the opportunity to measure abilities with the Macleod man. Johnny Franklin, a young Texan who had arrived in Alberta the fall before, secured work on the Strong Ranch upon his arrival, working for forty dollars a month and doing ordinary "helper" work. Hearing in the spring that Cleveland was paying seventy-five dollars a month for "busters" who would work his horses in shape for the Mounted Police market, Johnny asked for a chance, and was turned down, in an indefinite sort of way, Cleveland saying:

"Can you ride?"

"I guess so."

"Well, I'm not going to take a chance with a man who can't. I want these horses broken right, for they are to go to the Mounted Police."

Franklin "wanted to try," anyway, and finally they decided to take his bump of ambition away, bringing out a nasty old roan outlaw and telling him to "try" that animal. The lad from Texas did, and presented Cleveland and his men with a spectacle of the most graceful riding they had seen, thereby landing a job that lasted nine years and was a real sinecure, for there never were more than fifteen "raw" horses to rope, saddle and ride during any one day, and all for seventy-five dollars a month. Nowadays bronco-busters demand anything from five to ten dollars per horse, sometimes, in cases of really bad ones, requiring twenty-five.

Johnny Franklin was one of those horsemen who was a "natural" rider, riding on balance and using no more muscle than absolutely required. He was also a quick-thinking man who out-guessed his horse before it jumped, swinging instinctively into the correct position to withstand each jolt before it really came. Many "busters" get "churned" after a few years of "twisting," due to the terrific jolts from the bucking horses, weakening ligaments and interior muscles, but Franklin, the pride of the Macleod district, is as light and as firm and as unafraid of a bucking horse today as he was when he climbed on that outlaw of Steve Cleveland's in the spring of 1889.

Franklin enjoys the rare distinction of having never been thrown from any horse, which few of even the best rough-riders can assert truthfully. Some declare that John Ware never had been either, but Ware rode awkwardly, though with high efficiency. The great Negro would bounce all over his horse's back, holding by clutch of knees alone, bounding, rocking, laughing—but staying. Johnny Franklin rode with an easy grace that made him seem part of the horse, and as he rode he raked the fighting animal from tail to ears with his big spurs.

When the "Stampede" was held at Calgary in September 1912, where the best riders of all America were gathered to compete for the American championship, Johnny Franklin was chosen judge of the bucking, being acknowledged to be the master of that branch of horsemanship.

"Say, Johnny," said a friend to the sturdy old cowman, "how are you going to make a decision if two men should tie?"

Johnny was astonished that a sensible man should debate a single moment on such a matter when the solution was so easy.

"I'll just ride both them horses myself," he said. "That's the only way to find out which bucks hardest."

Returning to the events of 1889, we find that some changes had been made in the Cochrane Ranch, which, when not unfortunate in cattle, seemed to be bound to discover some other misfortune. Paul Lee Morrison, the successor to James Dunlap, who died in the fall of 1887, did not enjoy his new position long, he, too, going to the great beyond, his death occurring on January 17, being caused by eating poisoned canned meat. Morrison's successor, Charles Ramond, held the position for a few months, but he also met a sudden end, being drowned in the Belly River at Lethbridge that spring. After this the string of misfortunes seemed to end, for the new manager, W.F. Cochrane, remained in control of the ranch until it was sold to the Mormon Church a number of years later.

The Van Valkenburgs of the coast, a large retail organization with shops scattered through the Canadian cities on the Pacific, came into Alberta this year to purchase beef, George Lane, foreman of the Bar U, having severed his connection with that ranch and gone into the cattle-buying business, selling the first beef to the coast people, and thus opening up a big market for Alberta's surplus. Van Valkenburgs purchased about two trainloads that year, all of the stock coming from the herds of the High River ranchers, the Bar U selling the largest number. Alberta now had markets in England, on the Pacific coast, at Montreal, Winnipeg, and the local demands of the towns, Mounted Police and Indian reserves. The beef contracts for the Blood and Blackfoot reserves were let to I.G. Baker & Company, the Waldron Ranch securing the Piegan contract, W.R. Hull the Sarcees, and J.K. Leeson the Stonies. C.E. Geddes sent two carloads of horses into Eastern Canada, where they sold at good prices, and the Quorn Ranch imported a herd of Polled Angus.

In February Joe McFarland and Bill Parker left Macleod and went to the Kootenay River, where they built a cabin and corrals near the mouth of Scott's Coulee, a great boon to travellers in that unsettled country of long drives and long distances. In March the Pincher Creek Stock Association met, elected officers and arranged for a bull roundup, the Association deciding to take up all bulls in the district in an effort to improve the herds. A.M. Morden was elected president, W.F. Parker vice-president, C.W. Kettles secretary-treasurer, with R. Duthie, Louis Garnett, Alex McLennan, Chas. Smith and Captain Scobie as a committee of management. W.F. Cochrane and Jim Bell were appointed delegates for the eastern division and Duthie, Godsal, Brooks and Dan Bell for the western. The bull roundup commenced on April 10, and though it resulted in the corralling of two hundred and sixty-eight bulls of varying degrees of purity and blends, it was not thorough, for a number of animals were left at the Macleod end of the district. L. Hamond, a cowboy, met with a painful accident on this roundup when his horse fell on him, breaking his leg. There were a number of other accidents also, chiefly to horses, several being very badly gored by fighting bulls. Although there were a large proportion of scrubs among the animals that were collected, the worst and most were from the Macleod end, and became known as the "Macleod Appelloosies," which word must designate something, for western nicknames are not applied without reason.

Blackleg was bad in the spring, and large numbers of young stock died, Dr. McEachren coming in for considerable more criticism at the hands of the *Macleod Gazette,* which publication seemed to enjoy nothing better than taking a crack at the Dominion veterinary. A letter written at Pincher Creek on February 12 was published in the *Gazette,* and read as follows: "Another case of blackleg yesterday. A great many cattle have died of this disease since Dr. McEachren reported to the Department of Agriculture that the disease was entirely stamped out."

The disease was so bad that it crept into the adjoining Montana herds, causing considerable loss there. The ranch stock in the High River district escaped this season, though considerable loss was experienced by farmers there. Mosquito Creek country suffered, as did Pincher Creek, also. A strange fact in regard to this disease was that the districts that were affected the year before were free this year, Willow Creek, Belly River, and south and east Porcupine Hills being untouched.

Other troubles came to Dr. McEachren, though he managed to free himself of them by resorting to arbitration. "Dave" Cochrane, mentioned previously, had squatted on the Waldron range and seemed to be doing very well in his own peculiar way. Dave had a taking manner and had accumulated quite a number of household articles and stock. His kitchen stove was a splendid one, of which he was inordinately proud, though the mention of that stove to the Mounted Police Superintendent at Macleod would have precipitated a riot. Some few years before, about the time Dave commenced housekeeping, the Police at Macleod received a large amount of supplies, including a very fine range, which was put in the yard to await preparation for its reception in the mess-kitchen or wherever it was going. Dave, who was in and out of town all the time, "lifted" a lid, a door or a leg, every time he came around, the result being that before a very long time had passed, the article of kitchen furniture was denuded of all but the heavy frame. Dave could not very well smuggle this away unknown to anyone, so he displayed some more reasoning powers, slipping around after dark and emptying a pail of water on it and then waiting until it was red with the new rust. When his stage had been set to his entire satisfaction, he went to the officer commanding and asked permission to take that "old stove" home, seeing as it was not doing any good where it was.

"I used to belong to the force," urged Dave, "and they never gave me anything. I think you might at least give me that worthless old heap of scrap iron."

"Where is it?" asked the superintendent, who really could not remember any "old" stove.

"Here," said Dave, and he showed him the disreputable results of his craft.

"I did not know we had any old wreck like that around," said the officer. "Certainly you can have it." So Dave loaded it into a wagon and went home. Some time later this officer was in the neighborhood of Cochrane's place and stopped in for a meal. A bright, new, shining range, big, warm and most comforting, was fitted into the kitchen and doing splendid work.

"That's a pretty fine stove, Dave," said the superintendent conversationally. "Where did you get it?"

"Why," exclaimed the host in vast astonishment. "Don't you remember that stove? You gave it to me, you mind, and I've just polished it up a bit!" Tradition asserts that the officer simply scratched his chin and remained silent.

At any rate, Dave Cochrane had established a reputation by the time he had been on the Waldron range for a short time, and, as the years passed, the nervousness of Dr. McEachren increased, because he was mightily worried by Dave's presence. Dave Cochrane, located in the middle of a well-stocked ranch, with another man's calves rubbing right against his fences, tended to make that other man somewhat nervous, though he steadily refused to meet Cochrane's demands for restitution if he moved off, though the squatter even offered to be satisfied with a board of arbitration, instead of insisting on the full five thousand dollars he had originally asked of the doctor.

Sitting on their horses one day, and arguing this old, old matter, Cochrane lighted his pipe, puffed it into activity and vigor, held up the burning match, eyed it reflectively, then, turning to his companion, said:

"Did it ever occur to you, doctor, that a tiny flame like that could burn out your whole range?"

Just that, and nothing more, but Dr. McEachren moved with almost suspicious alacrity, agreeing without more ado to pay any figure that an arbitration board would fix, if Cochrane would move out at once. Lou Murray, John Herron and Billy Hyde were selected as the arbitrators, and they decided that twenty-seven hundred dollars would be perfectly fair for Dave, the latter moving out as soon as he received this sum.

Importations in summer and fall continued as usual, C.G. Geddes importing a band of general purpose and driving horses from Montana, Hodder of Calgary bringing in some thoroughbred Polled Angus stock, Samson and Harford of High River bringing six hundred and fifty beef cattle from Oregon. W.J. Hyde of the Trefoil Ranch imported a carload of livestock, including a Percheron stallion and two Clydesdales, while twenty-three new settlers went into the St. Mary's district with 288 horses and 1,877 cattle. Sir John Lister Kaye imported farmers, experienced Scottish and English agriculturists, to place on his great system of ten farms that he had established near Gleichen, J.R. Mollison, manager of the Mosquito Creek "76 " Ranch, which belonged to Sir John, being also manager of the farms as well. That fall there were many thousands of acres broken and under crop on these farms, two thousand acres on the farm at Balgonie, and the others averaging five hundred and six hundred acres. Sir R.W. Cameron purchased fifty thousand acres of land at the mouth of the Bow River to be used as a horse ranch. Sir John Lister Kaye had two thousand head of stock on his Mosquito Creek Ranch, the Oxley was running ten thousand, and there were a thousand horses and a thousand cattle in the Kipp district. All told, it is estimated that a hundred thousand cattle and many thousands of horses ranged the country between High River and the international boundary.

Lands in the south were very cheap, the North-West Coal and Navigation Company, which was working the mines at Lethbridge, trying to navigate the rivers, and preparing to build a short railroad into Montana, offering blocks of from ten thousand to forty thousand acres for a dollar and a quarter an acre, some choice blocks being held a shade higher. Mackenzie & Mann, railway contractors, purchased the Glengarry Ranch on Trout Creek from the original backers, retaining A.B. Macdonald as manager.

Prairie fires and other blazes were experienced in nearly every district, the fires often originating from locomotives. Brealey Brothers' ranch near Calgary was burned, a five–thousand-dollar loss being experienced, including a thousand-dollar stallion. A number of small prairie fires in the district did but little damage, and providential February rain extinguished a fire that had burned for a day near Belly River, just off the Stand-Off trail. Spring fires and June droughts promised a very bad year in the Lethbridge district, but late June rains stopped the summer blazes, though the fall fires, started often by the railroad, threatened severe loss, Dave Akers' buildings being saved only through heroic efforts of the Mounted Police and a number of settlers. One unfortunate incident happened on the Piegan reserve, where C.W. Kettles was then storekeeper. It appears that the Government sent inspectors through the reserves of Canada to check the supplies in order to keep track of expenditures. The official who came out to make the inspection in 1889 had a useless trip to the Piegan reserve, for the night before he was to check up the stores, the building took fire from some unknown cause and burned to the ground.

Someone killed some of Joe McFarland's cattle in June, and two suspects, Loudon and Fountain, were arrested but dismissed. A Piegan brave trimmed a Chinaman until even Confucius could not have recognized him, and escaped to the reserve and safety, and other interesting occurrences took place among the red men.

Star Child, suspected murderer, known thief and ex-convict, upon his release from the cells experienced a change of heart and averred he was for law and order, proving it shortly after he had been made a Police scout by descending upon two white men who were on the river bottom near Lethbridge trying to drink the contents of a ten-gallon keg of whisky. Strong in his new position, the Indian took the liquor and, despite threats, prayers and physical opposition, emptied it upon the thirsty soil. Other good deeds were performed by this Indian before he lay down that fall and died of consumption, one of the curses of the Indian tribes, though in this case it may have been contracted during the long years in prison.

The young Bloods spent a pleasant, if not a profitable, season in the Indian country in Montana, their peace pact with the Gros Ventres and Assiniboines being shattered to smithereens. Five young Bloods and one Piegan slipped over into the Crow country on the Big Horn and started back with a hundred head of Crow horses, hotly pursued by indignant Crows, and scattering exhausted horses here and there along the trail, the stolen stock being poor from a hard winter. Just as they swung into the Bear Paw Mountains, they were surprised by some friends of the Crows, a band of Gros Ventres, who attempted to interfere, and lost two scalps in the resultant opposition of the Canadian Indians. Holding the remainder of their stolen stock steadily in front, the raiders pressed into the mountains, meeting more excitement in the persons of an Assiniboine and another Indian, who also tried to stop the successful progress of the Bloods. The Assiniboine expired in a short time; his companion faded rapidly into the defiles. The Bloods stopped and rested, harvesting the spoils of war while doing it, Prairie Chicken Old Man taking the dead man's gun and scalp and another Blood confiscating the horse. Just as they completed this pleasing operation, they looked up and spied a troop of United States cavalry that had started out to stop their depredations, so, knowing well that the time for trifling and cheerful bickerings had passed, the raiders leaped on their horses and fled to the safety of the north side of the Medicine Line. With only five of the stolen horses left, they rode into their home camps, gloriously pleased with themselves and their three scalps, though they spent a few moments of nervous suspense when the Mounted Police corralled them and took them into Macleod, where a stern lecture was read to them and the five horses taken back to be returned to the Crows. The United States Government not seeming to wish to prosecute, the raiders were allowed to go back to their tribe, where their three fresh scalps aroused deep admiration and envy.

Desire for emulation burned fiercely in the breasts of other Bloods, and they longed so actively for opportunities to distinguish themselves also that they eventually found themselves in the Gros Ventres' country, with a strong herd of the latter Indians' horses before them, one scalp at belt, and the boundary in sight. Their vigilance relaxed a little when they reached Canada, so they were hugely surprised when a squad of red-coated Mounted Police swept down on them with guns ready. The Bloods scattered like chaff before the tornado, firing a few wild shots, but doing no damage. The Police pressed a hard pursuit, and one Indian in particular, mounted on a very weary little mare, was so crowded that he resorted to the desperate expedient of slashing the poor animal with his knife to spur her on. A few strokes, and a few spasmodic spurts on the part of the mare resulted in her dropping dead, the Indian then taking to a coulee on foot and escaping with the assistance of a providential nightfall.

Deerfoot, the Blackfoot who had been sought for so long, grew tired of a continual life of evasion and gave himself up, receiving a brief term in the cells at the Police barracks. The brewery at Lethbridge was closed by the Department of Inland Revenue, and whisky smugglers continued with their disreputable business, the result of which was one of the most cold-blooded murders in the history of the Macleod district, the murder being directly traceable to a debauch with the poisonous "rotgut," which possibly made the perpretrators really irresponsible. Four or five cowboys succeeded in securing a supply of whisky one late fall day and at once repaired to the river near the reserve to have an uninterrupted spree. An Indian buck and his squaw saw the carousers and, Indian-like, came to get what they could, being greeted with a hilarious welcome, and treated to the best in the jug. Morning saw the drunken party still there, but with empty jug and doubtless a certain feeling of remorse, their minds confused and blurred by the poison fumes. Finally the squaw and her companion started to their camp across the river, and one befuddled cowboy suggested that she might tell someone where she secured her liquor and thus bring them all into trouble.

"No, she won't," declared another, "gimme yer gun," and drawing that weapon from the other's holster, he shot the squaw dead.

"Now, by gosh, the buck'll tell," complained another cowboy helplessly, and he had no sooner put this possibility into the mind of the murderer when the revolver cracked and the defenceless red man toppled over. The white men, considerably sobered now, thrust the two stark bodies through a hole in the ice and left the country at once, though the bodies were not found until the next spring, when they were so badly decomposed that there was no indication of foul play.

Another cruel murder that occurred this year was the slaying of a half-breed squaw named Rosalie, by a gambler named Jumbo Fisk at Calgary. This particularly revolting crime was also the result of a drunken carouse. Fisk's first trial resulted in a verdict of not guilty, but Judge Rouleau refused to accept such a finding, the jury then retiring and finally disagreeing. On his second trial, Fisk was fortunate in getting a sentence of fourteen years. The Indians in the district were greatly aroused over this murder, and would have killed Fisk if the opportunity offered. The progress of the trial was also watched with great interest by the red men, for they wanted to see just how white men would treat another white man who had murdered an Indian, knowing well from experience how promptly and decidedly the white men treated Indians who had ill-treated or killed white men. The final verdict did not particularly please either the Indians or a good proportion of the white settlers.

A somewhat less gloomily morbid incident happened to Tom Purcel and a Mounted Police sergeant named Macdonald. Tom was the gentleman who had so grimly insisted on bringing "The Kid" before a court of justice for stealing his horse and saddle, Tom displaying a great desire then for enforcement of the laws of the land. On the Fourth of July, memorable all over the United States as the date of the Declaration of Independence, Tom was caught red-handed with a cargo of whisky, six well-filled five-gallon kegs, which Macdonald proceeded to spill upon the ground before his anguished eyes, no attempt at interference being made by the sufferer. Then Mr. Purcel accompanied Sergeant Macdonald to Macleod where his wagon, horses, and outfit were confiscated, and he himself was allowed to pay a heavy fine in lieu of a pretty stiff term in the penitentiary.

No story of the early times in Alberta, of its cowmen and their lives, would be complete without a description of the remittance men, the sons of the "good" families who were shipped into the new land, ofttimes because when once there, they could not come before the people of the old land so prominently. To the ordinary Western mind, a remittance man was a rich Englishman who had proven a failure in his homeland and had been shipped into the raw land to kill himself in quiet or to work out his own regeneration if possible. Usually the Westerner looked upon him with the scorn that a capable man feels for the unfit—the scorn that men who made their own way with nothing to start on but a sound constitution feel for the failures who commenced life with the proverbial silver spoon. For years the doings of the remittance men, their misfortunes, characteristics and errors of omission and commission, served as the sauce and dessert over many a campfire, for they did strange things. Young men who had tried the parental patience unto the snapping point were sent to the "colonies" to hibernate, young men who were cursed with drink, women, or idleness to the everlasting disgrace of the "family," so long as they remained within the scope of that exclusive circle, were given money and orders to go to Canada and the West. They went, leggings, monocles, caps, accent, and habits, and proved everlasting sources of enjoyment and personal gain to the hard-headed settlers and cowboys. A remittance man in any particular district was a local pride, and his doings were magnified and improved upon for the edification of the inhabitants of other districts, who, if they too owned such a person, listened in superior silence and then came back with more outrageous doings of their own man.

"Remittance man" is a household term in the colonies of Britain, and nowhere is it more thoroughly known and understood than in Western Canada, where the class established an enviable record for themselves and placed a smudge on the name of all Englishmen. There are good men in the isles of Britain, and a good Englishman is second to none in the world, but remittance men were all "Englishmen," and the early settlers took to classifying Englishmen as remittance men, to the great derogation of the men of England.

The remittance man of Alberta was as vigorous and busy as any in the land— busy and vigorous in pursuit of leisure, pastime, and mad excitements. He had

Remittance men were often regarded with scorn by Westerners and were frequently the subject of humorous anecdotes. Seen here are (left to right) Charles Sturrock, Mr. Figgis, Mr. Swartz, Wilson Jones, E. Buckler and Ted Lewis.

money— much money that was supplied by anxious "home" people whose anxiety was based on a fear that the undesirable might come home if he were not supplied with sufficient capital to keep things going—so he received sufficient and kept things "going," to the great edification of the pioneers. They saw him try to rip the interior from drinking places, they saw him paint the towns brilliant vermilion, they saw him waste money in sinful profusion, and they abetted him in this. They saw him the mainstay of the hotels, the proprietors of which boarded, gave sleeping rooms, and allowed the bar to trust him, for the regular quarterly cheque would cover any one man's drinking capacity, and the remittance men did not tote up the bar tallies very carefully anyway. They rolled in the gutters; they rode through the streets, they drank with everyone, and they made pitiful exhibitions of themselves. Sometimes, when they had been sent out to endeavor to make a new start in life, the disgusted folks at home learned of their incorrigibility and cut the remittance off short. Then the pioneers saw sadder things than plain waste: they saw these scions of good families become ragged and shiftless, drink beggars, hotel hangers-on, squaw men and worse, though sometimes a brighter result was shown when the latent manhood came to the surface and the remittance man became a man, an upstanding one to whom the opprobrious word "remittance" was no longer given.

The British training of feeling sure that all things learned in England were certainly better than any other teaching, the slow-moving brains of some, the lack of initiative, of originality, marked the general run of them as hopeless for Western life, where necessity is ever forcing and developing the inventive genius, the adaptability, and the power to make one thing do work for a dozen other articles.

A remittance man who wanted to buy some horses to start ranching visited a livery stable and saw two huge mules placidly munching oats from nosebags. The British mind, working hard, read the plain facts before his face and, believing he knew what he saw, was consequently not at all self-conscious when he turned to the attentive livery keeper and asked with dignity:

"I say, do you always have to keep them muzzled?"

He was one grade. Another is the slow thinker who takes time to consider what he hears. One time a party of lords who had come into the country to shoot ducks and other things employed a settler—an educated gentleman— to act as guide. One of the lords was a splendid man, and he warmed up to the settler when he learned that he knew and respected a branch of the latter's family in the old country, and they grew quite friendly. Ducks whirred up from a nearby slough, an aristocratic shotgun whanged and a splendid wing-shot brought a large drake flopping to the ground, a wire fence being between the marksman and his bird.

"Ow, my good fellah," called the nimrod to the guide: "Just climb ovah theah and fetch that bihd, like a good fellah."

The guide, amazed and surprised, promptly replied: "Pick up your own damn bird!" and walked on accompanied by the friend of the family branch, who remained very silent for nearly an hour, when he stopped, slapped his thigh and bellowed: "Haw! Haw! Served 'im bloody well right!"

He had been diagnosing the conversation between his friend and the guide, he had reasoned slowly and surely until he had arrived at the final and correct result that the guide had only done perfectly right in retorting as he did.

High River settlers remember a remittance man who went through one hundred thousand dollars in a few years and finally came down to driving a sawdust cart. This man had been sent out to try his hand at ranching, and he did it all—over the bar of the hotel—writing hopeful letters home of great success in the near future, of a splendid ranch, grand cattle, and a life of brilliant promise. One day this man received word that his father was coming to spend a few days with him on the "ranch," and he was greatly perturbed until he thought of the ingenious scheme of renting a ranch, stock, cattle, everything, for the week of the visit and "employing" the owner as hired man. This worked to perfection; the father came, a staid old English gentleman who displayed considerable pride in the success the son had made in the new land. Everyone in High River knew of the arrangement, everyone joined cheerfully in deceiving the old gentleman, treating the son as a man of weight and speaking often of the ranch. The father went out, saw the rolling plains, neat fences, snug buildings, goodly stock, and was so pleased that when he left for home at the end of a week, he tipped the owner of the place and gave his promising offspring another ten thousand dollars, which that worthy proceeded to dissipate in riotous living. Later, after a strenuous struggle with sawdust, the cart, and the cart horse, this man was taken ill and went to a hospital, where he proceeded to fall in love with a nurse and marry her. Fortunately for his wife, an affectionate aunt of the husband passed quietly away shortly after the marriage and left a large fortune, which drew the spendthrift back to England, where he became a respected and sober member of some high-class community.

Another and less cheerful case was of another married pair. These people were of "very good family," rich and respected, but had contracted the drink habit to such an extent that they disgraced the dignified relatives and were persuaded to try life anew in the far North-West. They tried, chiefly because it presented a choice haven where they could drink all they wanted to without bothering anyone at all, and they created considerable sensation. A remittance of one hundred and fifty pounds a month was settled on the pair, and they purchased some few hundred acres of land, furnished a house sumptuously with pianos and all comforts, purchased a thousand-dollar bull and a drove of thoroughbred hogs and filled their cellar with extensive shipments of choice spirits. Chosen friends flocked to them and they sorted out the most hilarious and kept them. They drank at night and morning, and their money went as soon as it arrived. One of their amusements was the intoxication of their seven-year-old son, whom they often filled with whisky for the edification of themselves and friends. One night when a party of the select were at this place, some brilliant spirit thought of the highly humorous idea of bringing the bull into the parlor and making it sing to the accompaniment of the piano, which would be played by the lady of the house, a most accomplished musician. No sooner was it thought and spoken than it was done. They dragged the animal from the snug stall, haled him to the house, prodded him into the lighted parlor and tied him close to the piano. The mellow chords had scarce commenced before the mighty brute had spied the window and realized that it was a route to freedom. Bellowing madly with fear, he surged to it, dragging the thousand-dollar musical instrument, ruining that as he crashed through the window, and ruining his own thousand-dollar neck as he broke it leaping forth. The company was quite downcast for a time, not because of property loss, but simply because the bull had refused to sing. They drank a few more rounds, and then the thinker was smitten with another thought, an original and stunningly appealing proposition. Everyone was hungry, there were some hundred-dollar pigs in the pens—why not have some fresh pigs' liver? By the light of lanterns they got it, cutting the warm livers from several quickly slaughtered porkers and throwing the remainder of the bodies on the manure heap. By morning they all went to town to replenish the stock of spirits, and those bodies of dead pigs lay and rotted where thrown, while the body of the dead bull ornamented the ground below the window. In time the woman died, the man went from bad to worse, and the boy, a stunted youth, became an early drunkard.

A remittance man who sought employment on a ranch was told to take his team and do some ploughing, the rancher later going into the corral and seeing his hired man standing irresolutely by the team and perplexedly eyeing one horse who seemed useless, there being only room between the plough handles for one animal! Another of the class was sent to town with a team of light horses attached to a light wagon. As it was a town trip, the animals were checked high, which the driver did not notice until after a long, hot drive, when he came to a creek and decided to water them. But how was he to get their heads down when checked up? A cowboy rode along just as the remittance person had solved this problem by standing knee-deep in the creek and lifting the rear wheels of the vehicle in order to let the thirsty horses get their heads down.

Weary and disgusted "home" people at times refused to send further remittances when they learned of the hopelessness of the particular subject they were interested in. In such event it became necessary for the unfortunate to either go to work or else deceive his people, the latter channel to relief being usually attempted first. One very successful deceiver lived at Calgary. His family scorned his pleas for money, saying he had nothing to show for the floods they had lavished on him, so he proceeded to show them they were mistaken. Borrowing a handsome horse, and hiring a uniformed fireman of the town, he took the two to a favorable position in front of the courthouse, where he photographed them. The low, substantial stone house, the sweep of open prairies behind it, the hills in the distance, presented a convincing proof that the house was the ranch home of the photographer, the commons and the hills his range, the horse his pet saddler, and the uniformed man his groom. More money was sent, and joy reigned until that too had vanished.

It can easily be seen from the foregoing just what basis the old-timers had for the solid contempt they felt for these useless sons, and the spirit of the old settler in the following anecdote will be appreciated:

There lived in Alberta at one time a man who was perhaps the finest liar who lived, a man whom Ananias would have sat at the feet of, a plain and fancy, straightforward and skilled prevaricator. Nobody believed him, nobody took any pains to keep him from knowing how they felt, for he really was an outrageous liar. Came a monocled remittance man, trustful, friendly, and the liar seized him as prey. All day they stood at the·bar rail and the settler talked while the new-comer paid. All day the victim listened, until at last even his confiding innocence could not believe all. Politely excusing himself, he left the untruthful one and went elsewhere, eventually meeting a settler to whom he confided his belief that the well-known liar was in his opinion an untruthful man. Indignation shook the frame of the listener, and shoving a horny fist in the remittance man's face, he demanded retraction and apology, quick and complete, which came promptly from the astonished and befuddled victim.

"What made ye do sech uh thing?" demanded bystanders in amazement. "Ye know thet ole larruper is the biggest liar in the world!"

"That may be all right," admitted the defender, breathing hard and looking grim, "but I ain't goin' tuh let no pin-headed remittance man say it!"

One of the class who joined the Mounted Police proved that there was in him the making of a man. When he first joined the force, he was receiving his remittance and proceeded to continue making a "bally ass" of himself until his superior officers bethought themselves of sending to the people in England and having the money supply cut off. This was done, and the subject was sent into the Yukon, where he made a name for himself that is enviable even among the members of the force. Accompanied by one man as reckless as he was and a team of tough dogs, he broke the first winter trail between Dawson and Fort McPherson, a tremendous undertaking that the natives had absolutely refused even to attempt.

The fourteenth or fifteenth heir to a dukedom, who had been sent into Alberta where he could not sear the eyes and the ears of his family anymore, spent his remittance so fast that he was always very tightly pinched before the next one arrived. This was so annoying that he sought and obtained employment on a ranch, and departed thenceforth for the hills, there to ride and chase the wild cattle. Two days afterwards he was back in town, and a curious native wondered audibly why he had quit work:

"Got a vacation?" he asked.

"I was dischawged," was the dignified reply.

"How's that?" sympathized the inquisitor.

"I cawn't just make it out," said the mystified ex-ranch-hand, "but my employah awsked me to go and get some cattle. I don't remember whether he awsked me to bring him two eleven-year-olds or eleven two-year-olds, but anyway when I returned, he was awfully angry and rude and dischawged me at once. I've just been thinking that I must have brought the wrong cattle."

They were the natural butt of the cowboys' jokes, and the mirth of most fore-gatherings. One man who was visiting at a ranch mounted his horse and started toward town, the rancher seeing him dismount about a mile distant, tie his horse to a fencepost, and walk back to the house.

"What's the matter?" asked the host.

"Nothing," came the reply in somewhat surprised tones, "I just forgot to take a couple of letters with me, so I came back to get them."

Once a party of cowboys were drinking in a hotel in company with a remittance man, and they had so pleasant an evening that when closing-up time came, they allowed the proprietor to push them outside and lock the bar before they remembered that there was a long, dry night before them and none had thought to lay in a supply of flasks. This calamitous occurrence darkened the spirits of all until one remembered that he had a flask in his room, number 40, right next to the apartment where a bride and groom had taken up quarters on the way from the farm to honeymoon in the city. The groom was big, wide, red-faced and thick, giving the impression after even a cursory glance that he was not only powerful, but was also impulsive.

"I'll go and get youah bottle," drawled the remittance man.

The tempter entered then, the cowboy thought of the groom, of his indications of impulse and strength, and he fell before temptation.

"Aw right, it's number 41," he said.

As near as could be learned by putting the somewhat disjointed episodes that followed together, the messenger stalked right into the room where the groom, who had just about prepared to retire, was taking a "nightcap" from a portly bottle. Usually when a groom sees a strange man boldly enter his bridal chamber he is not pleased, and this groom showed indications of being like all others. He glowered at the intruder, meantime holding the bottle in a loose and unconscious way. Without loss of any time, the messenger walked to the bottle, plucked it from the fingers of the owner, muttered perfunctorily, "Beg pawdon," and—

Downstairs the listeners heard broken words of expostulation, grunts, scrapings, and the body of the messenger was projected down the stairs, coming comet-like among the waiting cowboys. Picking himself up, he ruefully eyed the jagged remains of the bottle he held, thrust it toward the cowboy who had sent him up, and said apologetically:

"Theah was a bloody boundeh in youah room and he insisted on keeping youah whisky."

The sound of deep, compressed breathing came faintly from the head of the stairs, and then a great voice rumbled down:

"Bounder, be I? Wait jest a minnit en we'll see about thet boundin'." Then came a clump of feet descending the stairs, the creaking of the boards as some heavy weight pressed them, and the messenger left the hotel for the friendly blackness of the open prairie.

CHAPTER XIII

1890–1891

THE big ranches had been established for ten years, figuring everything from the date of the lease and the time of the first unfortunate Cochrane herds, and the livestock in Alberta had increased greatly, the herds now numbering scores of thousands. When 1880 came to an end, there were probably eight or nine thousand head of livestock in the Province, and since then they had grown by increase and importation into very large numbers, despite bad years and early bad management in some cases, and theft, disease and weather.

A census of the livestock in southern Alberta taken by the Mounted Police in 1890 showed over a hundred and fifteen thousand cattle south of the Little Red Deer and nearly two hundred thousand head of every kind of livestock. Though the Little Red Deer and the Big Red Deer formed the northern boundary of this district, there were not any very large herds in there, the majority being south of the CPR main line. The Mounted Police census was also somewhat general, though probably fairly correct, the figures being formed from estimates and judgments of constables on patrol, roundup figures, and similar methods. Roundup figures would be either too high or too low, and estimates of the big ranches would be low, while small ranchers would doubtless add a little to their real totals. Between Lee's Creek and Pincher Creek in southern Alberta there were 74,822 cattle, 8,729 horses, and 1,034 sheep. Between Mosquito Creek and the Little Red Deer, the Rockies and Crowfoot Creek there were 13,500 horses, 43,000 cattle and 18,000 sheep.

The British American Ranch, which was the property on which the Cochrane sheep were placed in 1884, was now a cattle ranch, this sheep venture proving unfortunate as usual with the Cochrane interests in the Calgary district. Disease, bad springs, unfortunate accidents, markets, thieves, and storms had reduced the herd to nothing at all, and cattle now ran on the range, which the sheep had not held long enough to spoil forever for the fastidious beeves, though it is well known that cattle will never graze in a country where sheep graze, partly because they do not like to and partly because the sheep browse so close that a steer or a cow cannot secure food. W.D. Kerfoot had apparently commenced ranching for himself, for in the following report of the stock in the Calgary district, his name appears while that of the B.A.H.R.C. is not in evidence:

Ranchers in Calgary district, 1890, from Mosquito Creek north to the Bow, with number of head of animals:

Name	No. Horses	No. Cattle	No. Sheep
McHugh Bros.	333	1,155	—
C.P. Cattle Co	83	112	—
Johnson Bros.	—	—	1,500
High River Horse Ranch	913	—	—
W. Iken	102	—	—
C.F.I. Knox	112	130	—
"CC" Ranch	—	900	—
W. Skrine	—	550	—
E.A. Cross	144	550	—
H.B. Alexander	—	1,400	—
N.W. Cattle Co	761	9,661	—
Sanson & Harford	50	2,800	—
Ross & Podgers	—	300	—
F. Brown	—	175	—
J.J. Sullivan	—	250	—
Ings Bros.	—	400	—
G. Emerson	—	750	—
Bow River Horse Ranch	505	—	—
Scarlett Bros.	—	250	—
D. Lynch	—	250	—
J. Robinson	275	550	—
Hull Bros	—	1,200	900
D. McDougall	275	500	—
J. McDougall	73	200	—
Leeson & Scott	—	615	—
Mount Royal Ranch	—	600	—
W.D. Kerfoot	53	500	—
W. Bell-Irving	91	150	—
J. McKinnell	—	240	—

Brealey Bros.	—	220	—
Shea & Madden	—	200	—
W. Cowan	—	200	—
Merino Ranch Co.	—	—	5,000
Quorn Ranch	1,212	5,000	—
Heald & Eustace	—	300	—
J. Fisher	107	160	—
J. Quirk	37	700	—

The business of breeding livestock on the range had by this time grown from a haphazard, go-as-you-please system to a fairly careful, methodical business, where most little items were looked into and taken advantage of if they appeared to be likely to be of assistance. The first big herds were thrown on the open country and left absolutely to themselves, cowboys being engaged to see that they kept within reach of their own ranges, to assist weak stock in the springtime, work on the roundups, and brand in spring and fall. But after the first ten years, various lessons had been learned, sometimes in very bitter form, and the ranching business entailed more detail work than previously. The best ranches were weaning and feeding their calves before fall, weak cows were being corralled for winter feeding, bulls were generally held at the home ranch, and the good stallions were usually kept up, though scrubs were still allowed on the range. Some ranches went even to the length of killing some of their poorest mares, and in some districts the worthless scrub bulls were being shot. The influence of more careful breeding was being felt on the horse ranches, and the reduction of poor bulls was also proving beneficial. Blackleg was appearing in such scattered instances that it could be said to be practically eradicated. Fifty-one head of fine Highland cattle were imported by the Bar U, who also imported a number of good stock from British Columbia, amounting to one hundred and fifty-two horses and seventy cattle. For the preceding two seasons, the Cochrane Ranch had stayed loyally by the British market, shipping in the neighborhood of a thousand steers each season, and they kept up their average this year, making as usual the largest shipments to that market of any ranch in Alberta. The Canadian Pacific carried 66 horses, 352 cattle, and 225 sheep to the British Columbia markets; and took east 3,155 cattle, 531 horses, and 1,822 sheep, the horses mostly going to Manitoba and the majority of the cattle coming from the Macleod and Calgary herds. During the year the ranchers imported from the East alone 250 horses and 1,421 cattle, including many valuable bulls, stallions and mares.

Robert Evans took a large number of cattle into the Macleod district from the United States, and Joseph Card and Harker of Cardston brought 2,227 sheep into that section.

Farming operations also showed a healthy growth, there being twenty thousand acres cultivated in the Calgary district, while the Macleod district claimed 6,313 acres cultivated, 73,822 Cattle, 8,679 horses and 1,004 sheep, with a human population of 3,800 souls, red, white and mixed.

The experiment with Ontario dogies proved a success, the animals fattening very satisfactorily and J.D. McGregor making good money on them even after paying an average of twenty dollars freight per head, in and out.

A great amount of hay was put up, as was now quite usual, though a few ranchers were becoming somewhat careless owing to being new and not really believing the tales of the mighty losses of "eighty-seven." Late rains damaged huge quantities of hay throughout the Province, but a spell of fine weather followed, and there was no hay fed until the day after Christmas, when the deep snows came and forced fairly general feeding. There was some winter loss, heaviest in the immediate vicinity of the Bow River between Calgary and the Blackfoot reserve, and severe spring storms, again worst in the Calgary district, did some little damage to the calf crop. Taking everything into consideration, the winter was fair and the entire year was a pretty good one from the cattlemen's point of view, though they were not feeling so jubilant as the sheepmen, who had at last encountered the kind of a year they prayed for—little winter loss, a prolific lamb crop, no spring losses, mutton and wool prices climbing. Sheepmen were happier than they had been for some seasons.

Conrad & Company (the Circle) moved three thousand head from the Medicine Hat district to a range near Macleod, because they felt it was a better district and one not so liable to drought.

The American ranchers in the states adjacent were in parlous straits, cattle starving from lack of range and a dry season, and markets very poor. Montana herds were hungrily pressing in on the fine, untouched ranges of extreme southern Alberta, and the Mounted Police officers stationed along the boundary were practically doing duty as herders, turning great throngs of Montana stock back across the line almost daily. One American rancher drove his herd seven hundred miles from the Musselshell to Canada, paid duty and settled. He said cattle in the district he had just left were then worth ten dollars per head, and horses were going begging at forty dollars.

Glanders appeared in the Macleod district, resulting in the shooting of a few animals, and wolves as usual committed wide depredations, the cry for bounty being still as vociferous and useless as ever before.

The liquor business maintained its activity, the Government's granting of permits to individuals having placed a strong defensive weapon in the hands of the hotel keepers, who secured sheaves of these permits and then were able to maintain a pretty extensive stock of wet goods. The whisky smugglers were busy and performed some very interesting exploits in which the police were sometimes the victims, while the lawbreakers themselves suffered quite often. "Dave" Cochrane, upon leaving the Waldron range, seemed to have embarked upon this adventurous sea of endeavor soon after, and handled his deals with his usual finesse, if a story of one of his ventures is true, an episode which old-timers in the district relate with enjoyment and detail. To lead up to this, it is necessary to first explain what happened to Lee (probably Tom) Purcel and his friend Al Dowser. These two men successfully crossed the boundary with eight five-gallon kegs of whisky and cached them for future reference, finally going to open the cache after an energetic Police trailer had discovered it and planted an ambuscade. The smugglers took their treasured kegs, put them in gunny sacks, swung the sacks on their pack horses, one keg to a side, and started down the long slope toward the St. Mary's, whereupon the police came joyfully down upon them. Dowser and Purcel displayed great mental perturbation and prompt action, talking vigorously to each other of the merits and characteristics of Mounted Police and cutting the sacks loose from the packs, while they themselves "drifted" speedily away before the approaching constables. Purcel was caught at Whoop-Up and sentenced to six months in jail, but Dowser was more fortunate, getting clear away, and then, in a spirit of revenge, stealing three Mounted Police horses, which he sold in Montana.

This episode probably taught Dave Cochrane a better scheme, for we hear of him with a cache somewhere near that old one of Purcel's and Dowser's, his liquor handily enclosed in kerosene tins. Corporal Spicer, of the Police, the man whom The Dog and Big Rib shot at some years before, discovered the hidden hoard and quietly assumed the duty of watching for the importers. In time they came— Cochrane, one companion and a string of pack ponies. The usual gunny sack packs were arranged and tied on, Dave took his string of laden beasts, bade farewell to his companion and proceeded down the grade to the ford on the river, Spicer appearing just as the pack train was taking the water. With a startled yell, Cochrane lashed up his animals and dashed across the ford, but the well-mounted officer overtook him as they emerged on the other side.

"I've been looking for you for a long time, Dave," exulted Spicer, "and I've got you now."

"I thought you was just tryin' to beat me out across the ford," replied Dave innocently. "Goin' my way?"

"I am, for you're under arrest," smiled the glad policeman.

"All right," agreed the philosophic smuggler, and they jogged along amicably enough until they came to the point where the trail branched off toward Whoop-Up, when Dave turned off in that direction.

"Where are you going?" demanded Spicer.

"Goin' to Whoop-Up, of course," returned the other in apparent surprise.

"You're going with me," was the grim retort. "I told you that you were under arrest."

"Aw, I thought you was jokin'," expostulated the suspect in an injured tone,"but if you say so of course I'll go along. Just wait a minute." He calmly dismounted and just as calmly proceeded to untie the cans from his pack horses and heap them on the ground. "I'll just drop it here and pick it up when I come home," he explained.

"You'll leave it on them and bring the whole thing to town," retorted the annoyed Spicer. "You're under arrest for whisky smuggling."

"All right," sighed the prisoner resignedly, and they ambled along in silence until they reached the Police headquarters, whereupon Spicer dismounted and commenced to untie the packs.

"What'r *you* doin' now?" demanded Dave in vast astonishment.

"Taking this whisky off," explained the policeman, without stopping his work.

"You're crazy," the owner of the tins asserted with conviction. "That ain't whisky."

A hasty glance proved this to be true, the cans all being empty, and Spicer glowered as he demanded the explanation.

"Oh," said Dave easily, "I'm allus afraid of prairie fires, an I'm takin' 'em home to sheathe my shack."

"Where's the whisky?" growled Spicer.

"The whisky?" repeated Dave. "Why, that's in Whoop-Up, three hours ago. The boy took it on right after you started home with me. Why didn't you tell me that's what you wanted, en' then I wouldn't uv had to come away up here out of my way."

The Police of the Calgary district were annoyed by an organized gang that was operating there, running off the stock of the settlers and holding it for reward. They did a lot of damage, but the leader of the gang was finally run out of the country and the organization was broken up. A man named Maclean raided the orderly room at Macleod and secured seventeen hundred dollars. He was given five years, though the money was not recovered. When sentenced, he made an alleged confession implicating two others, Robert Carter and Bertrand, but both were acquitted. The authorities of Montana sent word they wanted a cowboy named Tom Fallon, and the Mounted Police arrested him on the roundup, holding him at Macleod but finally releasing him, as the American authorities seemed to lose their desire for him as soon as he was located. Another American named Macdonald was arrested at Macleod and held there for a time, but finally escaped through the negligence of Constable Skelean, who was sentenced to six months and then dismissed from the force. Macdonald, who had broken out from an American penitentiary, escaped to the States and was never caught.

The Canadian Indians kept the Police quite busy, for they raided a little, stole a little, and shot a few cattle, though the police reports declare they did not shoot as much stock as wandering bands of American Indians killed on this side. Some gangs of white thieves made successful raids among the herds of the Indians also, and the Sarcees killed quite a few head of cattle in the hills, the country being so broken that it was impossible to catch them.

Crowfoot, the great old Blackfoot chief, died this year, and his loss was felt to be a blow to his tribe and the Indians in general, for he was a man of very superior intelligence, and held strong control over his people. In September a party of Police and cowboys went out after a band of cattle-killing Indians, but lost them, due, the police declare, to the impetuous cowboys giving the red men the alarm in time to escape. Nevertheless, the Police captured the Indians' tents and ponies, which brought at least some satisfaction. An Indian named Takes-Two-Guns shot some Cochrane cattle and escaped to Montana, and the two bad Bloods, The Dog and Big Rib, were finally brought to book in a very sensational manner.

The Sun Dance had just about passed into memory, as there was really not much need of warriors any more, and the teachings of the Mounted Police and Indian agents had gradually turned the younger men against it, even though the old men still longed for the thrills of the tests. Among the Bloods a great meeting took place, a sort of half-expurgated Sun Dance, and some two thousand Indians congregated, among them being The Dog and his partner. As usual with the Police when the Indians gathered in numbers, a squad was sent out to see that they did not exceed the limits of respectability and reason, this particular squad being under command of Staff-Sergeant Hilliard, a cool-headed non-com. The Dog and Big Rib were recognized, and the policemen promptly rode into that horde of half-fanatic savages to arrest the two criminals, who were the pride of the lawless element.

"Pull your guns, but don't shoot unless I tell you," commanded Hilliard as he shoved his horse into the glowering throng, and right then an incident occurred that nearly sealed the death warrant of the little squad. If trouble had started, there would have been no chance for the whites to escape, for the Indians had surrounded them in a dense mass. As the men drew their guns, one of them, Tim Dunn by name, snapped a cartridge with an unconscious nervous contraction of his trigger finger. Fortunately the shell did not explode, and the men, after a mental gasp, thanked their lucky stars—for the first time in their lives—that some cartridges were badly made, realizing that if shooting had commenced, they would not have lasted five minutes among those armed Bloods, half of whom were already clamoring to kill. Hilliard heard the click, and realizing what it meant, ordered Dunn to be disarmed.

"Take that man's gun away," he barked, and the order was promptly obeyed. Hilliard then urged his horse fearlessly into the press until he came before the impudently grinning quarry, dismounted, seized them, and was dragged hither and thither by the struggling renegades until Chief Red Crow and the minor chiefs arrived in a body and succeeded in establishing a temporary truce while the matter was discussed pro and con by the Indians. The chiefs wanted the two outlaws to be given up, the tribe in general demanded war if necessary to save their tribesmen, but the judgment of the leaders eventually prevailed, and The Dog and Big Rib went voluntarily to the police barracks under arrest, later taking a sentence of five years' imprisonment with the usual Indian fortitude.

Among the new men in the cattle business were P. Burns and Walter Wake, the former being now the cattle king of Canada, and the latter one of the best judges of beef cattle in the West, a valued employee of the great P. Burns and Company organization, and a prominent rancher in the High River district. Wake started punching cows in the South, and Burns commenced railway contracting, the CPR branch line to Edmonton having just started, and there being great numbers of men in the camps who must be supplied with meat. It is said that Mr. Burns' gigantic business was first started by his selling a team of oxen during his homesteading days at Minnedosa, Manitoba, and realizing such a profit that he bought another pair and fattened them, tried a few pigs and met with success, and then embarked boldly into the meat industry.

The year of 1891 was a remarkably good one for both sheep and cattlemen, the winter being fine and the spring free of bad storms. In the Calgary district, the increase among cattle was eighty percent, and horses sixty percent, which was a fair criterion of the other ranching sections. The only blackleg noted was one isolated instance when McHugh Brothers lost a few head of young stock. In the Porcupines and the foothills, the wolves were bad, running in packs of about twenty. A.E. Cross lost twenty colts, and other ranchers suffered severely. One old wolf, who hunted alone and who had been responsible for damage amounting to thousands of dollars, was finally roped and killed by a man who had followed him a month to secure the fifty-dollar reward that was offered for the destruction of this particular animal. No wolf bounty had been given as yet by the Government, and though the cry from the ranchers was imperative and the need most plain, there seemed no indication of such a thing coming to pass.

Big bunches of sheep were imported to the Cardston district from Utah, brought in by Joseph Card, Douglas, and a number of new ranchers, some of them old-time cowboys who had saved enough to realize the usual ambition and commence in business for themselves, running bunches of one hundred to five hundred head. John Ware had a ranch of his own now, with four hundred head of good stock; Miles and Minesinger had a band of three hundred; T.C. Langford started a small ranch with two hundred; D.S. McIntosh left the "44," with whom he had been for some years, and began business for himself, while other men in various districts did likewise. The Cochrane Ranch and some other southern ranchers shipped heavy consignments to the English market, and the ranches of the Calgary district sent twelve hundred of their primest steers to that country.

Two Clydesdale stallions from John Turner's stock farm were sent to the Winnipeg exhibition during the preceding fall, being the first purebred horses to be sent out of the Province for show purposes. Both returned with first prizes in their classes, this victory of local stock putting added vim in the careful breeders of livestock.

A fierce prairie fire raged in the Lethbridge district during the fall, threatening immense damage, but being at last extinguished by the ranchers and Police. This fire started on the Oxley Ranch on October 23 and swept across country to the Circle range, where it was finally stopped. McAbee's house on Sheep Creek was burned up in another fire, and a wide expanse of timber and winter range was burned over and destroyed.

An unfortunate tragedy occurred in the ranks of the Police during March, when Constable Herron committed suicide. This man was on duty in the Lethbridge district and was doing much patrol work. The bright spring sun's glare had brought on the terrible snow blindness. He lost his horse, and in the bitter weather of March 2 started out to find help, wandering aimlessly over the wide prairies, sightless, frozen by the cold winds, staggering along until about a mile from the Brown Ranch and succor, where he finally gave up in exhaustion and despair, and with his remaining strength placed his revolver to his head and killed himself to end his sufferings.

At this time the United States was engaged in their last war with their North-West Indians, but the Canadian tribes were running along in their usual rut of calmness, with rare outbreaks of exuberance among the young Bloods, as usual.

This tribe and their brothers, the Piegans, observed the Sun Dance ceremonial, but only the old warriors seemed to take much interest in it, though their reminiscences over the fires sometimes stirred the young men to emulation. Among the most successful of the Blood raiders this year was a young man who played a lone hand, and conducted his operations with Scottish canniness and success. His name was Running Fisher, and he went on an innocent visit to friends in the United States, returning through a well-stocked Indian district and travelling considerably of nights. A number of horses vanished from Indian herds in Montana about this time and their owners never heard of them again.

George Lane, who had been operating as a buyer for western markets for the previous two years, entered into a contract with P. Burns and the Government, and went to the Blood reserve to furnish beef to the Indians, where he remained a year, throwing up the contract then to go into ranching for himself.

A shooting affray between Mounted Police constables and Blood braves occurred on October 19, during which the casualties were even, one being wounded on each side. Red Crow was very sulky over this occurrence, and blamed the Mounted Police for messing up his men before they were sure they had done anything wrong, for it appears that the constables acted against two Indians who were either innocent or had so hidden their tracks that they could not be convicted. On the day in question, the two constables, Alexander and Ryan, were riding near Cochrane Crossing when they spied two Bloods who had their horses packed with the dripping meat of a fresh-slaughtered beef. On the impulse of the moment, the constables started after the pair, though there was really nothing, hide nor calf nor angry white man, to show that the meat had been illegitimately secured. The constables rode close, with every intention of arresting the pair, when one Indian, firing from the hip, nicked Alexander in the neck. Ryan at once opened fire and shot the red man through the body, the bullet entering the back and passing through lungs and breast. Neither of the Indians was arrested. Alexander rode back to town and the wounded Indian was carried on to his own people, refusing absolutely to have anything to do with a white doctor. Arriving at the reserve, he sent for the Blood medicine man, paid him ten ponies to cure him, and was riding around again within two weeks, apparently as sound and as well as he had ever been.

The king of the whisky smugglers was easily Gus Brede, a man who had tormented the Mounted Police for years, and who really carried on an open war with the authorities on both sides of the line. Year in and year out Brede had been smuggling whisky into Alberta and smuggling Chinamen back to the States, taking a load either way whenever he made the trip, and realizing a good, if precarious, income. At times he had been caught and heavily fined, but he looked upon this as a sort of legitimate expense account, and went immediately and cheerfully back into his old habits. Gus was the bugaboo of the officers commanding the southern detachments and divisions of the Police, and a heartfelt sigh of relief was given when the annoyance was removed by lightning. Every human agency had been called in to make Brede stop his nefarious traffic in Celestials and firewater, but a higher power than the NWMP took the matter in hand this year, and a bolt of lightning smote him and his trusty team down on the trail leading into Fort Benton, Brede and his horses being found dead by the wagon with a half-dozen frightened Celestials huddled beneath the canvas covers of the vehicle.

A great strain was removed from the Police force during the year of 1892 when the new license system went into effect, whereby hotels were licensed to sell spirituous liquor. The force had fought the liquor traffic to the best of its ability for years, but had found it impossible to stop it, owing to the unpopularity of prohibition laws among the people throughout the West. Westerners disliked prohibition because they felt they were capable of handling a liquor traffic in a reasonable manner, and the danger from whisky among the Indians was not worth thinking about, as the Mounted Police had the tribes in good control. Also, the Westerner often wanted to take a drink himself, and he did not believe it was component with his dignity to have to take his rye or Scotch surreptitiously. If he wanted a drink, he did not want to dodge a policeman to get it. The new license law was consequently greeted with general acclaim, and drunkenness increased considerably, though there was much better order. Poor Gus Brede had died just one year before the most lucrative branch of his business failed.

The worries of the Mounted Police were now confined to general crime, Indians, and the turning back of the great Montana herds, which were pressing in on Canadian territory in ever increasing numbers. On March 6, three Nez Percés brutally murdered an old Indian, and one of the trio, Nez Percé Sam, was given a life sentence. The extension of the Macleod-Calgary branch of the CPR was completed September 27, opening up the whole ranch country to railway service and settlement, farmers coming into the new district in numbers. The hotel at the end of the track on the north bank of the Old Man's River was the scene of an interesting travesty on justice that fall, when a Mounted Policeman named Burgoyne had an argument with an Indian. According to the narration of the old settlers, Burgoyne came down from Pincher Creek and was sent up to Claresholm by the officer in charge at Macleod to assist in some Police work there. Burgoyne started, but, arriving at the end-of-track on the north bank, discovered a thirst and went into the hotel to get a drink. Knowing the way to the liquor, he went down cellar to the place it rested, falling over a half-drunken Indian en route and so arousing the anger of that worthy that the application of the butt of the Police revolver to the skull of the native was deemed advisable. The Indian, or his friends, hastened to Macleod to make complaint, while the hotel proprietor worried, knowing how much trouble would come to his door if a case developed. Calling to his aid the Canadian Pacific Railway Company, which was represented there by a worktrain of freight cars and a complaisant and agreeable train crew, he suggested the solution of his and Burgoyne's dilemma. The train was backed to the loading platform, Burgoyne and horse were shoved aboard, the locomotive whistled and leaped northward on its errand, which was to deliver Burgoyne and steed at Claresholm in quick order. Success met them, the train stopping just out of town and unloading man and horse, the constable then riding into detachment, and immediately calling attention to the time by declaring that the detachment clock was wrong.

Meantime the officer in charge at Macleod was very much hurt to think that a member of the force would swerve from his path of duty to take drinks and bend a six-gun across an Indian's head. It was a flagrant breaking of rules—and heads—a disgrace to the force, and an example would certainly be made of Burgoyne if he was guilty. Investigation followed, and the hotel keeper loyally stood by Burgoyne in declaring he had not stopped there. The circumstantial evidence of Burgoyne's quick trip, by trail and on horseback, between Macleod and Claresholm was also most convincing to the men who conducted the investigation, the time transpiring between the hour he left headquarters and the minute he reported at the detachment proving conclusively that he could not have turned off his road, and in fact must have urged his steed to great efforts indeed to reach Claresholm in the brief time he proved he had! A perfect and entirely reasonable alibi was proven, and the police looked upon the afflicted and unrevenged Indian as a splendid type of Munchausen.

Though very fine horses were being raised in a few scattered instances in Alberta, the general run was not improving much, and big ranches could well have culled out goodly numbers of poor mares. The stock business was also hurt in all departments by a very bad spring that killed large numbers of calves, lambs and colts. Cattle prices dropped from fifty dollars to thirty-five for a good steer, there was no sale for common horses, and the outlook was pretty gloomy to cattle and horse raisers, though the prices for sheep were very good, and again the sheepmen did well. Timber wolves reaped big killings, being especially bad in the Milk River Ridge, where they slew a number of straying American cattle. The stock association, despairing of Government help, took the matter of a wolf bounty in their own hands and offered rewards for every scalp shown. This produced an immediate result, Indians, half-breeds and cowboys devoting more time to the pests.

The disastrous spring storm, the bitterest blizzard in twenty years, killed many cattle in every district. Early in the spring, Charles Craig purchased thirty-five head of dairy stock and took them to his ranch west of Macleod, where the blizzard of late April swept them away before it, neither hide nor hoof remaining, the conjecture being that the entire band was driven over a cutbank and perished that way. The same fate came to a small band of sheep and a boy who was herding them, the lad freezing to death and the flock vanishing, either in a river or over a cliff.

In the Lethbridge district, this storm came on April 24, rain one day, snow two days, with a bitter north wind. Numbers of cattle succumbed, drifting into the fence corners and there perishing miserably. The Circle Ranch, which had moved from Medicine Hat district the year before to escape drought, lost five hundred calves and thirty grown cattle, while it is known that other settlers lost an aggregate of at least twenty-seven horses, fifteen colts and one hundred and eighty-two cattle. Following this, in the Lethbridge district, a drought came, seeds planted in April not appearing above ground until August, and though the good hay on the dry sloughs in the Pot Hole and Milk River Ridge countries was cut in good season to assure much food for winter, the long-deferred rains came in sheets and ruined great quantities of it.

Seven prairie fires burned great wastes in the Macleod district, and one settler who was twice burned out was so unbalanced by his losses that he tried to shoot a passing neighbor, fortunately doing no harm, however. The Blood Indians, who had sixteen hundred horses on their reserve, lost a large number of colts, and considerable of their winter range was spoiled that fall by fires. In the temporarily unfortunate Lethbridge district, the fires created great havoc, burning for days, and not only taking the grass but also burning down deep into the soil. One fire, which was started on July 21 by a passing train, burned for three days between the Bow and the Little Bow, where Howell Harris, manager of the Circle Ranch, was then on a roundup. Harris saved possibly hundreds of head by the desperate expedient of slaughtering cattle and dragging the dripping half-carcasses along the edge of the flames by means of horses, fourteen beeves being thus sacrificed to save their fellows, and incidentally the horses and men present. The ground itself was literally on fire, and twenty-eight horses had their hoofs badly burned in the fire-fighting.

The spring storms did considerable damage in the Calgary district, but conditions improved immediately afterward, and despite low prices, the ranchers had a fairly good year, shipping many hundreds of cattle east, though the horse business was as dead there as it was anywhere in the West.

Macleod estimated a hundred thousand head of livestock on the district ranges that fall; Lethbridge claimed twenty thousand cattle and twenty-five hundred horses; Calgary, with a hundred new ranchers, counted thirteen thousand horses, forty-eight thousand cattle, and eight thousand sheep in the district including Fish Creek, Elbow River, High River, and the Bow.

Despite the bad spring and a literally blazing summer, the cattle entered the winter in good shape, and the ranchers hoped for better times the coming year.

Jim Christie, the pioneer horseman of Alberta, who had been ranching in the Pincher Creek country for many years, pulled out of there in the spring and settled on Nose Creek, north of Calgary, with one hundred cattle and one hundred horses. J.H. Wallace, the big Lethbridge rancher, arrived from Idaho with two hundred head of stockers and located in that district, where he has remained ever since, dealing always in beef herds, and not venturing into breeding at all. George Lane left the Blood reserve, where he was handling the beef distribution for the contractors and the Government, and purchased the Flying E Ranch in the Porcupine Hills, where he devoted his time from then on until his interests broadened toward their present wide scope.

E. Stanley Pinhorn, manager of the Oxley Ranch since John R. Craig gave up the reins of government, committed suicide this year, during an attack of mental aberration, and A.R. Springett, for some years an Indian agent in the south country, succeeded him on the Oxley.

The British market was closed to all Canadian beef on September 29, when the British Government placed an embargo on all stock from Canada owing to a fear of contagious pleuro-pneumonia, traces of which the British veterinary surgeons declared they found in the lungs of Canadian cattle off either the steamships *Hurona* or *Monkseaton* at Dundee. The suspected animals, or rather the particular animals in which the veterinary surgeons found the suspicions indications, were from Ontario. Despite the fact that the principal of the New Veterinary College at Edinburgh diagnosed the cases as the "corn-stalk disease," which was discovered in 1879, the embargo held, and resulted in a big flurry and thorough but futile investigations through all Canadian herds.

The annual bucking contest was held in Macleod for that district, and Billy (W.R.) Stewart won it handily over all competitors. Stewart was a very fine rider, though he was not considered as good as Johnny Franklin, who competed against him in this year, but had the misfortune to draw a horse that would not buck a single hop.

During the summer, Lou Murray and a stockman or cowboy named Leeper took umbrage at each other and threatened gunplay under the slightest provocation. Meeting one day on the steep cutbanks of the Kootenay River, Murray dragged out his gun, but Leeper was quicker and thrust the big muzzle of his six-shooter into Murray's ribs, galvanizing the latter out of his saddle and down the steep bank into the cooling waters of the river, where he swam to the safety of the other bank. Both were later arrested and fined fifteen dollars each for this violation of the peace, the consequence being that they looked more eagerly than ever to "meet up" with each other so as to wipe out the added grudge.

One day Macleod town was full of roundup cowboys who were certainly not worrying a bit about the serious side of life. Leeper was there, drinking in the Macleod hotel, while Murray was later discovered in the Queen's, each man utterly ignorant of the other's presence in the hamlet. The cowboys, full of the usual range deviltry, managed to keep them apart, one band devoting their attentions to Leeper and the other to Murray, holding each in the hotel in which he was discovered. During this time Billy Stewart had been the go-between, moving from bar to bar, telling Leeper what Murray had said about him when last he had met him, and then going back and telling Murray the terrible insults that Leeper had heaped on him at their last meeting. White-hot rage soon took possession of the two, and they talked darkly of dire happenings if only the other could be met with, Stewart and the two parties of cowboys sympathizing with each wronged individual and telling just how to kill most painfully and disgracefully. They drew Leeper's big gun from its holster over his hip, admired it, withdrew the ugly shells, and tucked it back in the receptacle on the innocent owner's belt; they treated Murray in a similar manner, and then Billy Stewart inveigled Murray from the Queen's to the other bar, followed, of course, by every cowboy there.

The door swung open in the drinking room of the Macleod hotel. Murray stepped in and swung his eye down the line of the counter, encountering the black gaze of Leeper just as the latter recognized the newcomer. The big guns leaped like magic to the ready hands, they stepped into the clear and advanced slowly, the "clickety-click" of the futile hammers on their weapons indicating just how earnest they were. Drawing almost body to body before they realized their guns were empty, they switched the heavy weapons, grip for muzzle, and commenced to bend the steel frames across each other's head, but here the hilarious spectators intervened and patched up a truce that remained unbroken, if strained.

CHAPTER XIV

1892–1894

MINUTE investigations throughout Alberta during the year following the British embargo resulted in the satisfactory conclusion that there was absolutely not a particle of pleuro-pneumonia or anything like it in the herds of the Province, though there was at one time supposed to be an outbreak of it at the ranch of McIlree, Gow & Struthers at Graburn's Coulee, south of Medicine Hat. Thirty head died there, and Dr. McEachren, the Dominion veterinary chief, made a personal investigation, finding no pleuro-pneumonia at all, nor anything like it.

In addition to this, the Government made exhaustive and thorough examinations of the samples of the supposedly afflicted stock killed in England the fall before, these investigations being made by Dr. McEachren and Professor Adami, the result of their examinations being that there were absolutely no indications of contagious pleuro-pneumonia in any of even these samples.

The sheepmen of Alberta, who had experienced two good seasons, were due for a check, which came promptly in the nature of scab, introduced among the Canadian herds by sheep from Montana. Dr. McEachren, while in the West in connection with the investigation of the suspected cattle at Graburn's Coulee, inspected the sheep on the South Saskatchewan and the Bow rivers, finding scab on the Saskatchewan and quarantining two districts there, but finding the Calgary ranches free of the disease.

Throughout the ranch country, the cattlemen were generally pretty well satisfied with things, prices stiffening a little and sufficient market for the surplus being easily found, there being a quick sale for steers at forty dollars. On the other hand, the horse ranches were unhappy, partly due to themselves, they having done some very careless breeding during the past years, the result being some few fine horses, a very great many poor ones, and a very weak market, even for good stuff. Several hundred horses were shipped to England for sale on that fastidious market, and met with a rather poor reception.

In March the Mounted Police were notified that a quarantine against United States cattle had been established, and that the enforcement was under their immediate supervision. Upon receiving this notification, the Police established a quarantine ground at the boundary between the north and south forks of the Milk River, where all stock brought into Alberta from the American side had to undergo ninety days' quarantine under Indian herders and Police supervision. The strict enforcement of the new regulations was made particularly onerous and disagreeable by the great herds of Montana range cattle that ever pressed northward along the line, and by the fact that a large number of Canadian range stock had drifted across into Montana under the influence of a number of northern storms in late winter and early spring. It was estimated that nearly three thousand head of Canadian stock crossed in this manner, the majority being Circle cattle, though three hundred head belonging to the Cochrane Ranch and branded "STV," with a dewlap, were known to have crossed near John Joe's spring, a short distance from Milk River Ridge.

Thus, when spring opened and the quarantine was enforced, there was this large herd of Canadian stock on the American side, and naturally the Mounted Police insisted on their passing through quarantine when they were brought back, driving the animals back into Montana again if the owners objected to the ninety-day delay. Finally Howell Harris, manager of the Circle Ranch, made statutory declaration that the Mounted Police had prevented men from rounding up Canadian cattle, and further that the Police were driving such cattle into Montana to the detriment of Canadian stockmen. Owing to the agitation thus started, the quarantine was lifted on all range cattle, the Canadian ranchers declaring that there was no need of enforcement of the regulations against Montana range stock, as that stock was free from every disease excepting, perhaps, a little blackleg. This lightened the work of the Police tremendously, though they continued to inspect and hold all domestic stock brought in by settlers and homesteaders, inspecting during the year four hundred and seventy-four animals belonging to families from Washington, Idaho and Utah.

Loud complaints of damage by timber wolves continued, though it was hoped the bounty would soon reduce their numbers. Sheepmen especially were suffering from these animals this year, and this, in addition to the affliction of scab on some of the wool ranches, caused the mutton men considerable loss.

Spring storms were disastrous to many cattlemen whose stock was below the Bow, thousands of dollars being lost through the death of early calves and weak cows. Walter Huckvale lost thirty to fifty percent of his stock this year, and other ranchers in that and adjacent districts lost very heavily. Huckvale pulled out of that district early in summer and went to Many Berries Creek, south of Medicine Hat, where he entered into partnership with S.T. Hooper, the latter running Bar N Bar stock, while Huckvale still used his fiddleback brand. This ranch grew in strength until, in a few years, the partners were running three thousand cattle and three hundred Clydesdale horses of a very good quality.

These April storms having occurred regularly for the preceding two or three seasons, William Pearce reiterated his belief expressed often before, that the Alberta stock industry would progress more rapidly if breeding was cut out and the ranchers went simply into feeding and fattening stockers.

In Macleod district there was grave fear that the winter would prove hard on the range stock, as the snows started early and came heavily, but the chinook winds of January took off the snow and saved hundreds, perhaps thousands, of cattle. Two men named Cook and Christie, who had located in the Boundary Creek country, were so appalled by the severity of the winter that they pulled out, bag and baggage, as soon as spring broke, returning to the States and remaining there. The settlement in the southern end of the Province was considerable, however, as is instanced by the following: There were nine families in the Kipp section, with 290 acres cultivated, and 250 head of stock. There were twelve families in the Mosquito Creek district, all ranchers, who shipped 1,376 head of beef to England and had 157 acres in crop. Two families arrived in the St. Mary's district, where the Brown Ranch was located. Fifteen families in the Kootenay district, mostly old settlers. They owned 756 cattle, 455 horses. There was a large increase in the settlement in the Big Bend, thirty families arriving from the States and Eastern Canada, bringing the total in the district to forty-six families. Five new families came in the Pincher Creek district, bringing the population to 440, and the stock up to 9,000 head of cattle and 2,500 horses. Scattered settlement went into the Milk River Ridge country, where Howe and Aldridge were the largest stock owners, the former running 400 cattle and thirty horses, and the latter twenty-five cattle and sixty-five horses. All settlers in this district were stockmen, because farming had been found to be unprofitable owing to the prevalence of summer frosts.

Captain Hockin located a horse ranch in the Boundary Creek section, and a rancher there named Olsen also imported sixty-three head of stock. Only two other families settled there, Jackson and Boledore coming in with the spring and locating. There were no new settlers in the Stand-Off district, and only three in the Porcupine Hills section. Throughout the Macleod district the crops were damaged by drought, and small irrigation schemes were planned and put into effect at Leeds and Elliott's ranch at The Leavings, at the Glengarry Ranch, and

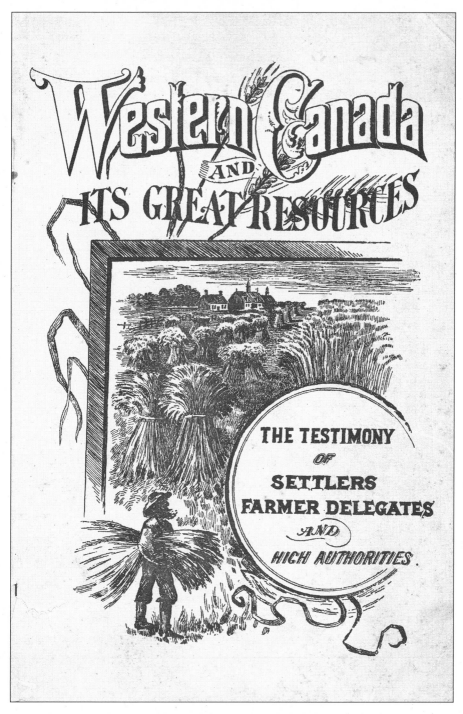

This cover of an 1893 federal government pamphlet encouraging settlers to discover the riches of Canada's West portrays a hard-working farmer reaping his rewards.

at C.A. Lyndon's. Owing to the failure of the crops in Big Bend, the settlers did not sow any grain, but decided to devote themselves strictly to stock-raising.

The Mormons in the Cardston settlement experienced a bad winter, a good spring, and a good year for stock-raising, while their agricultural efforts were so poorly rewarded that some seriously considered quitting farming entirely and going strictly into stock-grazing.

In the fall there were very destructive prairie fires in the Macleod district that destroyed much hay, the worst being a blaze that the Indians started near the Halifax lakes in September that burned up large quantities of hay that Mounted Police contractors had already in stack.

The dry season made the Lethbridge district again peculiarly susceptible to fires, and bad ones raged in the fall, doing considerable damage to stock and winter range. These fires were started in some instances by railway sparks, while some were set out by bone-pickers whose selfish aim was to clean off the grass so they could see the bones. Water in the springs and small streams was very bad, owing to contamination from the putrefying beef that died in considerable numbers in the coulees during the winter and spring. The Mounted Police made an official estimate of the stock in the district, which might be roughly correct, though it will be seen that they count Walter Huckvale's herd, which was moved to Many Berries Creek that summer:

LETHBRIDGE DISTRICT	NO. CATTLE	NO. HORSES
W.G. Conrad	6000	100
Conrad Bros.	5500	100
Cypress Cattle Co.	1900	75
James Pearce	1100	100
Walter Huckvale	650	40
Albert Whitney	400	100
Walter & D. Whitney	300	200
W.D. Whitney	150	25
E. Hasson	150	75
O.S. Main	150	12
John Ross	100	25
N. Walrock	80	150
J. Perry	70	—
Wm. Fixley	—	75
John Davis	50	60

Geo. Rowe	50	—
Russell	50	—
D. McNabb	35	—
Geo. Hauk	35	—
Josiah Davis	40	14
Ed. Holmes	35	10
John Duff	50	9
— Arnold	250	200
Browning & Maunsell	2000	25
Total	19145	1395

Four hundred and eighty-seven cattle and one hundred and thirty horses additional were owned by men not styled ranchers. Those owning less than twelve head were not included.

Cattle stealing and killing continued, the Indians doing their share, and not being very much blamed in some instances during the winter, as their reserves were overrun with ranging stock, and the temptation was always before them. One white man, James Bullock, received a sentence of three years in the penitentiary for stealing a calf from Maunsell Brothers' herd at Lonely Valley on the Milk River.

During the summertime, Walter Wake was working on the Oxley Ranch, but left before the season ended and went to work for E.H. Maunsell, riding the range south of Macleod in company with three or four other cowboys. One day they were attracted by the sight of a man riding rapidly south, and knowing that a person who was headed hurriedly in that direction was usually aiming for the boundary and its safety, they interested themselves enough to ride hard and intercept the unknown, who proved to be a stock-hand of the name of "Dublin." As soon as he saw he was among friends, Dublin drew rein and told of his reasons for haste, which appeared to be really urgent, for he said he had shot a Frenchman the night before in Macleod. As all the cowboys liked Dublin, who was usually a harmless, good-natured soul, this band decided to do what they could to ease his mind before he disappeared, for he was not just sure the victim was dead, though he had thrust the gun into his opponent's middle and pulled the trigger, after which he sobered up and fled into the night toward the southern refuge.

Caching Dublin and his horse in the brush, the boys proceeded with their work, one being alloted to ride into Macleod and learn just how much damage and excitement had been created there. Arriving, he discovered that Dublin had undoubtedly been as drunk as he had described, had attempted to shoot the lights out of a certain place of amusement, had caused excitement, had been interfered with by a Frenchman. The latter had stepped in front of the celebrator and tried to stop his fusillade, and Dublin had promptly shoved the muzzle of his gun against the other's stomach and pulled the trigger, whereupon the Frenchman crumpled up and fell to the floor in a sickly heap. The investigating committee of one further learned that Dublin's revolver had been empty when he pulled the trigger to fire the fatal shot, and that the French party, instead of being mortally wounded, had simply fainted, Dublin, befogged with drink, thinking the worst and leaving at once on the back of a fast, strong horse.

Another Macleod happening was a little adventure that befell John Cowdry, the banker, whose little office was known to every man in the south. One day when Cowdry was in this building, with the safe-door open, and all the money in the counter till, a man stepped in and presented the muzzle of a very useful revolver, demanding hands up and all the cash in the place. Cowdry turned, swung the safe-door shut, twisted the handles to lock the combination, and then coolly said, "Now shoot and be damned!" Not liking the possibility suggested by the latter portion of this sentence, the hold-up man showed indications of indecision, whereupon Cowdry stepped around the end of the little counter, gave his visitor some excellent advice, handed him twenty-five dollars and told him to leave the town before the police got him. The grateful gunman accepted everything, taking advice, money—and his leave to parts unknown.

A particularly ugly murder took place in the Pot Hole country early in December, when Lee Purcel shot and killed his old friend and neighbor Dave Akers at the latter's ranch on the Pot Hole. Many and many times to the personal knowledge of a number of settlers in the country, Dave Akers had befriended Purcel when he was sorely in need of a friend, going often to great lengths to do the favor. But bad blood had arisen between the two men, the direct reason being a discussion over some cattle. This hard feeling grew on Purcel until he decided he must kill Akers, so taking his rifle he went out and concealed himself near Akers' corral, waiting a considerable time until his victim came out of his house, when the rifle cracked and Akers dropped dead with the bullet through his brain. Great excitement prevailed in the district, and Purcel was arrested shortly afterwards, being tried in the following year on a charge of murder. It was a jury trial, and, as often occurs in these events, the jury came in with a peculiar verdict, when one considers the case as it appears from the reports of the time. They found Purcel guilty of manslaughter, and added a strong recommendation for mercy on account of the age of the defendant. The judge coincided with this decision and sentenced the murderer to three years in the penitentiary. It is said that at these same assizes a lad was given three years for stealing a colt.

During the year there was some horse stealing, but all complaints were not based on thefts, many ranchers complaining of losing their horses when they could not find them, and then leaving it up to the Police to look around. Thirty-three ranchers complained in one district of having cattle stolen, and Mounted Police patrols found twenty-nine of this "stolen" stock straying on the prairies. Wholesale slaughter by the Stonies of mountain sheep and goats, and of deer, caused considerable and often futile activity among game guardians in the mountains, the Indians being usually as successful in escaping the law as they were in killing.

Blood restlessness was rather pronounced among certain of the young men, and even the sedate Piegans broke out in a small disturbance. A Blood named Black Rabbit, who was wanted for breaking and entering, used the business end of his guns to persuade the police not to arrest him, subsequently escaping to the Blackfoot reserve in the States, where he was ordered to leave; so he came back and suffered himself to be arrested peaceably.

"Medicine White Horse," who escaped from the Police in 1891 after being sentenced to two years for killing cattle, grew weary of successful evasions, so he gave himself up and was rewarded with four additional months; Red Paint and Crazy Crow, a pair of Bloods, each received two years under a similar charge, though Red Paint was discharged shortly afterward on account of illness, the second time he had successfully accomplished this feat, though each time upon his release, his health showed a marvellous improvement under the sun and bracing winds of the southern districts.

Old Tooth, Painted Back, and Night Chief, three Piegan gentlemen, broke into and looted the store on the reserve, thereby causing a great deal of trouble for themselves, and incidentally also for H.H. Nash, the Indian agent who was active in tracing them down, and who, shortly after the incident, was attacked after night by unknown Indians, shot and stabbed, but left severely alone after he had put up a very fine fight and driven the miscreants off.

One day Constable Glave of Macleod swung up to the top of a "raw" remount and was carried in whirlwind fashion out onto the vast sweeps of prairie, the animal bolting toward the horizon as soon as it felt the weight of the constable on its back. Away, away the animal tore, with the man sitting tight and tugging frantically on the reins, the iron mouth of the frenzied horse giving no response to the tugs until he had dashed into the very centre of a band of range cattle. Constable Glave looked about him, felt his heaving steed between his knees, looked upon the wide and lonesome prairies, and into the bloodthirsty orbs of grazing "wild" cattle who were gazing at him with great intentness, their mouths slowly grinding on luscious cuds or succulent grasses. It was a very thrilling moment, wild animals all around, wild horse beneath, and escape absolutely up to the initiative of the man. Constable Glave did not hesitate a single moment. He drew his trusty revolver, killed a steer belonging to Thomas Holt, and escaped. Mr. Holt was quite wrought up over this occurrence and instituted proceedings to recover the value of his beef "critter," but the case was dismissed, as it was proven to be entirely accidental.

A Blood Indian named The Dog, and in all probability the partner of Big Rib, who caused so much trouble some years before, had by this time become so inoculated with respect for white men and especially the uniform and authority of the Police that he begged and secured a position as a Police scout and comported himself with considerable dignity. Big Rib and The Dog had received a five-year sentence in 1890, but good behavior or other exigencies might have turned this red man out before he had served his full sentence. At any rate The Dog was an Indian scout at Macleod when he entered into a controversy with Constable Currie and received a sentence of two years' imprisonment as a result. It appears that the Indian went into the detachment where Currie was in charge, and was lounging about when Currie, who, in common with many other white men, held all Indians in contempt, ordered him to "get out." From this moment, the actions of the red man brand him to be assuredly the real and original trouble-making Dog who annoyed the Police for so many years. Sullenly refusing to move, he waited until the incensed Currie seized him, whereupon he carved the white man very disastrously with his knife, thus leading to arrest and conviction.

The humor of cattlemen and cowboys has always been compelling, whether voluntary or otherwise, the spontaneous overflow of healthy animal spirits being often accountable for these outbursts. High River district and High River town, plodding along in its usual successful manner, taking good care of their stock, taking advantage of the Eastern and Western markets, importing good animals to improve their herds, yet had time to laugh over the adventures of a lank Eastern party named Perkins, who left the hilarious West shortly after he reached it.

Phil Weinard, who had been in the hills for about six months, arrived in town with a splendid crop of long black hair on his head, the result of being away from the barber shops so long, and Mr. Perkins came in about the same time and eventually wandered into the bar of the hotel where Weinard, still crowned with his luxuriant locks, had taken a temporary job of dispensing drinks. Weinard, lean, hawk-nosed, brown, looked the typical Indian, and Perkins eyed him askance, finally turning to Ben Rankin, who was sitting at a table, and asking him who and what the "Indian" was. In a low, respectful tone, Rankin assured him that Weinard was the rip-snorting kind of bad Indian who commenced killing as soon as he had two drinks of red liquor under his belt, and the Easterner's eyes assumed a worried expression, for the "Indian" was at that very moment swallowing the second glassful. After this second libation, Weinard shook his mane and, stepping from behind the bar, approached the two men with a half a lemon in his hand, saying to Perkins: "I'll bet I can shoot this off your head!"

Mr. Perkins edged toward the door, cleared his throat, and said: "Yessir, yessir," in a nervously hasty and agreeable manner.

"Stand still!" thundered the "bad red man," flashing his big revolver, but Perkins was outside and leaping through the snowdrifts with marvellous speed, arriving at Buck Smith's hostelry in a remarkably short time and telling that interested person how narrow an escape he had just experienced.

Times were slow and entertainment must be sought, so the cowboys and other local talent decided to exploit Perkins some more. A rancher named Hollan inveigled the shy victim back to the hotel, where Weinard was apparently in irons, the plotters having borrowed manacles from the constable stationed there. Though the "Indian" lay thus powerless on the floor, Perkins was not at ease, and so he was the first man to see the "outlaw" burst his steel bonds asunder and leap up ready for war. The nearest table being very convenient, Perkins dove beneath it, and Weinard fired a twenty-two calibre revolver beside his ear as he cowered. Screams of bodily fear came with great volume from beneath that table, cowboys drew their guns and blazed into the floor, pandemonium broke loose. One enthusiast fired upward, the bullet going through the ceiling and disturbing the rest of a man who at once hopped out into the corridor and demanded what was wrong.

"Shootin' scrape," yelled a joyous cowboy, "you better jump out of the window," which the man promptly did.

Perkins went back East in a very few days.

The long, uneven fight between the homesteader and G.E. Goddard of the Bow River Horse Ranch came to an end in 1894. Goddard, holding a lease that ran down close to Calgary, had been annoyed ever since he came, in 1888, by squatters coming onto the lease and settling on homesteads. For some time he endeavored to force them to leave, but finding this impossible, he at last in 1894 gave them permission to enter.

Medicine Hat district was the theatre of a lively discussion between sheepmen and cattlemen on account of the sheep herds grazing in great numbers on unleased Government land and thus rendering it unfit for cattle. Much fine grazing land was thus usurped by the sheepmen, and the cattlemen were highly incensed, though the trouble did not proceed past the talking stage, the land undoubtedly belonging as much to the sheepmen as to those who ran cattle.

Sheep scab appeared and dipping was done extensively, the herds being judged to be free of it late in the year, though this proved to be somewhat of a mistake. Sheep scab is a disease that shows up much more prominently and is far more pronounced among animals in a moist climate, so the disease with Alberta sheep was very liable to be active and yet not very apparent. This condition proved true in at least one instance, when Gordon & Ironsides, the great cattle dealers, purchased two large consignments of sheep in the Medicine Hat district, which they sent to England, the first proving a success, but the second, owing to scab breaking out among the animals while on shipboard, being a failure. The affliction of scab, this pronounced failure in export, and a low price for wool proved bad, and the sheepmen had a pretty hard year all round.

Investigations by veterinaries to locate pleuro-pneumonia continued, as also did the embargo. David Warnock, M.R.C.V.S., who had been appointed veterinary inspector for Pincher Creek district and the south the year before, made thorough examinations and inspections; J.L. Poett, M.R.C.V.S., also made extensive investigations, the results being that they came to the conclusion that the entire West was absolutely free from the disease.

Cattle ranches did remarkably well, and while prices for steers went down, there were a very large number sold off the ranges for cash. A few took cattle East, but the majority were purchased on the ranges by Gordon & Ironsides. In the spring there was a good calf crop, the cattle going into the winter in very fine condition, and the weather being steady with no excess of snow and no bad spring storms. The only loss was from timber wolves, which, the Mounted Police report, very seriously crippled some small ranchers, the animals pulling down such large animals as three-year-old steers, but being most destructive among foals. One man killed fifty head of wolves that spring. Mange was apparent this year for the first time in a number of seasons, being introduced, it was declared, by Montana cattle. Another bother to the ranchmen was a large number of homesteaders and squatters who settled in the south, seizing water, riverbanks, and springs.

Very few Canadian cattle besides those belonging to the Circle outfit drifted across into Montana in the spring, though a report was circulated to the effect that great numbers of Canadian stock had crossed the line during winter. The result of this report was that Howell Harris, manager of the Circle outfit, had his drifting stock rounded up for very little cash outlay, the Canadians joining the Circle men in their sweep of northern Montana, it being the only reasonable method of gathering all Canadian range stock. A thousand or two Circle critters were rounded up with about a dozen of other brands.

In March the Government passed an Order-in-Council establishing regulations by which school lands could be leased for grazing purposes, the leases to run five years with the right to cancel with one year's notice. Many ranchers took advantage of these new regulations to secure slight extensions of grazing acreage.

Horse breeding was not at all prosperous, due in some degree, Commissioner Herchmer asserts, to the low price of common horses in the East, which price, he said, was forced by the introduction of electric cars. A low-bred, common horse was unsaleable and a goodly proportion of ranch horses came under this classification, for, though a few ranches were producing some very fine animals, the majority bred poor stuff. A million wild horses were said to range in Washington, Idaho, and Oregon, and some of these brutes, scrubs of the worst kind, were being brought into Alberta. In addition to this there had been, up to a few years previous, great herds of such stock in the Kamloops district, and many of these animals had found their way by sale and trade to the Alberta market and the prairie ranges. An agitation arose among horsemen to have the Government establish a duty of not less than thirty dollars per head on all horses coming in, thus forcing the inferior stock of the States to remain there. As conditions now prevailed, the animals were bought for five and ten dollars, often less, and duty was based on a proportion of the actual cost. This enabled drovers to bring some herds across the line for one or two dollars per head, and made the Alberta market, with its struggling homesteaders, an especially desirable place on which to sell cheap horses. William Pearce, in an endeavor to solve this vexed question of poor stock, suggested that the Government take action and forbid the grazing of horses on the public leases. He further suggested that the horse ranchers be given three years in which to clean out their poor scrubs before the suggested regulations went into force, as the result of their enforcement would compel the ranchers to dispose of worthless horses and keep only their good ones. With range conditions as they were, it did not cost a man anything at all to range horses, for they rustled capably through the roughest winters. These suggestions were not carried out.

Thousands of stockers were brought in by the cattlemen, the Cochrane Ranch alone importing over two thousand two-year-old steers from the State of Oregon. American settlers came into the country in fairly large numbers, settling in the south or "trailing" up into the country north of Calgary. The livestock belonging to most of these newcomers was as a rule particularly wretched, few head showing any marked indication of breeding. The herd brought in by the Cochrane Ranch contained some good stock, though the great mass of range herds that crossed through quarantine seemed to be poor domestic stuff that had been picked up haphazard, and consequently was very mixed. The best range stock brought in, according to Superintendent Steele, NWMP, was one hundred and twenty-seven head, including one bull, for J.N. and P.W. West of Cardston.

At quarantine, all the domestic stock was herded at the expense of the Government, and the wild cattle at the expense of the owners; 3,043 animals entered quarantine, and 116 were lost, 39 by wolves at the Milk River Ridge station, 18 steers and 21 calves making up the total of this latter loss. The Cochrane herd tallied 2,121 cattle and four bulls, while the remainder of the stock through quarantine was largely for Mormons, coming from Idaho, Montana, and Utah, though some few head were from Wyoming, Nebraska and Washington. The cattle market in the States was at a very low ebb, and though Alberta markets were low, they were far superior to the American, as is shown by a few comparative "top" prices that were prepared this year: cows, three years and up, fifteen dollars in the United States and twenty-five dollars in Alberta; heifers, thirteen dollars in the United States and twenty in Alberta; yearlings, nine dollars in the United States and fourteen in Alberta; steers, fifteen dollars in the United States and twenty-five dollars in Alberta.

These were domestic prices. Range prices were: cows, fifteen dollars in the United States and twenty-two dollars in Alberta; heifers, ten dollars in the United States and eighteen dollars in Alberta; yearlings, seven dollars in the United States and fourteen dollars in Alberta; steers, ten and fifteen dollars in the United States and twenty-five in Alberta.

Four thousand two hundred and ninety-six head of sheep were brought over, three hundred for Lethbridge, three thousand four hundred and ninety-six for Beveridge & Ingles at Calgary, and five hundred head for G.W. Nickerson at Macleod. These were all Montana sheep. About fifteen hundred head of horses were brought through quarantine and taken into the Macleod district, a large proportion of this stock being scrubby and vicious, doubtless from the wild herds in the North-Western States.

A sad event occurred at Calgary on September 5 when Lieutenant-Colonel Macleod, C.M.G., passed away. Colonel Macleod was the commander of the first force of Mounted Police in Alberta and did wonderful work in establishing law and order in a lawless country, handling the warlike Indian tribes with rare judgment, and despatching every branch or his most difficult position with consummate skill. His loss was very keenly felt, throughout the southern portion of the Province particularly, where every old-timer was a personal friend of the grand old pioneer soldier.

During the same year, Jim Christie, pioneer horseman and stockman, who had removed to the Nose Creek district from Pincher Creek a few years before, was killed a short distance north of Calgary by being thrown from his rig.

Another fatality was the drowning of Saul Blackburn, a Waldron cowboy, at Legrandeur's Crossing on the Old Man's River on June 17, before the eyes of helpless comrades. Blackburn was driving cattle across the river, and, as is usual with cowboys when engaged in this delicate and dangerous work, he plunged into the river below the ford to hold the cattle from drifting below. The horse, half-swimming, had no sure footing, and was swept over eventually, rolling on Blackburn and then going tumbling down the current, both man and horse perishing in the swift waters.

In the southern part of the country, the farmers were edging in despite drought and bad seasons. Irrigation was being tried and found successful, and this encouraged the agriculturist. Blake & Miles, and Elton & Sons in the Porcupine Hills, built small systems, the Cochrane Ranch put in a ditch that would irrigate six hundred acres, and the Mormons in the Cardston district started work on an irrigation system that would supply water to eight hundred acres.

The fall was fine, and the cattle took the winter in splendid shape. Outside of the Calgary and High River districts, where no stock census was attempted, the numbers of livestock, horses and cattle in southern Alberta, exclusive of Medicine Hat, was over one hundred thousand head. Pincher Creek district had twenty thousand cattle and five thousand horses; the Porcupine Hills (south) had 15,515 cattle and 1,598 horses; St. Mary's, including 2,500 head on the Brown Ranch, had 4,871 cattle and 600 horses; Stand-Off, 14,095 cattle and 1,028 horses; Cardston, 2,000 head of stock; Boundary Creek, 773 cattle and 277 horses; Big Bend, which suffered another bad season of drought and high winds, had 12,915 cattle and 797 horses; Kootenay, 1,616 cattle, 406 horses; Kipp, 1,356 cattle, 302 horses; Leavings, including the Oxley and the Glengarry ranches, 17,150 cattle and 1,777 horses; Mosquito Creek, 6,838 head of cattle and 249 horses. An interesting item in these figures is that of mules and pigs and hens, which are carefully enumerated, the whole south country being able to bring a total of only 735 hens and roosters, and 141 porkers, a sure indication that though the farmers were settling in considerable numbers, they had not yet commenced mixed farming to any great extent. Six mules had also drifted into the new country, six hardy pioneers of the thousands now in Alberta. Their residence section was Kipp, and they were quite popular curiosities, every old-timer predicting positively that they would never like the country because of the smallness of their hoofs and the depth of the spring muds of Alberta.

The Indians were in good shape and seemed to be picking up white men's—and women's—ways to quite a marked extent, for one Piegan squaw was seen, shortly after treaty, taking a new sewing machine home in her buckboard. The Calgary district was pestered somewhat by the Indians—Stonies and Sarcees were suspected—killing cattle. A force of Mounted Police was stationed at Ings' ranch in an endeavor to catch some of these depredators, but found it impossible to secure a single red man, owing to the country being so broken and brushy that it afforded splendid natural hiding places for killing. This class of Indian hunting increased considerably in the south also, and at one time twenty Bloods were on trial charged with the offence. The Mounted Police declared that the practice could never be stopped so long as ranch stock was allowed to range on the reserves, or until the ranchers would pay rent to the Indians for the privilege of grazing thereon. The reserves were undoubtedly the best ranges in the south, and though the Indian agents and the Mounted Police frequently ordered ranchers to take their stock off, it was impossible to make them all do it. These frequent killings of cattle on the reserves led at last to a definite action from the Government, the Commissioner of Indian Affairs giving orders in September to allow no ranch stock belonging to white men on the reserves after that date.

The Indian Takes-Two-Guns, who killed Cochrane cattle and escaped a few years before, was captured on his return this year and convicted on the old charge, and an Indian named Mike, who was arrested in September for stealing horses in the Pincher Creek district, was so panic-stricken at the thought of a prison sentence and the loss of his freedom that he killed himself with a revolver. D.J. Whitney, a rancher in the Lethbridge district, found the body of an Indian in a badger hole in July, the bones scattered by coyotes, who had commenced to devour the body. A piece of paper and a necklace identified the body as that of a red man named Packs-Meat-On-Him, who had murdered his wife earlier in the year and then disappeared, doubtless committing suicide near the spot where his body was found.

In April the Mounted Police were the means of breaking up a very bad gang of half-breeds (to the number of about forty) who lived in the Sweetgrass Hills, and had been terrorizing the ranchers and settlers ever since the rebellion, at which time they settled there. They lived an absolutely lawless life, killing cattle when they wanted meat, stealing, bullying, swaggering through the country, absolutely strong in their knowledge of the fear which the settlers felt toward them. They openly boasted that they killed range stock when they wanted to; their leader swore to shoot anyone who tried to arrest him, and also kept the threat of burning stables and ranges before the people in the district as a reward for information or prosecution against him or his band. At one time a Montana ranch that had been losing many cattle in the hills stationed a man there to watch the breeds, and the fear of the gang was so impressed upon this man that he arranged to be out of the way when cattle were being slaughtered. Creditable information that was brought to the Police declared that the half-breed leader would ride to this range rider's door and tell him there would be killing on the morrow, and the range rider would keep himself scarce until the killing was done. In April this gang headed for the Canadian side of the boundary, where it was said they also killed as they pleased, and Corporal Dickson of Writing-On-Stone went to meet them, finding their camp just north of a cairn of stones that marked the boundary. Dickson acted with commendable judgment and ability, handling the ruffians with surprising ease, due to his own brains and cool head. The officer was alone, there was a goodly number of the half-breeds, and they were alleged to be ever ready to shoot a law officer at any time. Waiting until dawn, the lone policeman quietly collected their horses, led them away, and then when the vagabonds went to look for them, he entered their camp and discharged all their firearms, arresting the whole band when they hastened back to find the reason of the disturbance. The slaughtered remains of a Circle cow and calf were in the camp, and the Police thought they had a sure case against this gang; but later a surveyor, sent out by the Mounted Police to assure that the boundary cairn was correct, found that the camp was in reality twenty-two chains south of the boundary, the pile of stones being slightly off the true line. The breeds' lawyer then pointed out that his men could not be held, as they had been arrested and forced into Canada, while they could not be extradited because they certainly were not fugitive criminals. Naturally they were dismissed and started back to their old haunts, but the United States Government had at last taken action, and the gang was run out by a troop of cavalry.

During the autumn of 1894, prices rose slightly for prime beef animals, fat, dry cows bringing thirty-five dollars and four-year-old steers forty dollars. During the season, Gordon & Ironsides shipped 5,750 steers to eastern markets. Oats and hay went down in value, and the reports are that they were "very cheap," oats being down to forty-six cents a bushel and hay dropping to ten and twelve dollars a ton in the stack. These prices, particularly for the oats, seem today to be pretty fair indeed.

Winter fires, started by the C & M trains in late February, did a lot of damage, burning for two and a half days and sweeping nearly all of the country between the Little Bow and Willow Creek. Fall fires commenced just east of the Cochrane Ranch and swept the Blood reserve as far as Whoop-Up, burning a strip twelve miles wide and destroying much Indian hay. Two fires started from the Winder Ranch and damaged much Porcupine Hills grass, one burning over twelve miles in length and the other seven miles.

CHAPTER XV

1895–1896

THERE had been splendid winters and many of them up to this time (1895), but in the memory of the oldest rancher none had been better than this one, with its slight snow, warm chinooks, and cattle coming through winter to summer with good flesh. The calf crops were large and without loss, the lambing season was good and the foals on the range showed up splendidly. The police census of the Province of Alberta showed 268,000 cattle, 77,100 horses, and 112,585 sheep, the great majority of sheep being south of the Red Deer River, half the horses south of that point, and two-thirds of the cattle. In the Pincher Creek district alone, from that stream to Boundary Creek, there were 78,348 cattle, 10,740 horses, and 4,350 sheep, the latter being chiefly among the Mormons. George Hope Johnston's six thousand sheep on the Rosebud, Beveridge and Ingles' thirty-five hundred on Knee Hill Creek, and the herds of the Merino Ranch west of Calgary were in splendid shape, while the sheepmen of Medicine Hat felt that they had at last eradicated scab from their great herds. Twenty thousand fat cattle were shipped from Alberta that fall, ten thousand head coming from the ranges of Macleod, Gordon & Ironsides doing most of the buying for Eastern shipment, in fact shipping most of the twenty thousand head that went out, and paying slightly better prices than they had the year before.

The British embargo still held, but did little damage to the prices, and the horse market was pretty weak, the stock showing as yet little improvement. Shipments of horses that had been made the previous year to Great Britain brought unsatisfactory results, despite the great pains taken to send only the choicest stock, the average price being seventy dollars per head.

Alberta ranchers bought Ontario and Manitoba dogies in thousands, fodder being short in Ontario and the CPR cutting rates in half in order to enable the Eastern farmers to ship their stock to Alberta for feeding. Two thousand head were brought into the Macleod district alone, while the proportion was as large in Calgary and Medicine Hat and Lethbridge. From Calgary north, every pound of beef was used for local consumption, all exporting being done from the southern part, due partly to the very poor stock brought into the north districts in past years by American settlers.

Usually a cattleman hates sheep, and would rather die a slow and terrible death than have them, but some men—of the pure cattleman strain, too—are broad-minded enough to see that sheep are no disgrace if handled with judgment and moderation, as instanced by the fact that E.H. Maunsell and Conrad Brothers branched out into the mutton line this year, the latter organization importing twelve hundred head and Maunsell three hundred. When Maunsell's bunch arrived at Macleod, it was feared they were afflicted with scab, but after the most minute inspection by skilled stockmen and veterinaries who had wide experience in the cattle and horse industries, it was decided that the indications of scab were simply irritation from spear grass. So the ewes, the rams, and the lambs were turned out to gambol upon the green billows of the Macleod prairies, and Mr. Maunsell had everything a mixed farmer should have—horses, cattle and sheep—with the exception of pigs and hens.

Prairie fires that year were small and not serious in the Macleod district, but Lethbridge suffered extensively from fires that were caused in most cases by the AR&I and CPR engines, two thousand square miles of pasture being destroyed.

American cattle for some years had been edging north into the pastures of the unfenced southern prairies, but they were worse this year than ever before, the Mounted Police being kept very busy and administering several very substantial rebukes to the Montana stockmen who deliberately headed their animals north beyond the worn-out ranges of the state. In April the Customs Department, through the Police, seized two hundred and seventy-one head of this stock, which they kept in quarantine until the owners had paid expenses. Later in the year other herds were seized in this way, totalling in all 2,446 head, 1,200 being released to the American roundup hands, 795 upon payment of quarantine expenses, and the remainder escaping during a blinding snowstorm, through which the Police herders stayed in shelter, and it is thought some reckless Americans took advantage of the lapse, else the northern storm itself herded the cattle back to the home range.

Worthless mares were still inhabiting the ranges in goodly numbers, the ranchers being loath to part with them for nothing at all, though that seemed the only method of improving some of the herds, until a man arrived from the city of Buffalo and let it be known that he would buy all old, fat and useless mares to send to New York State, where they could be slaughtered and put into tin cans to be sold as good contract corned and canned beef. Whether this offer was ever accepted by any of the Western stockmen is not known, but it was certainly made.

A lynching that was apparently based upon some sort of merit was successfully negotiated in Lethbridge during the spring. Lethbridge seemed to be about as likely to spring original doings as any Western town, which had already been proven by the way they treated the mulatto Harrison in the early eighties. Since then the town had only indulged in the usual poker games, cattle deals, and other Western enterprises until the present, when it was aroused by the suicide of Charles Gillies on February 13. Gillies, who had been worried by the attentions shown his wife by a lodger named James Donaldson, finally felt his suspicions so well confirmed that he became disgusted with life and blew his brains out. The citizens of the town, simple, energetic men of action, decided that Donaldson was really responsible for this death, so they took him out one night and gave him a coating of tar and feathers, after which they escorted him uptown and shoved him into the entrance of the Lethbridge House, having led him there by means of a strong lariat that was attached to his neck. Sergeant Hare, a Mounted Police officer who was on duty that night, could not get the feathers and tar off his carbine, and was thus circumstantially and conclusively proven to have allowed his desires for strict justice to overrule his sworn word to uphold the law and see that peace was maintained. Another man who was most closely connected with the operation on Donaldson was a gun-gentleman named Charlie Warren, who led the gang upon request of other less reckless, but possibly just as enthusiastic, citizens. Hare was reduced to the ranks, and Warren was tried for inciting or some other such crime, the jury finally disagreeing, and the prisoner stepping over into the United States, leaving his bail to be paid by the bond-givers. Long-distance complaints then came into Lethbridge from Mr. Warren, who was deeply aggrieved because he had received only forty dollars of a hundred he had been promised if he would superintend the feather party.

Another affray which promised serious results arose in the hills out of a quarrel between John Lamar and Gilbert McKay, both cowmen, and the former the quickest and surest man with a six-shooter in the country. McKay became obsessed with the belief that Lamar was opening and reading or else delaying certain personal correspondence of his, and he voiced his complaint very loudly, working himself eventually to such a towering rage that he rode out to the Waldron Ranch with the expressed intention of shooting Lamar on sight. Arriving there, he called for Lamar, who stepped unarmed from the house, and was immediately made the subject of a most bitter verbal attack, which culminated in a demand that Lamar get his gun and shoot the question out to a satisfactory finish. Lamar tried to pacify the inflamed McKay, but only aroused him to more bitter invective, so bitter in fact that Lamar lost his temper also and repaired indoors, reappearing with his guns strapped on. McKay, seeing this, drew his revolver to shoot, but before he could level the weapon Lamar had drawn and fired twice, striking his opponent in arm and body and toppling him out of his saddle. First aid was rendered the unfortunate searcher for trouble by cowboys who came running up, and Lamar went to town to give himself up to the authorities, who, upon hearing the true facts of the case, decided that McKay had received that for which he had looked, and that no prosecution would be necessary.

Hoofbeats thundered into Macleod one day in summer, and some Piegan Indians dashed wildly to the Police barracks with the startling information that the slaughtered body of a red brave lay weltering in its own gore on the open prairies of the reserve. Quick action resulted, a small force of Police being sent at once to the spot to find the remains of the unfortunate, who, the informants had told the police, was shot through the chest. Hastening to the spot and the body, the Police dismounted—and awoke it, for it was a lone Piegan brave who, after painting his chest with red paint preparatory to participating in an expurgated Sun Dance, had fallen asleep, where his fellows had seen him and thought him dead.

This was only one demonstration of the fact that the ways of the red men were rapidly changing and that the brave days of old did not have the strong hold on the Indian nature that they once did; another being that Red Crow, war chief of the Bloods, whose father, and whose father's father, had dwelt in skin tents, and who had spent much of his own life beneath a tepee top, was now residing in a real house, fitted with carpets, linen sheets, stoves, and *clean* windows. And among all the ranchers of the south, the Blood Indians were making the best hay that was being put up anywhere. The only serious Indian trouble in the West was that which a Cree named Almighty Voice had aroused in Saskatchewan. This Indian, who was wanted for killing cattle, had been intercepted by Sergeant Colbrooke and a half-breed interpreter just as he was eloping with a thirteen-year-old squaw. Colbrooke stopped them, and told, through the interpreter, that he was going to arrest Almighty Voice, the latter replying that if such an attempt was made, he would kill the officer. Colbrooke fearlessly advanced upon the desperate man and was shot dead, the breed fleeing for his life and Almighty Voice escaping.

Several bad gangs of cattle thieves were broken up by the Police, one of the worst being an organized outfit that worked out of Calgary, carrying their depredations into Prince Albert. Three members of this gang were finally arrested and tried before Judge McGuire, the two who did the actual stealing getting sentences of three years and one year respectively, and the man who was suspected of being the brains and the money behind the operations being acquitted after a very interesting trial.

Constable L.C.J. Trustram, NWMP, won great praise for himself and his force on March 30, when he plunged into the high waters of the Milk River and saved an old man named William McEwen from drowning. A tragedy similar to that of the unfortunate Herron occurred in this locality late in April, when a man named Ross, a partner of Charlie Warren, who led the gang of tar-and-feather fame at Lethbridge, shot himself after becoming lost in a snowstorm near Baldwin's ranch. Ross and Warren had probably left Lethbridge, aiming for the boundary, and, becoming separated in the snowstorm, Ross preferred a sudden death to the cruel one of slow freezing.

An interesting sidelight on the comparative differences of Montana and Canadian administration of justice happened near Middle Buttes, Montana, in August. A Mounted Policeman by the name of Richardson, riding that country in search of stray horses, reached Middle Buttes and stopped there for a short time. While there, he met a rowdy named Long with whom he had experienced difficulty in Canada some time before, and Long, being pure bully, backed by friends, and in a country where the NWMP was not recognized, decided to pay off old scores. He had drifted into Middle Buttes, picked a quarrel with a man, pounded him cruelly, shot another seriously, and established himself as a "bad" man, which character he took pride in sustaining. On meeting Richardson, he struck him, and was at once seized and thrown by that husky officer, who proceeded to hold him helplessly on the ground until he promised to be good. This promise, wrenched from him in the presence of admirers, he broke as soon as released, picking up a large stone and hitting the Mounted Policeman with it. Richardson, not really knowing the right procedure in such cases in Montana, left the town and went out to a ranch five miles away, in order to keep from breaking the laws of order in Montana all to bits, but Long was drunk and enraged, and he followed the policeman, shooting him through the hip and incapacitating him. The rancher at whose place Richardson received the wound immediately harnessed his horses and drove to Macleod with the wounded man, arriving there about the same time as a wire from the sheriff of Middle Buttes district which read: "Long accounted for. Killed by Justice Brown resisting arrest."

This was sweet and satisfying, but did not give all details that led up to this most pardonable deed. Long, after he had disposed of Richardson to his own satisfaction, had longed for more blood and for more opportunities to impress upon the people of the district what a bad man he was. Brown, a very quiet, peaceable man, came to his mind as a perfectly safe representative of American law to mess up, so he spurred his horse to Brown's house and demanded that the latter come out at once and be shot. Quiet men often deceive bullies like Long, and this one proved to be of this sort, for he appeared very promptly with a shotgun at his shoulder and blew the belligerent Long into something resembling a colander.

In November of 1895 a very heavy fall of snow threatened enormous loss to the ranchers, particularly in the southwest, where it lay four feet deep on the level and the calves stood helplessly bound with only their heads sticking out, while the mature cattle could barely move and were entirely unable to rustle. The storm came so early that the cattle had not been rounded up properly, the weak cows and the "weaners" being still on the range. Pincher Creek and the country south of the Belly River suffered most, a few head dying before a chinook came and wiped off every vestige of the storm, leaving the range in very good shape. During the time this snow lay on the ground, a good many cattle were killed on the C & M branch of the CPR by getting in the right-of-way between the deep snowbanks and refusing to get out for the trains, and a number were drowned while standing on the rivers looking for water, the owners not having cut holes, and the thirsty cattle collecting in such large bands that their weight broke the insecure ice and threw them into the streams.

Bunches of wild horses scattered throughout the Province were in thriving condition and were considered a menace to the ranchers' ranging horses, because they were liable to take the domestic stock away with them, and they also contained a number of wild and poorly bred stallions. This led to the suggestion of impounding estray animals, but no action was taken. The Mounted Police, who were having great trouble every year from small ranchers crying over lost mavericks, suggested that such animals should also be placed under the Estray Act and be impounded. They said that fully a hundred mavericks were picked up every year at the fall roundup in the High River district alone.

There was a visible improvement in stock throughout the south, most noticeable among small herds, which were being kept closer and bred with more care.

The American stock, which had been poaching on Canadian ranges for some years, presented a very actual menace this year, great herds being clustered along the boundary from the White Mud River to the very foot of the mountains. For every Canadian steer ranging on the American side of the line, there were a hundred Americans on the Canadian side, and the luxuriant range threatened to be reduced to a very closely cropped stretch of pasturage. There was no doubt that thousands of head of cattle from worked-out American ranges were sent by rail to within a few miles of the boundary, unloaded and driven northward in search of food. Great herds of Texans, American "Circle" and "DS" stock crowded into Alberta, the steers, wild and vicious, proving very injurious to Canadian stock, worrying heifers, bulls and young stock, complaints coming from stockmen all along the line. The trespassing "Circle" stock belonged to the Benton & St. Louis Cattle Company, and the "DS" stock was the property of Conrad Kohrs & Company, both very large American firms. In their first spring sweep of the threatened districts, the Mounted Police collected and returned to the States a herd of three thousand head, two days later returning over the same ground and sweeping up five thousand, while the American roundup, which was working the boundary, took out over two thousand.

American cattlemen were also beginning to make advances to the Canadian Government, appealing through the attractive medium of more cattle, more wealth and more prosperity for Alberta. They asked for the quarantine on American stock to be raised, and said they would then register their brands on this side, putting in vast herds of steers, mostly Texans, hordes of which were waiting on the boundary line to overrun Canada. The American customs laws favored this proposal of the stockmen in so far as any of the stock brought in from the other side could be returned at any time, duty free. The Canadian authorities took no action, and the disappointed Americans continued to hug the boundary and hope for better days, which eventually came.

Horses remained low, though a good demand and fair prices ruled for heavy ones, the ranchers taking to these breeds because they were more easily handled and less liable to accident than the finer classes. Two carloads of horses shipped direct from Belgium to Knox & Hooper of Calgary were not a success, as they did not net the owners more than sixty dollars at that point. Sheep were away down, the market being forced to a low ebb by the great "sacrifice sales" among the Montana sheepmen, many of whom were being driven out of business by lack of range. Wolves continued fairly destructive among young stock, and the flies in the foothills in the north end of the ranch country were again very bad, causing cattle to worry, lose flesh and drift long distances. In the district between the Red Deer River, east of the track to High River, there were 268 ranchers with over 130,000 cattle, 52,000 horses and 32,000 sheep. Gordon & Ironsides wintered 1,200 head at Namaka and 3,000 at the forks of the Red Deer and the South Saskatchewan, while Conrad Brothers moved 1,200 from the Lethbridge district to winter near Queenstown. In the spring Walter Wake and George Lane, acting for the "44" and Gordon & Ironsides, shipped the first cattle out of Cayley station, this point becoming one of the greatest shipping points on the south line in after years.

The first official general roundup in the Medicine Hat district took place that spring, the Medicine Hat Stock Association having been formed in February, this being the first concerted effort of the ranchers of the district, J.H.G. Bray, an old rancher and an ex-Mounted Police officer of the vintage of 1873, being appointed secretary and stock inspector, the latter appointment remaining with him from then to the present, the Government taking over stock inspections in the following year and confirming Mr. Bray's appointment by the Association.

Later in the year, on December 28, the Live Stock Growers' Association was formed, embracing all the district stock associations of Alberta, the meeting being held in Calgary and D.W. Marsh being elected president, W.F. Cochrane first vice-president, F.W. Godsal second vice-president and R.G. Mathews of Macleod secretary-treasurer. Representatives elected were F.W. Cowan for Bow River, John Ellis for Medicine Hat, W.H. Andrews for Maple Creek, T. Curry for Lethbridge, E.J. Swann for Sheep Creek, F.S. Stimson and G. Emerson for High River, A.B. Macdonald for Willow Creek, D.J. Grier for Macleod, Charles W. Kettles and R. Duthie for Pincher Creek. At this first meeting, the assembled stockmen urged upon the Government the necessity of wolf bounty, and recommended the appointment of stock inspectors at various points of shipment, indicating individuals who would be suitable at each point. Those recommended were P. Osborne of Calgary, Sam Sharpe of Cowley, A.W. Fishe of Pincher Creek, C. Sharples at New Oxley, J. Johnson at New Oxley, D. McIntosh at Nanton, R. Urch at Kipp, and C. Brown at Davisburg. These appointments were not particularly popular with some of the inspectors who represented small shipping stations, the emoluments of office being five cents per head inspected, and each inspection of one or a hundred or a thousand head necessitating a long ride and the loss of a day's time.

The new association did not take up the matter of new branding laws, though the Police and many stockmen considered this quite necessary in order to be able to administer the law satisfactorily, one instance of this necessity being cited:

A cow bearing the distinctive "WR" (Waldron) brand was seen on the range lavishing huge affection on a calf that carried a "77" hair brand and returned the affection of the mature mammal with great vigor, wiggling his tail and gamboling around in the excess of great family joy. Anyone seeing these two beasts would have jumped to the conclusion that they were undoubtedly mother and child, despite the man-made evidence to the contrary in the shape of dissimilar brands. A brand, judged the Police, can be placed on any animal, but affection is born in them. The result of these quite natural deductions was that John Mitchell, who owned the "77," was arrested and tried on the charge of stealing the calf, but was acquitted. Hence the desire of the Police for more satisfactory branding laws.

The markets improved, and the completion of the Crow's Nest Pass branch, which was expected the next year, would open up a great market, not only for beef but for thousands of tons of hay, which would otherwise waste on the prairies. Mines and lumber camps must be fed when the railway went through the rich valley of the Crow, and the prairies were the natural source of supply. In fact, the demand up the Pass was already a strong factor in the calls for Alberta stock, any kind of beef being acceptable, and P. Burns supplying most of the call with fat cows, four hundred of which he was killing every month for this trade alone. A shortage of fat steers was also evidenced on the ranges, due to the good demand and fair prices of the year before, when a large number of three-year-old steers were tempted off the range, consequently leaving a void in the ranks of the prime four-year-olds of the next year. This demand for fat cows was the first step toward cutting into the breeding capabilities of the ranchers' herds, a condition of affairs that grew and in time hurt the industry considerably.

The Medicine Hat district shipped 906 steers for export, and the Lethbridge district sent 5,287 head. Great care was urged in these shipments, and stock inspections urged in Lethbridge particularly, as one American steer in a car could hold up an entire trainload at Winnipeg; 298 horses were sent from the Lethbridge ranges for the Manitoba market. Imports into the Province by trail and train totalled two thousand horses and seven thousand cattle, the latter mainly "stockers." Considerable lump-jaw was found in some herds, and a number of cattle were killed on this account.

Prairie fires did but little damage, the ranges being very fortunate in this respect. One interesting fire took place in the Macleod district, and Frank J.P. Crean, then a cowboy on the range and later a civil engineer and Government explorer, called Inspector (Colonel) Sanders on the telephone and asked him if he would get half the fine in case he told who set the fire out, the even split of fines being the practice in such cases, as it helped the Police to find culpable parties through the greed of neighbors.

"Certainly," the inspector informed the informer.

"All right, I did it," confessed Crean. "I'll be right in," and he was as good as his word, as Sanders also proved to be with regard to his own.

Fugitive Crees who had fled to and located in Montana after the rebellion of '85 were this year ordered to return to their native country, the United States Government sending troops of cavalry to escort the natives to the boundary, where the Mounted Police were to take them over. One hundred and ninety-two regretful and reluctant Crees were herded northward under strong escort, the number having been reduced by one when Day Bow, a Cree participant in the

Frog Lake massacre, committed suicide when convinced he must really return. Inspector R.B. Deane and one constable met this horde of savages and placidly took charge of them, much to the astonishment of the American officers and men who composed the cavalry escort to the boundary.

All of the Canadian tribes were behaving well, and the Blackfeet had even commenced coal mining, one of their sub-chiefs, Buckskin, having discovered coal in a coulee while hunting stock, and at once proceeded to become a coal baron. Today his mines net him twenty to thirty dollars a day, and he banks the money at a Calgary bank so he will not be tempted to get drunk. There was much trouble from Indian stock thieves in the south, but these were mostly United States red men, the Blackfeet, Piegans and Bloods being very quiet. Cattle stealing by white men resulted in a few convictions. George Colby stole a steer from a man named McLean, a resident of the Springbank district, and was convicted and sentenced to a jail term; a man named Mott at Medicine Hat contracted with a local butcher named Adsitt to supply him with beef at a very low price. Though Mott had only a few head of stock in his herd, he supplied large amounts of beef, much to the surprise of local people, who instituted inquiries that led to police investigation and conviction for stealing cattle. In November J.P. McHugh was charged with stealing a Circle steer from Conrad Brothers' herd at Queenstown and was acquitted, Judge Bouleau, in giving the decision, making some very frank statements concerning the methods of the cattle industry. He said that for some years it had been the habit of the Conrads to take up and brand all unbranded stray cattle south of the Bow and east of the Blackfoot trail; and that it had been the habit of the High River Stock Association, represented by B.C. Rankin, to take up all unbranded strays west of the trail and south of the Bow. If the ranchers thought they could make a law of this kind to suit themselves, it was time their minds were disabused. Further, he said, he would severely punish any one of such concerns if brought before him. Undoubtedly, Judge Rouleau's one idea was to have a law that would be fair to rich and poor, big and little rancher alike.

Hull Brothers, ranchers and meat dealers, dissolved partnership this year, W.R. Hull assuming the Alberta end of the business and his brother, John R. Hull, taking over the Kamloops and British Columbia end.

Inspector (and interpreter) Jerry Potts, the best guide and scout in the North-West, and the man who piloted the Mounted Police into Alberta in 1874, died at Macleod on July 14, after twenty-two years' service on the Police force. Potts did splendid work in assisting Colonel Macleod in his handling of the tribes in the early days, and he was a loyal, trusty soldier. His death was deplored by Indians and white settlers throughout the south.

Jack Graham, now a rancher on the Belly River, and then one of the best cowboys in the country, won the roping competition at the annual Macleod show that fall. All spring, Walter Wake had been riding an ugly, homely pony while roping calves in the hills, and when Wake went up into the Crow's Nest Pass, the horse came into possession of Graham, who decided to try his qualities in competition. Calves are quick and tricky, steers are slower on the turn; the pony was in fine condition for calves, so he found the steers easy, Graham winning very handily with a horse that had been scorned before by the other cowboys. In the line of exhibitions and competitions, Alberta cowboys kept their end up very well this year. The territorial exhibition was on in Regina when Duncan McIntosh and Billy Stewart reached there on a fall hunt for stock, and they promptly entered for the sports, McIntosh going into the roping competition and Stewart the bronco-riding. Both captured first prizes and kept the reputations of their districts up to their usual high notch.

The one lone Indian trouble in the south was that of Charcoal, which has been written and told in prose and poetry a great many times—always with a little touch of art—but it is given here from the staid reports of the different Mounted Police officers and private citizens who actually took part. Charcoal killed two men, shot two or three others, and frightened hundreds before he was finally caught and put in jail, the story of his pursuit and elusions and eventual capture being one of intense interest.

On October 12, Charcoal, alias Bad-Young-Man, a Blood Indian, discovered that another red man named Medicine-Pipe-Stem had debauched his home, so he promptly and thoroughly killed Medicine-Pipe-Stem and went on a lone war trail, his hand against every man's. Poor Charcoal, being but a simple savage, did not know that he might not be punished very severely for the deed he had done, for he had been told that white men hung Indians who killed, and he was assured he had signed his death warrant when he drew his rifle sights down on his victim. Taking his family, four squaws and two children, he vanished in the hills, first shooting Indian Agent McNeil through the arm, and then leaving word that he would come back some time in the near future and kill both McNeil and Red Crow, the chief of the tribe. Charcoal certainly ran amuck when he started. Thirty Indian scouts and all the Mounted Police in the district combed the country in search of their quarry, but failed to gain sight or sound of him excepting in one instance, when he stole an overcoat from a settler named Henderson, much to Mr. Henderson's indignation. A few days passed and the forces of law and order scoured the land, one band under Inspector Jarvis finally cornering the renegade red man and his family in a cluster of timber in the hills. Charcoal met them with a shower of bullets, directed with such good aim that he almost hit Inspector Jarvis. Just why this should be considered good shooting the Police records do not say, but they assuredly mention that the red man "almost" hit the inspector. After quite a while had passed and the silence had lasted some time, the Police charged and captured the camp, two squaws and one child, Charcoal in the meantime slipping around with a couple of squaws and one child and stampeding the horses of the police, the animals being found later at Legrandeur's ranch, forty-five miles away.

Again Charcoal vanished, being heard of once when he shot at a Piegan while endeavoring to secure another horse, and the Police went to great lengths of coercion and persuasion to secure some trace of him. Charcoal's other boy, the one who escaped with his father while Inspector Jarvis was capturing the camp in the hills, was arrested on the reserve, and he consented to lead the police to the spot where Charcoal's two squaws were camped in the Porcupine Hills—after lengthy persuasions. Long Man and Red Horse, two half-brothers of Charcoal's, were arrested but released after proving they knew nothing of the man, and then new tales began to drift in concerning the doings of the lone Blood whom the Police were endeavoring so hard to cut from his base of supplies. On the 25th of October the fugitive killed a steer belonging to Hatfield on the north fork of the Kootenay; on the 30th he stole a horse from a Blood Indian and went south.

A reign of terror seemed to exist, particularly among the tribesmen of Charcoal. Red Crow, who was a warrior and a chief, and who had adopted white men's ways to the extent of a house and furniture, eschewed a bed during the time the outlaw ranged, and slept nervously upon the floor, knowing that Charcoal knew the position of the bed and might shoot through the window and kill him if he slept on it. Red Crow also worried much over the possible fate of his son, who was attending school at High River. White Calf, another Blood warrior who apparently had reason to dread Charcoal, took to sleeping in the loft at his home and pulling the ladder up after him.

About the end of October, the Police made still further efforts to bring the nerve-racking strain of the chase to a close, arresting two of the fugitive's brothers, Bear's Backbone and Left Hand, with twenty-two relatives. These two heads of families, with all their wives and progeny, were taken to jail and told they must find Charcoal. Left Hand was turned loose with the information that the Police would hold his family as hostages, and that he must catch, or inform the Police where to catch, Charcoal before the tenth of the month, or he would rest in jail with the remainder of his family. Bear's Backbone was given his liberty with similar information, leaving his mother, wives and children in the hands of the Police.

In the meantime, Charcoal shot at and narrowly missed Corporal Armer at Cardston, taking a shot at him from behind a water trough, and then heading up Lee's Creek to Lamb's ranch, where he broke in and stole some food.

On November 10, Constable Hatfield and five Indian scouts found his camp on Beaver Creek, and Hatfield sent two Piegans to Pincher Creek to warn Sergeant Wilde. The latter at once set out, with Constable Ambrose, Scout Holloway and the two Indians, Wilde telling his men to shoot the outlaw as soon as they got within good range. There was a slight snow on the ground, and the party sighted Charcoal, riding one tired horse and leading a pack horse, on the north fork of the Kootenay near Thibaudeau's ranch, Wilde reiterating his orders about shooting, saying he wanted them to commence shooting when within range, and in no case to go nearer than fifty yards.

Holloway rode to within one hundred and fifty yards of the fleeing Blood, dismounting to shoot, only to have his rifle miss fire. Charcoal had increased his lead to a half mile by the time Holloway climbed back on his saddle, and the Police horses were tired, so some of the pursuers dismounted and plodded ahead on foot, though Wilde pressed on and came up to the Indian about fifty yards from where an unarmed rancher named Bratton was herding some cattle and saw the whole tragedy. The officer fired no shot, but rode fearlessly up until he was alongside the red man, reached over to seize him and was shot through the body, the bullet entering his right side and being found in his left gauntlet, Charcoal firing his rifle as it lay across his saddle. Wilde fell, tried to rise, tugging weakly at his revolver, whereupon the desperate brave rode back and shot him in the forehead, thus making sure of his work. Dismounting and leaving his own weary pony, he took Wilde's horse and carbine and pressed on, being out of sight when the other constables and scouts arrived. Bratton offered his horse to the upcoming pursuers, but the only man who continued on the trail, an Indian named Many-Tail-Feathers-Around-His-Neck, refused it, saying the animal was too nervous and wild for him. Leaving the rest of the party with the slain constable, Many-Tail-Feathers-etc., proceeded in pursuit, riding Charcoal's deserted steed.

Next day the country was filled with Police, grimly determined to capture or kill the man who had slain one of their most popular brothers-in-arms, and the hills and slopes were searched with a thoroughness that left nothing unseen. In the meantime, Many-Tail-Feathers had followed Charcoal and come upon a party of ranchers composed of John Herron, Thibaudeau, Craig and Foster, and thus reinforced hastened on, eventually seeing Charcoal standing in a neck of timber. The only member of the ranchers' party who is reported as being armed was Herron, who had his revolver, but as Charcoal was two hundred yards away when seen, this weapon was of little use, especially so since the red man had a very serviceable rifle. Tail-Feathers shouted to Charcoal to surrender, and was answered by such a shower of bullets that the entire party took up a less exposed position a hundred yards further back, Herron keeping up a futile fire with his revolver. Darkness came, and with it went Charcoal, the ranchers taking his trail up again in the morning and pressing doggedly on toward the reserve, whither he was evidently heading.

Weak and spent, exhausted from sleepless nights and endless flights, Charcoal had about decided to give up and kill himself, after one more visit to his reserve and his people. His family was his last thought, apparently, and he went to them, arriving long after dark at the house where his two brothers sat disconsolately thinking that the tenth of the month had passed and they and their families must remain in jail. Charcoal opened the door and looked in, but something seemed to warn him he should not remain, for he turned and started back to his horse, but was overpowered by the two men, who leaped upon him as he turned to go. Taking their prisoner into the house, they gave him a smoke, and then noticed he was bleeding badly, stoically dripping to death from holes he had punched into the veins of his arms with a moccasin awl, so they plastered the wounds with flour, bound them up, and kept their prize alive for the Police, who arrived in the morning, the arrest being made on November 12. With Charcoal chained to the floor in the Macleod guardhouse and guarded by five policemen, the body of unfortunate Sergeant Wilde was laid to rest with military honors, the Blood chiefs and scores of settlers following the body to the grave.

The following ranchers and settlers assisted the police in the pursuit: John Herron, G.J.B. Jonas, G.B. Ryckman, A.E. Dempster, N. Nash, B. Bolster, F. McKenzie, W. Reed, M.D. Gray, Legrandeur, S.S. Fraser, Thibaudeau, Foster and Craig.

CHAPTER XVI

1897–1900

THE appointment of stock inspectors, which had been agitated for some time, came to pass in 1897, and though some districts were satisfied, there was some dissatisfaction felt at Macleod, and a cry was raised to have the Mounted Police take up the work over again, as they had done in the past. This did not last very long, though, and matters settled down to a satisfactory basis.

A bad November threatened heavy loss, which did not materialize on account of good weather following promptly, though the spring calf crop was light in most districts, and the cause of this was laid to the one bad month. Pincher Creek suffered the heaviest loss.

Charcoal was tried by jury at Macleod on January 21, with Mr. Justice Scott presiding, and was found guilty of murder and sentenced to hang on March 16. He maintained his dignity to the end.

The Western Stock Growers' Association continued their wolf bounty, paying ten dollars per head for every female wolf killed. The matter of water reservations for range stock was also a live issue, and the Association endeavored to hare some action taken by the Government in this regard, appointing a committee composed of A.B. Macdonald, D.H. Andrews, F.S. Stimson, D.W. Marsh, and A.E. Cross to interview Mr. Frank Oliver, the member, and have some effort made to establish the much needed concessions. They failed to get any satisfactory assurances.

Straying American cattle were numerous along the boundary as usual, most particularly in the Milk River country, where the great herds of the Bloom Cattle Company and the cattle of the Shonkin and Marias ranges thronged in thousands. Texas mange was beginning to appear among this stock, but the ranchers did not worry, and neither did the Canadian Government, the belief being that this disease simply arose from poverty of the blood, and would disappear as soon as an animal took on fat. The Spencers of Montana, who ran thousands of head of stock just across the boundary and needed the Canadian range very badly, went down to Ontario during the spring and purchased six hundred head of dogies, which they brought West and turned to range on Many Berries Creek, branding this stock with their American brand, 3CU, thus starting a ranch on the Canadian side and making it impossible to tell whether the animals there were the original stockers or herds from their American ranges. Though they had not received the permission of the Government to do this, they arrived in Alberta with the cattle branded, and promptly turned them out to rustle.

Lump-jaw appeared in small numbers in some of the herds, and it was discovered that some stockmen were selling animals so afflicted to canneries, though the meat had been adjudged unclean, even Indians refusing to eat it. Glanders were also found, sixteen head being killed for this disease in the Calgary district, along with twenty-three lump-jawed cattle.

Though the calf crop was small in the Macleod district, the cattle were fine and good business was done. William Morgan, a rather "intensive" breeder who had come from the States on August 6, 1895, showed the best calf crop in the south, and had quite an argument over it with the Mounted Police. Mr. Morgan came, as already stated, on August 6, 1895, and settled on Halfbreed Creek with eighteen cows, four calves, and eighteen steers. On November 2, 1896, Mr. Morgan had a herd of eighty-nine head, by police count, including eighteen calves. These remarkable indications of breeding possibilities of Halfbreed Creek astounded the Police but brought no other results, because in a prosecution, the Crown must prove ownership before they can make a case. A little later on, Morgan and a man named Conger, both of whom had been under Police surveillance as a result of the prolific increase in the herd there, were arrested and charged with stealing a calf from a cow that had a "W" brand, but the case was dismissed, the learned judge holding that the prosecution had not proven that the cow's brand was unknown. Morgan was then tried for stealing a bull, convicted, and let out on suspended sentence.

The story of Morgan and this bull is amusing, showing various Western idiosyncracies. The animal was known throughout the range as "the blue bull," every stockman knowing him and his owner. One day the blue bull vanished, the Police were told, and investigation resulted in his being discovered in Morgan's corral. Leaving a plainsman of a sort to watch the corral to see if any branding developments followed, the Police constable handling the case returned to headquarters, intending arresting Morgan the next day. But someone had handed a bottle of whisky to the plainsman guard, he became drunk and drunker, he went to Morgan and in a spirit of spirituous effervescence told him why he lingered. Late that night the owner of the bull dashed into the police detachment and awoke the officer, exclaiming in great excitement:

"The blue bull's going over the bridge, the blue bull's going over the bridge!"

Sounds from that structure indicated that this was quite true, for there were muttered, suppressed oaths, the clump of hoofs, and the swish of a stinging whip. In trousers and undershirt the policeman rushed out, mounted his horse and galloped down to see what was the trouble, finding Morgan profanely urging the "blue" and reluctant bull to return home under cover of night. Swearing in hushed tones, putting the repressed vigor of his voice into his sturdy arm, Morgan lashed the obstinate beast along at a snail's pace, eagerly anxious to get the animal where it belonged before sun-up. Consequently his trial on the charge was more or less of a reminder, and his suspended sentence was quite understandable.

The Police "fell down" more delightfully in another case that was tried about the same time, and concerned a stub-tailed calf and his devoted mother. A Mounted Police constable riding along the Milk River patrol in springtime often noticed a particularly striking calf, with a stub tail and bright red and white markings, running with a Circle cow. Finally the constable missed the calf and instituted an investigation, the result of which was that he found it in a strange corral, whose owner's brand was certainly not a circle. Just to make himself absolutely sure, he drove the bereaved Circle cow to a spot where the stub-tailed calf could see her, and the little animal ran to her with joyous, mumbling blats. The corral owner was consequently arrested, tried, and dismissed, the learned judge ruling as he did because there was no witness who had expressed an opinion that the calf was the offspring of the cow.

Horses experienced a sharp advance this year, partly due to the construction camps in the Crow's Nest Pass, and more decidedly to the great numbers of people who were heading for the Yukon gold fields via the interior route across the mountains west of Edmonton.

Five thousand men and a thousand teams commenced construction on the Crow on July 13, the contractors bringing many horses from the East, which took sick of influenza of thoracic form, while the native horses were almost immune. Only prime "tops" were shipped east off the ranches this year, as all the "rough" cattle went into British Columbia points, P. Burns alone killing six hundred a month. Four-year-old prime steers were bringing forty to forty-five dollars, and three-year-olds from thirty-five to thirty-seven and a half. Old, fat cows and inferior steers easily drew twenty-five to thirty-five dollars, and the ranchers cleaned out their herds, going so far as to spay their heifers and thus reduce the breeding possibilities of the ranges.

Heavy horses between thirteen and sixteen hundred pounds went up to seventy-five and even a hundred dollars each, which was considered a good price and absolutely satisfactory. The sheep industry was at a standstill, wool being low, seven to nine cents, and the English market regulations of immediate slaughter tending to discourage exporting.

Almighty Voice, the Indian who murdered Colbrooke in the fall of 1895, returned this year to his own country and was finally located on a hill where he had built a rifle-pit and cut radiating lanes through the dense poplars in order to better protect his position. The Police attempted to storm this redoubt, but were repulsed by the murderously accurate fire of the Indian, who was assisted in the pit by a boy of fourteen. Inspector Allen was shot in the arm, Sergeant Raven through the thigh, Corporal Hockin was mortally wounded, Constable Kerr was killed, and a settler named Grundy also met his end before the Police and civilians surrounding the place withdrew to await reinforcements. These arrived in the shape of a stronger force of Police and two guns, a seven and a nine pounder, and the besiegers then shelled the Indians' position while Almighty Voice's tribe sat on a nearby hillside and watched, listening meantime to the doomed murderer's mother chant his death song. After the shelling, they found Almighty Voice and a lad called Little Salteau lying dead in the pit, and another dead Indian in the brush down the hill.

A cowboy named McKenzie, who had come into the Willow Creek district and won the liking of the whole countryside by his cheery good nature and exceeding skill in riding, came to a sad end in Willow Creek during early August while attempting to ford in high water and being swept off into the current, where he drowned. Charlie Sharples was himself nearly drowned endeavoring to save the young man.

Ben Rankin, a well-known cowboy who had been with the Bar U Ranch for a long time, also came to his death this year. He had left the plains and gone into the mountains to work, locating at Sandon, B.C., where, in a moment of temporary insanity, he committed suicide by shooting himself in the head.

Unfenced railway tracks were responsible for considerable damage to range stock, a number of head being killed or maimed by the trains, fifteen head perishing at one time on the tracks near Willow Creek.

Crow's Nest Pass construction was leading into the heart of the Kootenay Indians' country, and the Mounted Police who had been sent in there to maintain order during the introduction of the railway were informed that there was a bloodthirsty Kootenay who had expressed the intention of shooting Fred Kanouse, the old-time trader, on sight. Kanouse, it will be remembered, had quite an active argument with a band of Kootenay Indians at Fort Warren in 1875, wherein a number of good warriors were sent on to their happy futures. This warlike young brave, who was now talking of death and shooting, was a descendant of one of the deceased warriors and wanted to wipe the disgrace of defeat from his family escutcheon by killing the cause of it. The Police officer in charge in the Crow sent word to Kanouse that blood would doubtless flow if the young brave met the trader, and told him he had better remain away if he did not want trouble.

"But," protested Kanouse, "I've got some business at Cranbrook, and I want to go up there. Under present conditions, would it be unlawful for me to go armed?"

The Police gave the required permission to carry a gun, and Kanouse strapped a six-shooter on and went up into the mountains, arriving at Cranbrook in good shape and not at all worried. There were a number of Kootenays in the place, but the one particular buck was at that moment on the reserve, so Kanouse paid another Indian a dollar to go out and inform the avenger of family that he was in town. Any message like this is always cheerfully and speedily delivered, and the man in search of Kanouse's life was promptly notified. When an Indian hunts he likes to hunt, and is often very much put out if the hunted hunts him, especially if it be white men. This young brave was all Indian, and he experienced an immediate change of heart, casting his guns to one side and speeding unarmed to Cranbrook, where Kanouse placidly waited with the intention of giving the red man the first chance. This self-sacrificing resolution was happily unnecessary, the Kootenay greeting his family's depopulator with pleasant words and indications of peace. In fact they held a nice conversation, during which the Indian told Kanouse that his tribe had resolved never to fight white men behind logs again when they returned after the disastrous fight at Fort Warren. They had tried it several times but always with little success, the last time being the most convincing.

By the time the year 1898 came around, the class of cattle in the range country was said to be inferior to earlier times, due to a certain extent to the ranchers reducing their herds of Shorthorn bulls, which had previously been used very extensively, and introducing Herefords and Polled Angus indiscriminately. Many wretched bulls were still on the open range, and some of the imported ones were little better, their only superiority to a fair grade bull being a long pedigree. Ontario breeders of purebred stock had by this time given themselves a black eye in this regard, it having been learned that many of the so-called thoroughbred bulls from there were the culls of the herds, and that purebreds bought from American stock farms were much better, the American dealers taking more pains to please their Western Canadian patrons, knowing well that they understood a good animal when they saw it.

The stocker business received a setback, and feed ranches found themselves hampered through lack of steers. The Manitoba, Ontario, and other outside sources of the supply of stackers, which had become pretty well established during the preceding years, were very much reduced, and the supply of two-year-olds on the Alberta ranches was nil, the breeding ranchers preferring to keep them and realize the full profit themselves.

The question of feeding all stock throughout the winter again came up and was widely discussed, many smaller ranchers taking it up and some of even the larger outfits seriously considering it. On the whole the ranchers did pretty well, the winter being very good, the calf crop fair, and steers bringing a good price. Many people who had commenced spaying heifers discontinued this and resumed breeding, a healthy indication of prosperity and confidence.

Many of the best steers marketed in the fall were raised by the Indians, who, assisted by the Indian Department, were gradually getting some very fine animals, the Department keeping the bulls moving from reserve to reserve and thus doing away with the dangers of inbreeding. Sheepmen did better than cattlemen, prices being up, with a very active demand, the grade of Alberta ranch sheep having improved rapidly during the preceding three or four years. Wolves were bad, and coyotes were particularly destructive among the sheep, destroying sixty head in one corral alone. In 1896 a sheep could be bought for a dollar and a quarter, while this year they were worth three and three-fifty. Wool jumped correspondingly during the same period, from seven and eight cents to fourteen and fifteen. Indeed, the sheepmen were more happy than the cattlemen, even though the latter did not have anything but the shortage of stockers to really complain about.

About this time, there was an outbreak of mange noticed among the cattle, particularly along the Little Bow, but little attention was paid it by the Ottawa authorities, the belief that it disappeared when summer grass fattened the stock still being sufficient to convince the Government and many ranchers of its unimportance. American stock difficulties along the boundary were as pronounced as ever, and the ranchers of Medicine Hat were making loud complaints of Spencers' "3CU" herds, which were becoming a nuisance to everybody.

The territorial Department of Agriculture took over the handling of the brand issuances, and the Alberta Stock Growers' Association made another effort to have the Government pass a law making a recorded brand on a beast *prima facie* evidence of ownership, memorializing the Minister of Justice the third time to this effect. Old brand ordinances, as was being seen more clearly every year, were so confusing as to be comparatively safe for thieves.

In the Lethbridge district, irrigation was being developed, the Mormons starting fairly extensive systems and the AR & I Company preparing to put in much larger ones. The only damage recorded against any of the herds in this district during the year was a loss of thirty head of young stock from blackleg on Sir Roderic Cameron's ranch. Two thousand head were exported from the district for the British market, the shippers being A.J. McLean, of the "CY," and Gordon & Ironsides.

In Macleod district an equally large number of steers were exported, and a good year was experienced, though the spring promised badly, late March storms cutting the calf crop down considerably and threatening damage to the herds.

The Calgary district was untroubled until fall, when numerous prairie fires burned off great plots of country, the worst being one that started in the Gladys district on election day, November 4, and burned off nearly all the country between the Bow and the Little Bow, killing a number of cattle.

The horse business picked up markedly this year, and indications pointed to a continued strong market. George Lane, of the Flying E and formerly of the Bar U ranches, bought a good herd of the famous Percheron "Diamond O" stock from James Mauldon of Dillon, Montana, and placed them on the "YT" Ranch, which he established that year. This herd was the nucleus of the splendid stock that Lane now owns, prizewinners at every big horse show on the American continent.

A sure indication of the pulse of the horse market was the fact that large numbers of horses were stolen during the year, thieves keeping as delicate a finger upon the strengthening throb of business as the legitimate dealers. Scores of complaints of thefts were made, and in many cases were based upon fact, though sometimes they were not. During the times that horses were cheap, the owners had not bothered to look after some of the herds, feeling they were no temptation to thieves anyway, but now that the market had picked up, they were commencing to round up their animals, and, finding many gone, they jumped at once to theft, when as a matter of fact they might be either dead or strayed. Both American and Canadian horse traders were still carrying on the business of bringing American scrub stock to the market, this year in greatly increased droves, for prices in Alberta had soared fifty percent. Heavy horses were bringing a hundred and twenty-five dollars, and ordinary animals ranged around seventy-five and one hundred. The traders with the poor stock from the States made big money, buying the stock off the range across the line for five to ten dollars and grading probably up to thirty, declaring a value of twenty dollars per head, paying the twenty percent duty demanded, and then selling the animals to the settlers in the Province for prices ranging from eighty to a hundred dollars.

The Canadian Pacific Railway Company, in order to facilitate the handling of cattle shipments from the range country, appointed a livestock agent this year, whose duty it was to handle all the work with the cattle shippers, arranging for cars and all the details so essential to speedy and careful handling of this freight.

Prairie fires were not dangerous, and did little damage except along the Red Deer River. The haystacks and sheds of the Circle Ranch at Queenstown caught fire and were burned entirely, incendiarism being suspected, for there was snow on the ground at the time.

Medicine Hat district exported 4,131 head of cattle, importing forty-four thoroughbred sheep and twenty-seven purebred cattle. The customs station on the St. Mary's River passed 3,144 horses into Alberta, and 312 cattle, the horses being largely the poor American stuff spoken of previously. Ranchers in the south country imported forty-nine thoroughbred bulls from Iowa and Utah.

Twenty-seven head of lump-jawed cattle were killed on the High River roundup, and it was felt that the last of this disease was in sight.

The Yukon gold rush still continued, and Superintendent Steele, who had been at Macleod for some years, was sent to the Yukon to take charge of the Mounted Police there, Superintendent R.B. Deane succeeding him at Macleod.

"A great year for all livestock," these are the words with which the condition of the ranchers and farmers of Alberta was summed up for 1899, but this did not mean that every kind of domestic animal experienced a good market, as the sheep industry was stagnant. The price of beef and of horses went up, there was plenty of grass, few fires, and no snow until December 10. In certain localities, the appearance of mange was beginning to cause a little nervousness, but it was not considered serious at all.

March was variable, and April was unusually severe, though little loss was suffered, the only big outfit losing anything of any account being the Circle at Queenstown, whose big herd of stockers had been sent there to winter just before the supposed incendiarism of the fall before destroyed seven hundred tons of their feed. These animals were essentially dogies of the most pronounced type, and when their supply of food was thus abruptly taken away from them, they did not know how to rustle for a living, the result being that many strayed and succumbed.

There was a big increase in export to British Columbia, the price in April and May being exceptionally high for those seasons, buyers paying four and a half cents on the hoof and taking all the good animals that were offered. The stocker supply was deplorably short, and the "beef" ranchers were short in their herds. William Pearce again urged that more attention be paid to the feeding branch of ranching, declaring that while Alberta was not ranking very high as a breeding country, it was admittedly unsurpassed for feeding and maturing. He did not mean that the country was not a good one in which to breed stock, but simply that the system then in vogue was generally too indiscriminate and careless to place the native stock in a very high grade.

The ranchers, as usual every year, imported a number of good animals, with the aim of gradually working up the quality of the beef, and the CPR also became more active in encouraging improving the breeds, supplying certain districts with stud stock on very good terms. A number of pedigreed bulls were imported, but were not of the highest quality, if one is to judge by comparative prices, the values of the imported stock running from one hundred to two hundred and fifty dollars, comparing unfavorably with the six thousand dollars that an Argentine Republic rancher paid for one bull this year in an endeavor to improve his herds. If price was any criterion of the value of bulls, the Argentines were assuredly trying harder to improve their stock than the stockmen of Alberta were.

The mange question was under discussion during the entire year, and though no loss had as yet been experienced and none was actually expected, efforts to obliterate the disease were made, more or less energetically. The Government authorities were inclined to believe that Montana was a very live source of danger from infection and contagion among stock, as there were half a million cattle and twenty million sheep in that state, and the health conditions of the herds were attended to by only five veterinary inspectors. Inspections there would naturally be merely superficial, and in all probability many large herds were entirely neglected by the State officials, no matter what work was done by private veterinaries.

In April the Alberta Stock Growers' Association held a meeting at Calgary to discuss the mange and the best method to handle it, dipping stations being decided upon, to be located in districts most easily reached by all. In order to facilitate matters, W.F. Cochrane offered to pay for the erection of a dipping station at Kipp, on the NWMP land there, permission having been given by Commissioner Herchmer for such use of the land. It was agreed that Mr. Cochrane should go ahead with the construction of the station, and he proceeded at once, advancing all money required excepting three hundred dollars, which the Government gave. In the meantime, the Government had laid Alberta from the main line of the CPR to the international boundary under quarantine for mange, and efforts were made by both the Government and some of the biggest ranchers to have every infected head treated, but many ranchers, not realizing the seriousness of the menace, entered half-heartedly in the work and the result was that not half of the affected stock was dipped. The disease was appearing most strongly in the herds of the Lethbridge district, and from there north to Mosquito Creek, and many head even in these sections were allowed to escape the dipping station on the expressed belief of many ranchers that there was no danger, and the stock would soon be clean, good blood and good feed doing more than the dipping solution. Consequently, mangy cattle went into the next winter.

Wolves seem to have been worse again, the animals appearing to have gained in boldness since the bounty was removed the year before. At the same meeting that was held to consider the mange situation, it was decided that the Association would resume paying a bounty for these pests. The quarantine regulations went into effect on July first, and by that time there had been dipping stations commenced at Goose Lake, Spring Creek in the Medicine Hat district, and at Kipp. In the fall, the stock association refunded to W.F. Cochrane the monies he expended in the erection of the Kipp station.

At the third annual meeting of the ASGA, which was held that fall and was very largely attended, the question of mange came very prominently into the discussions, the general feeling being that if the affected animals were dipped, it would be a sufficient safeguard.

The American herds were neither so numerous nor persistent along the boundary during this year, the grass in the States being generally very good and the range being sufficient for the stock. A few hundred strayed over, but all that were collected by the American roundup that came on the Canadian side to clean up the strays was a total of two thousand and twenty-five head, many thousands less than had been seen in Canada during the preceding two years.

As has been said, the sheep business was slack, and on the top of this there were outbreaks of scab on the Rosebud, on Buffalo Lake, and at Macleod. It will be remembered that E.H. Maunsell had branched out into the sheep industry a year or two before, and stocked with a herd of Montana muttons, the animals showing indications of "spear-grass irritation." Unfortunately, this irritation developed into a very bad attack of scab, which so disheartened Mr. Maunsell that he decided to have no more dealings with such treacherous livestock, voluntarily butchering every head affected. Such heroic methods put a very speedy stop to danger from scab in the Macleod district, and Maunsell deserves credit for his summary method of handling the plague. Selling the unaffected stock, he quit the sheep business and directed all his energies again to the cattle, that having been the branch he had been thoroughly studying and developing during the preceding twenty years that he had been ranching. The disease in the other herds, which came to both the more northern districts from Manitoba, was treated by the slower and not-so-sure process of dipping.

Summer rains of exceeding frequency did a great deal of damage to farmers, and in some instance killed numbers of stock. Heavy rains in May, June, July and August kept the rivers booming, and retarded crops until after frost. The toll of death in the rivers was considerable, four drownings being recorded, the two-year-old son of Alfred Wilson being drowned in Pincher Creek; two half-breed girls perishing in the St Mary's on May 30 while attempting to ford; and Albert McFarland being swept away to death by the waters of the Kootenay on July 17.

Prairie fires were few and harmless, the rains ensuring much of this safety, and the ranchers went into the winter with the range rich in a splendid supply of good grass.

A considerable number of stock were drowned in the river bottoms and coulees of the southern rivers when a cloudburst in the hills on September 13 flooded the valleys and the coulees, nearly taking out the Pincher Creek bridge, fifty feet of the approach going before the rush of the flood.

The export trade to England kept up, Medicine Hat district alone shipping 5,458 head of stock, a good proportion of which went to the British market. Incoming stock by trail was not so extensive, most of it being sent in by rail now. The total inspected at Coutts during the year was 644 cattle, 86 horses, and 164 sheep.

P. Burns & Company, whose business had grown to tremendous proportions during the ten years Mr. Burns had been in business in the Province, established an abattoir and cold-storage plant at Calgary, the greatest of its kind in the West. From a small beginning, the great Burns Company had grown to a tremendous organization, with a string of shops across the country from Alberta to the coast, and on up in the Yukon, and with great ranches scattered through the ranch country, where the stock was fitted for the market.

An unsolved tragedy occurred on the Milk River Ridge in September, the body of a sheep herder named Oliver Jenkins being found on the range, with the neck broken. A jury hastily called from among the ranchers of the neighborhood to endeavor to learn the cause of the unfortunate sheepman's demise listened to the brief evidence, and then decided with little loss of time that the shepherd had died "from a visitation of God."

Beef sold off the ranges during the winter and spring of the first year of the twentieth century indicates that the start was auspicious, though the prices did not reach within half a cent of those of the year before. Nevertheless, it was a grand year for ranchers, from the beginning of the winter to the end of fall, though mange and a few prairie fires worried the stockmen, especially the latter, which caused heavy loss in certain districts. The biggest beef prices were four cents on the hoof, dropping to three and a half cents in the fall. A grand total of nineteen thousand head of cattle were exported, and the market was strong. In the Macleod district there was a light loss in calves, but a bountiful crop. Early March chinooks broke up the rivers, and ice jams were formed, the streams rising to great heights, and the Willow Creek railway and traffic bridge going out on March 11, forcing the transferring of passengers in baskets. Calgary district also experienced a mild year and a big crop of calves, with good prices and ready sales. A large number of Manitoba dogies were placed on the range in that section. Medicine Hat had a fine, dry year, the cattle putting on firm, hard flesh and being in very prime condition. Rains in August, September, and October made haymaking difficult.

A good price was paid for stockers, and there was a splendid supply, Manitoba selling large numbers, and British Columbia commencing to take advantage of the prairies as maturing areas, sending big herds of young stock.

Dr. McEachren, Dominion Veterinary-General, in reporting on the condition of the mange, stated he believed it to be well under control, and that it would be easily handled in future. This belief he backed with the reports of the secretaries of the Stock Growers' Associations, R.G. Mathews of Macleod, and J.H.G. Bray of Medicine Hat. Mr. Mathews reported to the Department of Agriculture, stating that the roundup had discovered only one hundred and seventy-four cases of mange and that all had been treated; Mr. Bray reported that the mange in that district was almost gone. Based upon these reports, Dr. McEachren was not to be blamed for believing as he did, and to strengthen this was the fact that only one steer was rejected at Montreal that fall on account of mange. But on the other hand:

Mange was so serious that the ranchers were beginning to have live worries about it. At a meeting of the Western Stock Growers' Association, it was moved by Howell Harris and seconded by F.S. Stimson that all cattle found affected at the roundups should be dipped, the motion being unanimously carried. Further, A.B. Macdonald and W.R. Hull moved that the secretary write to the Montana Stock Association asking co-operation in this method of handling the disease. Mange was getting bad, and there were mangy cattle from the boundary to the Red Deer River, the disease growing steadily worse and spreading rapidly. The Montana Stock Association agreed to assist in the attempt to eradicate the disease, but the right methods were not adopted in either country, both dipping affected stock only. American stock was very bad along the boundary in the Medicine Hat district, thousands of head crowding northward and mixing with the Canadian herds.

Horses strengthened again, and there was a good demand for animals of eleven hundred pounds and heavier. The ranchers had again commenced to make careful roundups of horses, and again found that the increased demand showed as strongly among thieves as it did among legitimate buyers. The Mounted Police, in commenting on this state of affairs, declared that the only horseman in the Calgary district who had lost none from this source was R.G. Robinson, who rounded up his stock regularly, whether prices were good or not. The contention of the ranchers that thieves stole more when the market was good was not supported by the Police, who took advantage of the opportunity to reprove careless methods. "Lots of colts have been stolen before branding," is the report, "though horse stealing is not on the increase. Owing to the increased value of horses the ranchers are looking more carefully after their bands and find their losses more regularly. Often they stray, after long neglect, and the ranchers at once jump at theft as the solution."

Dr. I.J.F. Burnett, V.S., NWMP, goes into the question of saddle horses rather exhaustively, and pays a high compliment to Frank Strong of the Strong Ranch at Macleod. Up till the time of his death, Mr. Strong had been breeding carefully to produce good saddlers, and of his work, Dr. Burnett says:

"Frank Strong bred along good lines for saddle horses. Some of his horses are still in the Mounted Police force after sixteen years' service." Burnett further expressed himself as against the stabling of young horses in southern Alberta, declaring that open sheds were all that was required. He suggested crossing cayuses with saddle horses, then the offspring to a thoroughbred, stating he thought that would produce a very fine line of saddle stock. Of the stock in the Province, he commends the stock of Eagle's Plume, Acrostic, and Silk Gown, though the latter was not a thoroughbred. Yorkist and Preston, two other well-known imported stallions, had proven unsuccessful in breeding good saddlers. There is probably considerable truth in Dr. Burnett's remarks about saddle stock, which branch of the stock industry had not been so carefully developed as the heavier breeds, though the Quorn Ranch had imported splendid mares and some good stallions. But the Quorn was ever unfortunate, and the stock never seemed to meet the sales they were expected to. In fact, there came a time when the handsome Irish hunter mares that were shipped in to breed remounts for the army were sold wholesale on the local market, and it was not uncommon for the fine-bred animals to be seen toiling in construction work and at other plebeian labor.

The stock association endeavored to arrange a new schedule for compensation from the CPR. Some time before, the ranchers had met the railway representatives and had drawn up a general schedule of prices on which basis cattle killed on the railway tracks would be paid for. The railway agreed to pay half, and the ranchers were to stand the other half. The original schedule, ranging down from forty dollars, was not thought to be quite equable, and the Association wished a general raise in the set scale. They suggested forty-five dollars for four-year-old steers, forty dollars for threes, twenty-eight for twos, and thirty, twenty-six, and twenty dollars for cows. Meeting with the CPR authorities to discuss this matter, they received the assurance that the railway would endeavor to be fair with all stockmen, but they did not meet with an acceptance of the prices proposed.

The Association set the wolf bounties at ten dollars for females, five for males and three for cubs, adding a coyote bounty of twenty-five cents, payable to half-breeds and Indians only. In October, they raised the female bounty to fifteen dollars.

The fourth annual meeting of the Association saw the retirement of D.W. Marsh from the presidency, which honor he had held since the formation of the organization four years before. Upon Mr. Marsh's retirement, he was succeeded by D.H. Andrews.

As an indication that the Government was not paying the proper amount of close attention to Western stock affairs and to their Indians, the Association members at this meeting endeavored to have the ordinance amended so that all cattle killed for Indians should have proper inspection.

The irrigation ditches in the Lethbridge district were opened this year, the water being turned in on September 5. Settlement in the Cardston, Pincher Creek and Calgary districts had grown rapidly, and wealthy rural communities had sprung up, this being especially true in the south, where the Mormons had settled and where irrigation had been introduced some time before with great success. A large number of good bulls were brought in, most of them being Shorthorns, and Medicine Hat district exported eighty-three purebreds—sixty-nine Galloways and fourteen Shorthorns.

Stock thefts seemed to be on the increase, and a number were brought to book, including one man who neglected to pay duty. This man, Joseph Pocha, smuggled, and was not even content with that, for he smuggled stolen horses, and had reached as far into the interior as High River before the Police nabbed him and put him through an experience that resulted in three years in jail.

The early winter saw two interesting thefts in the south country, the culprits escaping in both cases, once, indeed, with the loot. On the 19th of January, C. Scott sold three head of stock to E.H. Maunsell, one of them having a broken jaw. This animal Scott suggested killing at once, and Mr. Maunsell followed the advice, finding upon skinning the carcass that the block "J" brand, which appeared on the outside of the hide, was simply the "C" of the Cochrane Ranch on the inside, Scott having worked a hair brand on the animal. Maunsell at once stopped payment of the cheque he had given Scott, and this was sufficient warning to that individual, who vanished into the thin air of the boundary.

The other case was more mysterious, and is a mystery to the present day. Just before Christmas of 1899, James Daly bought twenty-four head of gentle stock from W.I. Joll of Moosomin, shipping them to the nearest railway point, and Joll and Daly taking them thence to Daly's ranch, arriving too late in the evening to brand, the animals being put in a pasture overnight. Next morning every hoof was gone. Now note how Providence helps some men. The thieves and the weather seemed to be working in absolute harmony, for there was not a sign of the trail of the band, a snowfall having settled over the prairie after the herd was moved. Later this stock was located in a bunch of wild range cattle, but when Joll, who had returned home, had come back to identify them, they had again been moved, this time it was said to Montana, where all trace was lost. The Police, who had been handicapped in all their work by the Boer War, which had taken many of the constables, were so short-handed that they could not spare a man to ride herd on the suspected bunch and watch it until Joll arrived, this so arousing Mr. Daly that he sought to secure damages from the Government, which attempt, however, proved to be a failure.

Prairie fires caused considerable damage to stock in the Lethbridge district, and a few people were fined. One murder occurred, which has remained unsolved, that of James S. Huggard, who was found dead on Nose Creek, north of Calgary, on November 7, the man having been most brutally killed with rifle and axe. The murderer escaped scot-free, there being absolutely no evidence pointing to the perpetrator.

An illustrated lesson in chivalry and etiquette was given in the town of Cardston during the year, and the teacher had to pay for administering it. A man with an unreliable tongue and a mind bred to scandal and meanness circulated a considerable series of the usual kind of slanders about a young lady there. This young woman, American and essentially Western, neither went into hysterics nor retirement, but on the contrary stepped right calmly into the eye of the public. Seizing a trusty revolver, she went forth on the highways and sought a meeting with the man in question, finally succeeding right where she most desired, in the streets of the town. The steady, gaping mouth of this revolver, the dull sheen of the leaden pellets in the cylinder, the eminently business-like glint in the eyes of the lady and the ring of her voice, produced the desired effect upon the scandal-monger, who dropped to his knees and apologized most humbly, declaring himself a very Ananias. Escaping after this satisfactory display of himself, he sought the Mounted Police, laid complaint, or caused it to be laid, and the lady was fined ten dollars for enforcing the gallantry upon him.

R.G. Robinson moved his stock this year from the old ranch on the edge of Calgary to its present location at the Big Springs on Nose Creek. E.H. Maunsell secured a splendid lease this year, arranging with the Piegans and the Indian Department for the grazing privileges of the reserve, Maunsell agreeing to pay the Indians certain rental, and putting up the cost of a four-strand fence, sixty-five miles long, around the range, the Indians doing the construction work. This lease was for a ten years' term, and the "IV" herds had as good range as any in the whole of Canada.

The greatest blackjack game ever played in Alberta took place about this time in High River between Captain Iken and a man named Todd. Iken was an English gentleman, honorable and upright, popular with every rancher and settler in the country; Todd was a sort of cowhand who played games of chance by preference, and usually had but few dollars to back his play. By the time the game was completed, Todd had won a quarter-section from Iken, the property being worth about a hundred thousand dollars, as it lay on the very edge of the town and, in fact, the High River school was later erected on it. Iken, though he was pretty sure that he was playing in a game where the most he could hope was to get his own losings back, had stayed until this land was gone, and he prepared to get the deeds and turn it all over. But the men of High River rather like to see an "even break," and they did not consider this was one, as Todd had not been able to come into the game with the backing that Iken had. Todd had gone in for gain, pure and simple; Iken had played because he loved the excitement, and that was all. So High River arose and urged Iken to pay nothing at all, but this he would not do, allowing himself, though, to stand for a compromise of forty acres and a team of horses. Todd accepted, and sold the forty acres in a short time for forty thousand dollars.

CHAPTER XVII

1901–1903

HEAVY beef shipments were made to the Klondike during 1901 and a big export of prime beef to England and Eastern points. Great herds of Manitoba and Ontario stockers were shipped in, a total of twenty thousand head of these animals being on the Alberta ranges by fall. The efforts to improve Alberta herds were becoming centralized, and the Government and stock associations commenced to work together with the aim of introducing better blood in the livestock-throughout the Province. The materialization of this united effort was the first provincial auction sale of bulls, which was held at Calgary, where sixty-four head of good stuff was distributed at an average price of eighty-one dollars and a few cents. This sale was arranged to be held every year, and was to form a basis of introducing pure blood and also of exchange in order to obviate inbreeding.

Homesteaders and squatters poured into the country, thousands and thousands of them, the range was fenced, the waterholes and springs taken, and the ranchmen were worrying about the eventual end, which they had seen some years before. With range and water gone, they would be put out of business, and again they endeavored to have the Government reserve water in suitable places, but without avail. This was the second effort of the kind that had been made by the Association in three years. Complaints came in from settlers along the Milk River concerning great herds of trespassing American cattle that were again forced into Canada by the burning of all the range north of the Marias and east of the Sweetgrass Hills. The cattlemen of the Marias had driven sixteen thousand head north of the hills, posting six men to line-herd, and bands of these cattle drifted the few remaining miles north to Canada and mixed with the Canadian herds. Complaint being made to W.S. Prenitt, the secretary of the Montana Stock Association, the American ranchers sent three additional men to ride the line. The feeling between the Canadian and American stockmen was of the best throughout these discussions and difficulties, as is evidenced by the Americans turning out in the spring to help round up Canadian stock that drifted over the line during a hard January snowstorm.

Eight thousand and fifteen head were exported from Macleod district, and 5,954 from Lethbridge. Over two thousand horses were sold out of these two districts also.

It was felt that mange had been about cleaned up, only five cases being found and treated on the roundups at Pincher Creek, Mosquito Creek and the Oxley Ranch and Willow Creek. At the April meeting of the Western Stock Growers' Association, John R. Craig moved, seconded by D. Warnock, that as mange had been practically stamped out, the Association should request the Government to lift the quarantine. Dr. McEachren also reported that mange was entirely eradicated from certain districts, and R.G. Mathews reported to him that the roundups only found one case at Willow Creek and a few at High River. He reported that there was no mange in the Porcupine Hills, none on the Cochrane Range on Kootenay Lakes, and a few in the Medicine Hat district, one of these few cases being an Indian Department cow. Against these statements, which were doubtless correct as far as Mr. Mathews knew, J.C. Hargreaves, Government veterinary surgeon, reported that the ranchers refused to admit the prevalence of mange, that seventy-five percent of the herds along the Red Deer were affected, and that only thirty head had been treated. In the extreme south, Jesse Knight found several cases among his herds and treated them, reporting good results and clean cattle after the treatment.

Sheep prices went up to eight cents for carcasses, and as usual the industry appealed at once to a number of stockmen, who brought great herds into the country and turned them on to the range. On November 5, 6 and 7 Inspector Burnett of the Mounted Police inspected 41,565 head imported by the Knights of Raymond, and on November 13 he inspected 17,731 head for C. McCarth. When fall came there were 80,000 sheep in the Lethbridge district, and there were lively threats of real clashes between the cattlemen and the sheepmen over the range that was being rendered useless for cattle.

The Police reported that cattle and horse rustling was on the decrease, but that calf-stealing was growing in favor, the extreme south country being especially favored by the wielders of the "running-iron." A considerable difficulty was being experienced by the Police in ferreting out many cases, because the ranchers who lost stock were often afraid to make complaint, fearing retaliation in burned crops or buildings, and explaining this timidity away by a carefully careless statement that it was worth a cow or a calf to find out just what their neighbors were like, anyway.

One of the most disastrous fires that ever swept the cattle country occurred south of the Blackfoot reserve during the fall, and caused a loss of forty or fifty thousand dollars to stock and pasturage. Dan McNelly lighted his pipe south of the Blackfoot reserve, and the flame spread south and west with extraordinary rapidity, travelling from forty to sixty miles an hour. Horses, cattle and wild animals were helpless before this terrific speed, and hundreds either perished outright or were so badly maimed that they were killed or dragged themselves off to some secret spot to die lingeringly. Whole bands of horses and cattle were burned to death in the bottoms along the Little Bow, the majority of them having their eyes burned out, or else their hoofs and legs burned off, many being still alive when the range riders found them after the fire and mercifully shot them down in scores. Fifty square miles of country were destroyed and the losses were tremendous, George Lane estimating his at fifteen thousand dollars, while others placed it at twenty-five thousand. George Ross lost two thousand dollars' worth of animals, and P. Burns, three thousand dollars' worth.

A very barefaced attempt at cattle stealing took place in the Okotoks district in June. Gould and Hubbard of that section purchased thirty-one head of yearling and two-year-old dogies and placed them in Gould's corral upon their arrival, when they were at once earmarked, though not branded. On June 5 they were run off, being taken right from the corral. Staff-Sergeant Dee of the Mounted Police traced them to Crooked Coulee and thence to Tongue Creek, where he found the stock near a corral, where they had been branded and their ears cut off, the ears being still in the corral. Three men, James Lineham, James Fisher and Alexander McDougall, all of Okotoks, were charged with the theft, strong indications pointing to them, including the plain and simple truth that the brand found on the animals by Dee was James Lineham's. In spite of a reward of two hundred dollars for the capture of each of these three men, they all succeeded in getting out of the country.

Down in the Medicine Hat district, a more satisfactory case was disposed of, one in which the innocent calf and the devoted cow again played prominent parts. The American roundup came over in August picking up strays, and one cow with the "WW" brand of William Wallace of Montana seemed very loath to leave Canadian herds and territory. Escaping from the American cowboys, she dashed madly back into the ranks of the Canadian stock, where she found her calf just as wildly, but more blindly, searching for her. This interesting state of affairs led the Police and Canadian stockmen present to make an investigation, which brought to light the interesting information that the cow also carried the "2PD" brand of a Canadian rancher named Henry Marshall. Wallace was arrested and tried, being fined seventy-five dollars and costs.

Sam Larson, a bad man from Montana and an escaped fugitive from American justice, came into Canada and evaded the customs to such effect that he was caught, tried and sentenced to a month in jail. Meantime the American authorities were communicated with, and a sheriff came over to await Mr. Larson's freedom, but the prisoner did not appreciate this, and disappointed the waiting official by escaping two weeks before his time was up; he was never caught.

During the winter in the Macleod district there was little loss of range stock, even dogies doing well. The spring saw some deaths among the herds, principally from two-year-old heifers that died calving. A heavy September snowstorm visited the Cardston district and did considerable damage to stock.

The prices of horses went up, and the ranchers throughout all the Province commenced to again pay great attention to this branch of the business. The Boer War considerably aided the strengthening of this market, as remounts were much in demand. Colonel Dent of the Imperial service purchased one hundred and five head of remounts at Pincher Creek in June, and other bands at other points. In August he inspected six hundred more head at Pincher Creek and purchased one hundred and twenty-four. The demand for horses was becoming so heavy from the army, the homesteader, and the railway contractor that the Alberta ranches could not supply the market, and a profitable business was again done with American horses, these imported herds being somewhat better quality than the earlier ones. Every bunch that was driven in contained some first-class mares, and as the purchasers were mostly farmers, these animals would be bred and produce good stock. Mares were greatly in demand, ranchers buying all the good ones obtainable and holding on to their own, even the Indians refusing to sell their breeding stock. There were fourteen cases of horse stealing in Pincher Creek district and six in the Calgary district. A story of a horse-stealing gang having a cache in some secret niche in the Livingstone range caused the Police to send a patrol there under Sergeant-Major Generaux, who found the tale unfounded. It was an old one that had simply reappeared after lying dormant for some years. In October, thirty head of stolen Canadian horses were sold in Montana, and a band of forty head were stolen from a man named Lindquist on Boundary Creek, the thieves in this case being caught, the ringleader, Percy DeWolfe, receiving ten years in the penitentiary, and Frank Smith, his helper, getting one year.

Indians were now a fairly respectable portion of the community, and the Bloods, who once were the best warriors and thieves in the south, had turned most of their energies to more respectable channels and were making good money in various ways. That fall they made eight hundred tons of hay under contract for the Cochrane Ranch, four hundred for the Brown Ranch, and a lot of smaller contracts, while they sold a thousand head of ponies to Eastern buyers. They also were weeding out their ponies and breeding larger horses.

In July John R. Craig, who had been ranching on Meadow Creek since he left the Oxley Ranch in 1885, found six of his horses gone from the range, and traced them to Millarville, where he claimed to find them among some horses belonging to a man named Joe Fisher, the animals then having Fisher's brand on them. Informations were sworn out against two employees of Fisher's, Cutting and Stagg, and they were tried before a jury. There were two three-year-olds, one two-year-old, and three yearlings in the bunch, and the defence claimed that they had all left the home range and gone south to Meadow Creek, the three-year-olds migrating in 1899, the two-year-old wandering to the same range all alone in 1900, and the three yearlings starting out to support themselves in the spring of 1901, all going to the same place to sort of form a colony of themselves. One man swore positively that he had seen the three-year-olds when they were yearlings, that he had never seen them since, but that he certainly did recognize them now at three years of age. Says the Mounted Police report of that date in telling of this remarkable case of migratory horses and powers of recognition: "The Crown made a clear case of horse stealing against these men, the judge summed up strongly against them, but the jury brought in a verdict of not guilty!" This was not the first time that the Mounted Police had learned that they should not place their trust in juries, particularly in stock cases. The aftermath of this case is doubly interesting, after the finding of the jury and the acquittal of the men. Upon Mr. Craig's return home, he declares he found the horses—brands and all—upon the home range, and the acquitted parties never tried to get them, nor were the animals ever returned to the owner of the brand that they carried.

A whisper of early days, a gleaming glimpse of the long ago, was presented to the citizens of Pincher Creek that summer when a resident of that thriving district, Sampson Jackson by name, took umbrage at something that a lawyer, A.C. Kemmis, had done, and proceeded to seek satisfaction. Entering the town with two heavy revolvers sagging his belt, he repaired to the legal party's office and in a matter-of-fact but very seriously intent manner suggested that they "shoot out" the argument at sixteen paces, Jackson chivalrously offering to let Kemmis have choice of the weapons. These methods appeared so crude and messy to Kemmis, educated as he was in the finer system of the law, that he hesitated and managed to stave off the consummation of the meeting, though Jackson was very urgent and insistent. Mr. Kemmis, or someone else, told the Police of what was going on and what was proposed, so a red-coated officer came and restrained the eager gunman until he had thought a few times and finally decided that shooting to settle an argument was obsolete anyway.

Up to the present year, by rules established by the livestock associations, mavericks on every range in the whole continent had been anybody's stock, and the roundup associations took them for their own benefit. This year Howell Harris and George Lane suggested another improvement on this, they moving at the April meeting of the Alberta Stock Growers' Association that in future all blotched brands found by roundups should be sold by the roundup captains the same as mavericks, unless ownership was proven.

Settlement of the southern districts, though not so pronounced as in Calgary and High River, where thousands of farmers had gone onto the land, was fairly healthy, most of the settlers bringing in goodly herds of stock. Sixteen new settlers went into Kipp's Coulee with 2,500 head of cattle, 300 horses and 20,000 sheep; three settlers went into Chin Coulee with 500 cattle and 60 horses; fairly large sheep ranches were established at Verdigris Lake and Tyrrell's Lake, and three other settlers with 500 cattle and 30 horses located there. On the Milk River, between McIntyre's camp and Kennedy's Crossing, there were forty-one new settlers with 11,104 head of cattle and 855 horses. The principal stock owners were Spencer Brothers, with 5,900 head, and G.P. Ashe, with 1,140 head. Sam and John Spencer of the "3CU," who had located a few years before this in the country south of Medicine Hat, had by this time been forced to adapt themselves more to conditions, their American brand being allowed to be registered in Canada, but on the opposite ribs to the position it was carried in the States. Hence they could ship no stock through Canada unless it had either a brand on each side or else on the ribs allowed by the Canadian brand, and this condition also helped the Police to look through the herds and find any American "strays" from the Montana ranch. Before this new order of things came into being, the Spencer herds increased marvellously on the Canadian side.

The day of the open range had passed when 1902 arrived, and had been passing for some years, though not keenly realized until this year, when the ever-growing fences of the homesteader and the grain rancher closed the country in. Wire fences seemed to have actually sprouted of their own accord, and whichever way the rancher turned his herds, they rubbed themselves against the barbs, with the exception of some big leases in the country south of Medicine Hat, a few in the hills and on the Red Deer. Superintendent Sanders of the Mounted Police at Calgary summed conditions up when he reported: "The days of the big rancher are numbered, and unless he purchases enough land outright to run his large herds on he will have to seek pasturage elsewhere. Yet, though the methods must change, the stock industry is bound to be the principal one in this district for many years."

The Provincial Government, in its efforts to forward immigration and to make things as easy for new settlers as possible, gave permission for the fencing of old trails that had been in use for thirty years, and this caused considerable commotion, especially in the Macleod district. Fences and homesteaders had grown so thick there that not only was there strong discussion about the closing of the trails, but the ranchers were seriously considering giving up the holding of roundups in the future, as with fences everywhere this was no longer necessary. Some of the large stock owners had to cut their herds down owing to the vanishing range, and some prepared to move to the more open country on the Little Bow, the Red Deer, and the Cypress Hills. As is usual in cases of crowding ranges, there was bitter feeling between the sheepmen and cattlemen, especially in the Lethbridge district, where great herds of sheep were being run. Many large bands, averaging thousands of head, had gone into that section, and the range in places was utterly destroyed for cattle, it having been proven fairly conclusively that cattle cannot eat over a sheep range until two years after the sheep have left.

Generally speaking, it was a good year for stock, the copious rains bringing splendid and early grass, and the stock rounding into good shape very early. The majority of ranchers fed their weak stock and calves as usual, but the sheepmen of Lethbridge country made little or no provision, which, due to the mild winter just passed, caused them no damage. Many thousands of sheep rustled all winter, and yet the average loss was not over seven percent, cheaper perhaps than the cost of putting up food, though not so humane. There was a loss of young calves, colts and lambs, the stock perishing from exposure in the cold and incessant rains of May. Cows and mares drifted before the storm, the prairies were mere quagmires, and the young stock lagged and dropped from exhaustion. The worst floods in the memory of the oldest inhabitants prevailed not only in one section, but in every river of the Province, the streams rising to tremendous heights and destroying cattle, sheep, horses, property and people, fortunately few of the latter. All over Alberta the rivers rose and swept corrals, houses, stables and stock out of the river bottoms, bridges went out like paper structures, ferry boats went down on the crests of the floods, and the whole vast country was a well-soaked sponge. Men in boats rescued settlers in towns and on ranches, miles of railroad tracks were washed out or under water, and one of the wonders that has come down from that wet year is the few lives that were lost, for the ranchers and homesteaders performed prodigies of recklessness in endeavoring to save their livestock from the bottoms and streams, desperate chances being taken by settlers everywhere to save their animals, and yet the only drowning recorded during that time in the southern country was the death of Billy the Kid (William Nash), a very popular young cowboy, who met his end in the raging waters of Pincher Creek while working to save some stock that was imperilled.

There was little complaint of loss from wolves, less, in fact, than any preceding year, and there was practically no disease in evidence, the rains and the early grass soon wiping off all signs of mange and persuading many ranchers that, in spite of the cattle that had died of the disease during the preceding winter, it would never be very dangerous now. Others of the larger ranchers did not feel so confident, and the veterinary branch of the Dominion Department of Agriculture was nervous, too. Dr. J.G. Rutherford, V.S., one of the greatest authorities on livestock in the world, had been appointed chief of the branch that year, succeeding Dr. McEachren, and he considered the threat of the disease so serious that he arranged to meet the members of the Alberta Stock Growers' Association at a meeting in April, when he could discuss with them the best method of eradicating the disease. Mange had appeared that spring, not only in cattle but in horses as well, High River and Gleichen being the districts where affected horses were found, this being the first time horse mange had appeared in Alberta for years, it being introduced this year by horses from Montana. There was mange among some of the cattle along the Milk River, some along the Red Deer, and the ranchers belonging to the Association wished to have the herds cleaned thoroughly. Dogies and Eastern cattle were proving very susceptible to contagion from mange, taking it far more readily than the tougher range stock.

The livestock association met in April, and Dr. Rutherford, after discussing the situation, decided to leave the matter of treatment in their hands, feeling satisfied they would handle it satisfactorily. George Lane and A.J. McLean spoke of the necessity of getting after the disease and stamping it out, and the discussion then took a general form, it being finally decided to dip all infected animals, as in the past, but to make a much more thorough search for such beasts, any showing the slightest indications of mange to be put through the vats. W.F. Cochrane, manager of the Cochrane Ranch, was elected to the honor of president of the Association.

The suitability of Alberta as a finishing area had now been proven by years of experience, and the stockers of Ontario, British Columbia and Manitoba were being shipped in. This supply seems not to have been sufficient, as feeders had extended their purchases down as far as Old Mexico, considerable numbers of Mexican dogies being tried on the ranges during the season. There was at first a fear that these animals would bring disastrous results by introducing the hated tick among the herds, but this fear proved to be groundless, no bad results following the shipping in of the descendants of the first range cattle ever run on the American continent.

The second annual provincial bull sale was a success, much more marked than the first, widespread interest being displayed and prices ruling higher, while the demand was much stronger. Two hundred and twenty head were sold at an average price of $95.81.

There was little or no complaint concerning trespassing American herds, this being partly explained by the fact that great herds of cattle from that country were brought in, the owners securing leases and paying duty in order to do this, establishing great ranches in the Province south of Medicine Hat. The American roundup that came over to pick up their strays found not over a thousand head. A glance at the leases let by the Government during the year is plain indication of the reason why the American herds did not crowd the boundary any longer—they were across. The Spencers secured two leases, over 90,000 acres; the Creswell Cattle Company of Trinidad, Colorado, over 65,000 acres; C.E. Hallward, 60,000 acres; J.A. Thatcher, 65,000 acres; A.G. Day, over 65,000 acres; D.A. Richardson, nearly 11,000 acres; James O. Born, about 10,000 acres; John Jenson, 4,500 acres; J.W. Taylor, 47,000 acres; George M. Cannon, about 10,000 acres; J.D. Wasesha, 22,000 acres; Wallace and Brown, 100,000 acres; a total over half a million acres.

J.H. Wallace, whose lease was obtained this year, had been in the country and engaged in the "beef" ranching business since 1892. Tony Day, manager of the Creswell Cattle Company, was an old Texas "Panhandle" cowman, having been in the stock business forty years when he first came to Alberta, following the range from Texas north as the newer states opened, and finally being crowded into Canada this year with twenty-two thousand head of stock, which he had shipped to Billings, Montana, and trailed from there to the new lease, seventy miles south of Medicine Hat. Day paid the Dominion Government forty thousand dollars in duty when he brought his herd over.

The Montana Spencers, who had, as has been said, many cattle on the Canadian side, and who secured a great tract under lease this year, experienced the discomfort of being caught "with the goods," which is always embarrassing. On March 7, Mr. Bourinot, of the Canadian Customs Department, went, in company with Superintendent Deane of the Mounted Police, to the Milk River country to interview the Spencers regarding certain cattle that had been said to have been driven into Canada without payment of duty. On May 14, Deane and

Bourinot went with a roundup party and looked over Spencers' stock, with the result that six hundred and six head of their cattle were seized, as no duty had been paid on them and the animals bore the brand of the American ranch. The Spencers raised a strong objection to paying this, and upon making a deposit, their objection was forwarded to Ottawa to be threshed out, with results not said to have been eminently satisfactory to the owners of the "3CU."

Another somewhat similar case occurred in September when Special Customs Officer Studen asked the Mounted Police to assist him in holding certain cattle that were being gathered by the Circle and Brown ranch companies. Both these companies had ranches on each side of the line, and the customs officer was assured they were endeavoring to elude duties. The result of his roundup was that three hundred head of American stock was detained, two hundred belonging to the Brown and Circle outfits, and the remainder to small American stockmen whose animals had simply strayed across the boundary. The Circle and the Brown companies paid sixteen hundred dollars' duties, while the other stock was turned back into their own country.

The cattle market was very lively, and large numbers were sent out for export, 1,418 from Macleod, about 4,000 from Lethbridge, and somewhat equal numbers from Calgary and Medicine Hat. A considerable number of "top" steers were shipped over the Great Northern to the Chicago market, concerning which departures from usual channels of trade Inspector J.V. Begin, RNWMP, Lethbridge, says: "Strange to say, after paying duties of eight dollars per head, from five to eight dollars more per head was realized on these shipments."

The horse industry continued to be splendid, two thousand horses being imported for the Alberta homesteaders, and about six thousand sold off the Alberta ranches to points outside the Province. Owing to the large numbers of animals being shipped to South Africa, the supply of saddle stock was cut away down below the average, and the result was that this class of animal went high in price, the Mounted Police and even some ranchers experiencing considerable difficulty in securing mounts. One rancher went over into Montana to get a cow pony, and paid the hitherto unheard-of price of one hundred and fifty dollars for it. The spring rains and bad weather caused some loss to one horse rancher who had just entered the Province to commence in business. This was a man named Dowse, who had located in the Porcupine Hills with seventy-five Percherons that he trailed in from Montana, losing thirty-five head en route, due to the mud, the rains and the rivers.

About this time the Bar U Ranch changed hands, George Lane and Gordon, Ironsides & Fares purchasing it. The ranch was the second oldest in the Province, the Cochrane leading it in age, but not in success, for the Bar U, or the North-West Cattle Company, as it was officially known, had been splendidly managed and handled from the start. It had been founded by the Allans of Montreal, and their first manager, F.S. Stimson, had remained in charge until the ranch was sold to Lane and his associates for a sum of nearly a quarter of a million dollars. With the Flying E, the Bar U, and half a million acres of land on the north side of the Bow River, east of the Blackfoot reserve, that they leased from the CPR, they had as big a range as any cattle concern in the Province, and brought their herds up to about thirty thousand head, increasing to this figure by breeding and importing.

The Mounted Police experienced a fairly busy and quite effective year, teaching a considerable number of the lawless that the stock business was a poor industry to attack. Fires were not dangerous anywhere but in Lethbridge and the Cochrane range on the Waterton Lakes, and on the Little Bow. Twenty thousand acres of grazing land burned on the Cochrane range with no damage to stock; the Little Bow country was badly scorched by a three days' fire that killed some stock there and in the Lethbridge district; and January fires swept fifty thousand acres of Jesse Knight's range, destroying also much grass and some stock in Kipp's Coulee, the north side of Milk River, and Chin Coulee. The most serious fire was on October 29, starting between Kipp and Blackfoot Crossing, with a heavy wind blowing. Despite the extreme efforts of the ranchers and settlers, this fire burned all the country between there and the Little Bow, east of the trail, destroying much hay, a number of ranch buildings, and considerable stock.

Stock stealing continued, and a number were caught, but the Police still claimed they were hampered in this work by the ranching public themselves. "Cattle and horse stealing are the crimes I believe to be most prevalent, and they are most difficult to detect," reports Superintendent Sanders. "The way the cattle and horse business is carried on in the range country is such as to tend to assist this class of crime. One cause is the holding back of information in fear of retribution and revenge. It is quite likely from letters received from British Columbia that a good many horses stolen from that Province are brought into Alberta by the Crow's Nest Pass. Several large bands have come in that way this year, and, after being disposed of between Macleod and Edmonton, we get complaints of thefts."

Junius Young of Cardston was given four years by Chief Justice McGuire for stealing horses. Other cases of white men were treated more leniently, but all Indians who were convicted were made examples of. Fred Brouillette, charged with stealing a horse from a Piegan named O'Keefe, was dismissed by Judge

McGuire on account of his parents; Joe Vare, found guilty of stealing a colt from Henry Cwoonty of Gillingham, was let off on suspended sentence by Judge McGuire because he was young and had invalid parents. Hugh Brewer, a Blood Indian, was convicted and given three years; Ugly Head, another Blood, convicted of stealing horses from the Brown Ranch, was given three years; a half-breed Indian who stole four horses from a Blood named Chief Moon, and who had taken these animals to the Cochrane Ranch, where he had a job hauling logs, was given two years by the Chief Justice.

Alfred Barton, a rancher near Whisky Gap, in the Milk River Ridge, pleaded guilty before the Chief Justice, and his attorney presented a largely signed petition for leniency, the signatures being those of ranchers and settlers in the district where he lived. This caused the judge to remark that if the people who were the principal sufferers from this class of crime chose to petition leniency, they were the best judges. Barton then gave bonds and was released, to come for sentence when called upon.

The Indians were showing improvements in many ways, some of them not being entirely desirable. Many of them were going in for a sort of mixed farming, running small bands of stock and cutting a lot of hay, generally of a very good quality, they seeming to take more care in the making than the general run of white ranchers did. They continued to steal considerably from one another, and their clannish spirit would crop up always in case the thief was caught by the administrators of law and order. In such event, the owner usually declared the animal had not been stolen at all, that he had given it to the thief to sell for him. Thus, and often, were the wheels of justice blocked by the tribal brotherhood of the red men. They sold a considerable number of horses, mostly to Eastern dealers, who shipped the animals to points in the Eastern Provinces. With regard to their drinking propensities, they remained as thirsty as ever, but found it very much more difficult to secure liquor. Perhaps it was due to gratitude, and perhaps it was due to the selfish desire to retain their known channels of supply, that they had learned by this time to always declare they had "found" the liquor whenever they were discovered intoxicated or with the stuff in their possession.

A new horse ranch started this year in one of the most historic spots in Alberta, the Kootenay Plains of the early traders. Tom Wilson of Banff, the best-known guide in Alberta, and an ex-Mounted Policeman of the vintage of 1880, placed a band of horses there and has bred them up to a very large herd. When the North-West Fur Company was endeavoring to get the Kootenay trade a hundred years ago, they went to the Kootenay villages and solicited it, receiving the reply that the Rocky Mountain House, at the junction of the Clearwater and

Saskatchewan rivers, was too far away. This post was the nearest one to the tribe, and yet it was some two hundred miles from the Kootenays' main camping country. In order to overcome this obstacle, the company manager agreed to go every year at a certain date to the plains at the headwaters of the Clearwater, taking merchandise and trading stock for the tribe, if the latter would journey up there and meet him. This was agreed upon, and for years the traders met the Indians on these flats, which in time came to be known as the "Kootenay Plains," from their association with the annual trading pilgrimages of the Kootenays. Surrounded by mountains, rich in grass and winter feed, Wilson's ranch is one of the most picturesque in all of the Province, and is one of the best horse ranges, deep though it be in the mountains.

Range conditions prove to many a good man that the veneer of civilized polish is not more than skin-deep, and that a man is a man no matter how he wears his clothes. To many people this knowledge never comes, but the men of the range judge a man for himself and not for his attire, and this gift is often acquired by those who live among them. Big ranchmen handle money in huge amounts, treating the large sums they receive for their yearly shipments with a familiarity that only comes to men of big dealings. Not long ago, a Macleod rancher sold many hundreds of steers which he must deliver at Winnipeg, so he accompanied them to see that they were handled right. Registering at the palatial railway hotel there, he spent his days in the stockyards attired in high boots and workaday clothing, thinking nothing of it, for men wore clothes that were suitable in the country where he lived. Upon making his delivery, he was presented with the price of the animals, a certified cheque for ninety thousand dollars, and, business completed, he decided to start for home. The manager of the hotel knew him and valued his patronage; large financiers dropped in to see him, and several accompanied him toward his waiting train. The party, including the hotel manager, strolled leisurely toward the iron gates over which a skilled and mighty police Cerberus presided and sifted the wheat from the chaff, for only Pullman passengers were admitted within the sacred precincts. The hotel manager and the other gentlemen of broadcloth passed in safety, but the rancher, who had loitered a few steps behind, was promptly and sternly stopped.

"Here, you can't get through here," declared the limb of order.

"But my train is waiting there," protested the amazed cattleman.

The policeman-guard cast a skilled and judicious eye over the attire of the other, read the unmistakable signs of work, instantly catalogued the wearer, and insisted that he knew not of what he spoke, that the homesteaders' excursion did not leave until the next day.

"But I have my ticket for this train," objected the stockman, fidgeting nervously, because the five-minute gong had sounded.

"Let's see it," gruffly commanded the officer, and the rancher reached into his inside pocket, felt a long, narrow, familiar piece of paper and flashed it toward the policeman, who seized it, scrutinized it keenly, gasped, and then stared with hard and suspicious eyes at the toil-marked man before him.

"Where'd ye get this?" he demanded in steely tones.

"From the manager of this hotel," replied the innocent.

"Well, I guess ye'd better—" began the officer, using the usual "come-with-me" tone of his kind, but the hotel manager bustled back at this moment and intervened, and the staring, goggle-eyed guard silently passed the ninety-thousand-dollar cheque back to the rightful owner, for the cattleman had drawn that valuable strip of paper out in mistake for his ticket.

When Hon. A.L. Sifton, Premier of Alberta after the downfall of the Rutherford Government in 1909, was appointed to the honor of Chief Justice of the Province in 1903, a very important change took effect in the lives of cattle and horse thieves, as well as the stockmen in general. The Chief Justice brought joy to the hearts of the Mounted Police and the ranchers in general by the strict and impartial way in which he handed stiff sentences to convicted rustlers and brought consternation of a kind to the stock associations by a ruling that made it unlawful to take up mavericks at the roundups and sell them for the general good of the Association. Ten-year sentences for stealing stock grew quite common, and it was felt by the range public that these examples of harsh punishment would produce a very beneficial effect, and the thieves at liberty would be much more careful in future.

His lordship's decision regarding mavericks was made on November 20, at Medicine Hat, and was brought around as follows:

A German settler near Josephburg, who could not speak English, had a fourteen-month-old steer that stayed on his place until driven off by the May snowstorm and lost. He heard in the course of time that the Plum Creek roundup had gathered it in, and he went to the captain to claim it. Not being able to make himself understood, he went away and shortly returned with his English-speaking son, who made an intelligible claim and was scorned by the captain, who pooh-poohed the requests and declarations of the owner, made no attempt to find if the claim was well founded, and eventually sold the animal at auction as a maverick, getting nineteen dollars for it. Another German settler in the same locality lost a heifer under somewhat similar circumstances, though he did not know it had been taken up by the roundup and consequently made no claim.

The first German brought his troubles to the Mounted Police, who, knowing there was no law that forced owners to brand stock, laid a criminal charge of theft against the captain. The Medicine Hat and Maple Creek associations engaged P.J. Nolan of Calgary to defend him, and the defence was that unbranded animals on the range were incentives to theft, and held that they, as associations, did no wrong in selling them. Also, the mavericks might as well belong to some members of the associations as to any other man, being easily made by cows becoming separated from calves. The practice of selling mavericks on roundups to help to pay expenses had been an established custom of some years' standing.

Judge Sifton's decision knocked this custom sky-high, and established a ruling that was really more satisfactory to ranchers, settlers, and farmers than the former uncertainty had been. He said that people had a perfect right to allow their stock to range the prairies unbranded if they so desired, and that a roundup had no right to gather cattle that did not belong to the members of the Association without consent of the owners. Holding that the punishment in this particular case did not necessitate imprisonment, and considering that the requirements of the case would be met with in a formal conviction, he added that the secretary and other members of the Association were equally guilty with the captain of the roundup. They were well aware that the sale of mavericks was unlawful, for they had been trying for years to legalize it. Orders for the return of both the heifer and the steer to the two Germans were made and carried out, and the maverick ruling of the stock associations had received a severe blow.

Other theft cases that were heard before Chief Justice Sifton were disposed of with equal promptitude and vigor, as is shown by the following ones. In May two horsemen from Montana arrived at the ranch of Eugene Hasson on the St. Mary's River with their horses played out by a hard ride. These men, Bert Edsall and Webb Galbraith, left these animals at Hasson's and proceeded to Lethbridge on borrowed mounts, and Galbraith, upon returning to the ranch, asked Hasson to allow him to continue riding the animal he had borrowed, Hasson to keep the horse that Galbraith had ridden from Montana. Hasson agreed, and Galbraith went away, selling Hasson's animal at Cardston, while Hasson placed his own brand on the new horse. Acting on a complaint from John LaMott of Piegan, Montana, the Police found a horse at Hasson's that proved to be the one stolen from LaMott and ridden in by Galbraith. Information was therefore laid against Hasson for having stolen property in his possession, knowing it to be stolen, and he was convicted and given five years.

Joe Hill, "Mexican Joe," stole horses from Piegan Indians on the reserve, took them to Morley, was caught and sentenced to five years.

Through Sheriff C. Wallace Taylor, of Choteau, Montana, the Canadian authorities learned of the whereabouts of James Fisher, wanted for participation in the theft of the thirty-one head of dogies at Okotoks. Taylor arrested him, and he waived extradition, coming voluntarily to Alberta and standing trial, receiving a five-year sentence.

A most cold-blooded theft occurred in the Macleod district late in May that was as cruel a performance as had been heard of. Fred Broulette, the lad who was dismissed the year before by Judge McGuire on account of his parents, was riding with a man named Pike, the latter meeting death in the swollen waters of the Old Man's River in crossing. Broulette never stopped, never even rode down the bank to see if he could be of any assistance, never even told what had happened. On the contrary, he went to the place of a man named Johnson, got the latter to row him back across the river, picked up Pike's pack horse and saddle and went on to the ranch of a man named Whitney, to whom he sold the pack horse and tried, but without avail, to sell the dead man's saddle. Upon learning of what had happened, Whitney posted at once to the Mounted Police at Macleod with the information, and Broulette, or Brouillette, was arrested at Lethbridge, receiving a sentence of four years.

Three suspected—and in some cases known—thieves evaded the law and went to distant lands, Old Mexico, Argentine Republic, and other points, to escape the trials that were before them. A stock detective employed by the Ranch Association had been living on the ranch of Ed. Johnson near Kipp, when Ralph DuMaurier drove one of J.H. Wallace's steers in and slaughtered it. Foster, the detective, took this information into Macleod, and E.H. Maunsell laid information against DuMaurier, who was arrested on the Kootenay River. His hearing being remanded, he was allowed out on bail and promptly hastened to Old Mexico, where he effectively concealed himself.

Through the efforts of another stock detective of the Association, C.E. ("Hippo") Johnson was arrested for wholesale cattle stealing in the Macleod district and was cornmitted for trial, R.B. Bennett for the accused asking for a postponement to the next sitting of the court in order to secure important witnesses. This was granted, and Johnson was held in the cells until he decided to get out, which he did, found a horse waiting, and "drifted" on it across into the United States. The Police constable who was on guard at the time of the escape was later convicted of negligence by the Police authorities. A thousand dollars' reward was offered by the Alberta Stock Growers' Association for the capture of "Hippo," but it availed not and he was not heard of again for some time.

One of the most bizarre cases of probable wholesale stock stealing came to a head in the Calgary district this year. Carl Schultz, a German, came into the town of Calgary a couple of years previous and proceeded to make himself very prominent through the newspapers, endeavoring to interest the citizens in a meat-packing plant similar to Armour's great establishment. He held public meetings, delivered long orations, found the scheme did not take particularly with the townspeople, and switched to another proposition, an up-to-date ranch. This was something that Calgarians knew some little about, and the plausible Schultz placed his scheme before them in such vivid colors that capitalists backed him in the project. He bought, fenced and stocked two or three sections and conducted the business for a couple of years. Owing to his prominence and personality, the roundups believed him to be above suspicion and never rode through his fields, until he resorted again to the newspapers to let the people know what a success he had made of ranching. Bad weather or not, storms, snows, hurricanes or earthquakes, said Mr. Schultz through the public press, his ranch was immune because of his extremely efficient management. His cattle never died, his herds were always filled up, his stock came to market in the primest condition, all owing to Mr. Schultz's up-to-date methods. Canny cowmen read this and snorted, especially as they had seen dead cattle in Schultz's fields, and an investigation was undertaken, resulting in the opinion that the modern methods of this German rancher were nearly as old as the brand of Cain. Once the secret of his success became known, any old cattleman could have improved on his system. On August 2, two well-branded steers were found in his fields, bearing, beside these marks of well-known ranchers, the newer brand of Schultz. This was enough, and Schultz was arrested, given preliminary hearing, and sent for trial on three counts. Bail of two thousand dollars was given, and so the German had an opportunity to leave, which he did without ostentation, speeding as far as he thought necessary, never stopping until he reached the Argentine Republic. Sad Calgary citizens who gave the bail bonds paid two thousand dollars when it was learned without doubt that Schultz was never coming back.

Fires were rampant throughout the Province, and vast tracts were blackened and destroyed, houses burning, stock buildings crumbling and stock perishing before them.

April fires destroyed as much property as did the fall blazes. On April 6, a fire burned the barn of a settler named Traverser, near Claresholm, and the next day two wide districts were swept, one starting from engine sparks at Slide-Out bridge and burning to Whoop-Up before a terrific wind, the other burning from north of Macleod to the Little Bow, this fire being extinguished on the 9th by a snowstorm. A CPR engine started a fire at Woodpecker on April 23, and burned a large area. Fires at Writing-On-Stone raged for two days before the desperate settlers had it

A group of men beat down a prairie fire at Ghost Pine Creek, near Elnora, Alberta, *circa* 1906.

extinguished. In the fall, a lot of winter range was destroyed on the Blood reserve, and on October 15 a man named Bruen threw a match on the ground near Nine-Mile Coulee and caused a widespread conflagration. That same day a furious blaze started near Tilley, west of Medicine Hat, burning off all the country north of the Red Deer and as far east as the junction of that river with the Saskatchewan. A fire that started from a CPR engine spark destroyed the house, stable, grain, and feed of a man named Green, and all the grain of another settler named Saxe, on October 17 in the Macleod district. Settlers and police fought these fires with every means available, horses, wet blankets, ploughs, drags and backfires, and saved much property, often working night and day, with the women cooking and carrying provisions to the weary firefighters. Whenever possible, the police laid charges against those responsible, and a number were fined, including the railways.

The ever-recurring plaint of the Canadian ranchers that American herds were stealing their range produced results this year, and never again was the same trouble experienced to any great extent. Canadian ranchers more than suspected that the American herd owners deliberately drove their stock to the boundary so they could drift to the more plentiful shelter, water, and grass of the Canadian prairies. They also declared that American roundups came into Canada, gathered and branded their young stock and turned it loose again, and that in their beef roundups they were not too particular whether they always took their own fat cattle. In the spring John McDougall, Commissioner of Dominion Customs, issued the following bulletin, which had great effect:

"Round-up parties entering Canada for the purpose of taking out cattle or other live stock are required to report at the nearest Customs House after crossing the frontier, and obtain a permit from a Customs Officer in the regular way. This permit will be subject to the conditions that the round-up while in Canada shall be accompanied by a member of the Police Force or by an Officer of the Customs, whose duty it will be especially to see that Canadian cattle are not disturbed on their ranges and that American livestock are taken across the line and properly reported outwards, and also that foreign livestock unentered for duty are not branded on Canadian territory.

"The pasturing of foreign live stock is not permitted without duty entry at the Customs House. The owners will be held strictly accountable after June 1st, 1903, for keeping foreign live stock out of Canada, under penalty of seizure and forfeiture. Casual estrays will be especially dealt with from time to time, but the onus of proof that the cattle are estrays in any instance will rest with the owners."

After the issuance of this bulletin, the Montana Stock Association grew quite busy and not only employed line riders, but also built twenty miles of fence along the boundary, six miles south of the line. Owing to a big snowstorm in May that prevented roundup operations, the Customs Department gave the American owners a month's grace, and the cattle were picked up and taken out during June. One outfit that went on the ground early in May was so completely broken up by the May blizzard that they could not reorganize until the first part of June, their horses scattering all over the country as they drifted before the furious storm.

A pretty nasty winter and spring was experienced by the Alberta stockmen, the severe weather of February forcing general feeding, and March and April being bad, with the great May blizzard to make good measure. March was severe, snow crusted heavily, hay ran low, and cattle lost strength and flesh. April opened with snow, and the stock had little opportunity to recuperate, and the May blizzard in the south was very damaging, stockmen sustaining severe losses among calves, lambs and weak "critters." Hundreds of Eastern dogies perished from exposure, many of them succumbing during the May blizzard, in stockyards, railway cars, and on the open prairie; thousands of sheep and thousands of lambs died during the winter and the bad spring. Speaking of the Lethbridge district, Inspector Begin, RNWMP, says: "The big snowstorm in May somewhat dampened the spirit of some of our largest sheep ranchers. Knight and Son have decided to go out of business. The largest owners are: Knight and Son, McCarty of Raymond, and Peterson of Grassy Lake. They have now about thirty-five thousand head. The lamb crop was poor, and little or no provision was made last year for wintering sheep. Sheep cannot be successfully raised without hay and good protection, such as sheds, being put up for them."

By the end of the year, all that remained of the forty-odd thousand head of sheep that Jesse Knight had imported a few years before was about three hundred, he having disposed of them during the summer and fall, as he had not at all relished the experience of the preceding winter. Other sheepmen, though following his example, were selling out more gradually.

For the cattle, the spring and fall fires destroyed much grazing, and early frosts killed the grass before it matured, thus hampering the beef herds, fifty percent of the "beef" on some ranches being held over because they had not put on sufficient flesh to make them marketable, and the price of beef was low—the cattle not being in shape, and markets being weak, a great many of the ranchers would not sell, preferring to run the chance of another winter season in hopes of better prices and better animals.

Considerable mange was found among some horses and cattle, particularly in the herds of ranchers named Arnold, Roy, and Rodgers. These herds were quarantined, treated, and the disease pretty well stamped out. Large ranches commenced dipping their stock, Jesse Knight erecting a dipping station for his herds on the Milk River, and other ranchers having already had them for a year. Some small ranchers who could not afford construction of dips were treating their stock by hand-scrubbing, which was really the best treatment, though impossible for owners of great herds.

In spite of the fact that many ranchers did not sell half of their mature stock, the export figures were slightly larger than the year before, the totals being 16,555 cattle and 2,767 horses, as against 14,035 cattle and 1,126 horses. Eight hundred and fifty-one head were shipped from Milk River for the Chicago market.

Two hundred and sixty-eight thoroughbred bulls were sold at the third annual bull sale at Calgary, the prices ruling fairly strong, being $96.61 per head.

The Alberta Stock Growers' Association held their annual meeting in Calgary and elected A.B. Macdonald, manager of the Glengarry Ranch, as president, Thomas Tweed of Medicine Hat, vice-president, and George Lane, second vice-president. At this meeting the Association sought hard to get relief from crowded ranges and a certain ignoring of their interests in favor of the many thousands of farmers who had come and were coming in. Settlements were crowding nearly every rancher, and the only self-protection was to buy or secure closed leases. They memorialized the Government, begging the right to purchase their leases or else have closed leases for given periods of years. Further, they asked to have certain districts, which were unfit for agricultural purposes on account of summer frosts, poor ground, and such natural disadvantages to farmers, reserved exclusively for ranging, but they failed.

Horses were hard to get, and good horses of all classes were scarce. The demand had increased a hundred-fold during the past four or five years, coming chiefly from homesteaders, and the British Army. "If it had not been for the much-abused American horse," declares a Police report, "things would have been in a parlous state." The four largest lighter horse ranches that were in the Territories in the earlier days, the Quorn, Oxerart, Bell Brothers, and the Strang ranching companies, had gone out of business and left no one to take their place, most of the remaining breeders going in especially for the heavier classes, Clydesdales, Percherons, Shires, etc. Some of these animals were mediocre, as Veterinary Inspector Burnett, RNWMP, says. He writes: "Two specimens of Percherons which I saw had been purchased by syndicates, one for three thousand and the other for four thousand dollars. Five hundred dollars would, I think, be a good price for the two."

At exhibitions held in different districts at which prizes were given for the animals best suited for Police work, there were only a few head brought out. Three horses were entered at Calgary, and none came up to standard; five were shown at Pincher Creek, where A.M. Morden won first and L. Bell won second prize; at Macleod Mrs. J. Grahame won over six competitors, Johnny Franklin winning second. In the three centres of stock breeding, only fourteen horses were brought out to compete in these saddle classes.

The Indians, with the exception of the Sarcees, had become nearly self-supporting, and the worst done by any red men was horse smuggling and a little rustling. Mounted Police patrols seized several bands of ponies that were being run into Alberta by Indians who desired to evade duties, the natives having improved in white men's ways sufficiently to realize the pecuniary advantages of such actions. They sold great herds of their inferior cayuses to Eastern dealers, the Bloods and Piegans selling ten thousand dollars' worth of such stock. These tribes were also supplying their own beef, purchasing among themselves, the Bloods having a total of 3,500 cattle, and the Piegans 1,750.

Glanders broke out in certain districts, the worst outbreak being in the Calgary district, where forty-two head were killed, thirteen belonging to the Eau Claire Lumber Company.

CHAPTER XVIII

1904–1906

THE ranchers of Alberta at last acknowledged to themselves and the public, in 1904, that the mange was bad, that they lost cattle every year from it, and that there was strong necessity for drastic action. Before this time, they had hoped that the disease was confined to merely a few head each year, but as seasons went by and the mangy stock grew in numbers, they did not hesitate to admit the presence and prevalence of the parasites. In the spring Dr. J.G. Rutherford, V.S., Veterinary Director-General of the Dominion Department of Agriculture, held meetings with the ranchers at Medicine Hat, Lethbridge, Macleod and Calgary on the subject of the proper method of taking care of this big problem. His belief in compulsory dipping of every head of mange stock, whether infected or not, was met with approval, as evidenced by the stand taken by the Alberta Stock Growers' Association at their annual meeting, which was attended by Dr. Rutherford. At this meeting, May 13 and 14, D.H. Andrews and Dr. McEachren moved that the Association endorse the compulsory dipping of all cattle in infected districts. The motion was unanimously endorsed, and on June 27 the regulations were passed by the Government, making dipping compulsory between September 1 and October 31.

Dr. Rutherford was commended by the ranchers for his active participation in the attack on mange; 160,928 head were dipped in Macleod, and that was not all on the range; in the Calgary district, out of 193,600 head, there were 100,000 dipped. In all districts the ranchers were pleased with the result of this treatment, and in the Cardston country, they erected fifty-one dipping stations, putting many of their horses, as well as all their cattle, through. In preceding winters, there were many ranchers who had not noticed how many cattle they lost through this scourge, the disease growing worse in winter, and stripping the hair all off the affected animals, thus leaving them unprotected and poor for the winter blasts to work their will on. Ranchers, convinced at last of this, and shown conclusively wherein they were losing money steadily by neglecting the scourge, worked eagerly at the dipping, continuing work long after the expiration of October 31. The treatment not only removed mange, but also proved an effectual check to lice, worms, and other skin diseases of the stock.

The only danger remaining to be feared by the Canadian ranchers lay in the American herds, and this was much reduced by the Montana stockmen, who extended their boundary fence to ninety miles in length.

Swarms of homesteaders came into the districts, and the herds were more constricted than ever. Markets were healthy in demand but low in price, so much so in fact that many small ranchers kept their stock, this being the second year they had not sold on this account. Lethbridge and Calgary districts shipped over 30,000 head for outside points. At the Calgary bull sale, nearly 300 head were sold at an average price of $100.

Very few American cattle found their way to the Canadian side as compared to former years, the action of the Customs Department putting such a damper on the American ranchers who had settled near the boundary that some sold out and others removed their herds to points further in the state. The Chonkin Pool roundup tried to evade the quarantine laws with about 1,800 head they had collected on the north side of the boundary, and succeeded to a certain extent, though not entirely. Staff-Sergeant Allen, of the Mounted Police, went to the roundup and endeavored to arrest the foreman, but that worthy fled to the refuge of a few yards on the south side of the Medicine Line, from which vantage point he scorned the law. Allen, though hugely disappointed that the man had escaped, turned his attention to the cattle that the Americans had not yet shoved back into the States, called upon some Canadian stockmen who were present for aid, and turned back 450 head, all with mange. The mange order, which provided that no cattle should leave a quarantined district without the permission of a veterinary surgeon, had been violated and the stock was kept and treated. The expenses of treating and holding these animals amounted to $1,268, which the owners paid, expressing at the same time a deep regret that their foreman should have behaved in so undignified a manner—but keeping that loyal henchman still on the payroll.

In November the Police, acting under orders from Special Customs Officer Studen, seized 768 head, 750 head of which belonged to one American firm. It was alleged that it had been the habit of this firm during the past three or four years to gather their female stock every fall and take them back to Montana, leaving the steers to rustle on the Canadian ranges. This seizure cost the owners $1,060, duties and expenses. Having thus legally entered Canada, they decided to ship to Chicago, shipping from Maple Creek, Another blow was met at the boundary, where the United States custom authorities, considering that the animals were now naturalized inhabitants of Canada, insisted on another duty being paid before the herd could go on to its eventual destination in the Chicago abattoirs.

Mormons were growing very strong in the south, hundreds of families entering the Province to locate. The range was cut and split by fences, the country was overrun by settlers, and the great Cochrane Ranch Company, the pioneer big ranch of Alberta, decided that the day of the big cattleman was over. They had 10,000 or 12,000 head of cattle on their great range, they owned 500,000 acres of the finest land in southern Alberta, and they placed it on the market. The Mormon Church purchased this half-million tract this year at a price of $6.25 per acre, the land rising within two years to ten, fifteen, and a few years later to twenty dollars an acre. It was a splendid business venture for the Church, whereby millions were made. The cattle on the ranch were not then sold, the Church giving permission for them to be kept another year on the land. But before that year had passed by, a considerable number of Mormon farmers were already on the range preparing it for the farms, which sprang up with marvellous rapidity, fenced, ploughed and seeded.

Another project that brought tens of thousands of people to Alberta within the few years following was started in the district east of Calgary during this year, the great irrigation reclamation scheme of the Canadian Pacific Railway, whereby 5,000,000 acres of dry land was to be made fertile by irrigation. J.S. Dennis, who had charge of this tremendous work, opened by it a vast empire that promised to become the richest mixed farming district in all of Alberta.

An extraordinary disease, said to be the first of its kind ever found in Canada, was discovered in the spring among the horses of W.T. McCaugherty in the Macleod district. This trouble, which was finally pronounced to be maladie du coit, was reported to the Department of Agriculture by Inspector Burnett, V.S., and Dr. Rutherford immediately went to Macleod and examined about forty mares that were affected. In May, Dr. Rutherford examined them again in company with United States Veterinary Surgeon Knowles, who concurred with him in pronouncing it maladie du coit. Over one hundred animals were found to be affected and placed under stringent quarantine. On the last day of August, Dr. Rutherford and nearly all of the veterinary surgeons of the western part of the Territories met at the quarantine field for the purpose of examining the animals and destroying those found incurable. A thorough examination resulted in the discovery that some of the mares were in better condition than when placed in quarantine, so Dr. Rutherford decided to hold them in the pasture through the winter at Government expense. By fall a number of other affected animals had been picked off the range, and when winter came, there were 280 head in quarantine.

A peculiar and fatal accident happened to one of the Mounted Police horses this year, the second known case of the kind in the Province. The animal was bitten by a rattlesnake near Lethbridge, and after lingering in agony for a month, was shot to put it out of its misery. The other instance of the kind happened some years before, when a horse that was lying down was struck on the shoulder, dying within a half hour.

Prairie fires burned wide expanses of country and destroyed a lot of buildings and property, a number of people being fined for setting them out, and one man, Thomas Flack, living near Stavely, being fined for refusing to help fight one. Every district was damaged to more or less extent, but Lethbridge suffered most, its usual high winds, and wide, open sweeps of prairies making it peculiarly susceptible to this danger. Children playing with matches at Lethbridge set out a fire that was driven for miles by the high winds, finally dying of its own accord in a big coulee, but, en route, burning the haystacks and outbuildings at Ashcroft's ranch, and the hay and corrals of T. Leadbeater. Good winter ranges were ruined in many sections of the south.

The strict, stern sentences given by Chief Justice Sifton and expected to put a feeling of fear in the hearts of stock thieves had not resulted as yet in any marked reduction of boldness or frequency. Twenty prisoners, sentenced to terms varying from one to ten years, had been sent up from the Macleod district alone, but the docket for 1904 was just as populous as before. Joe Hill, the "Mexican Joe" who had been sent down the year before to do a five-year term, was taken back to Macleod, tried, and convicted on another charge, this added punishment being given because of his brazen effrontery in denying the charge and allowing suspicion to point to an innocent party.

Charles McLaughlin, of the High River district, was sentenced to seven years for stealing a horse from Eckford of that district. The animal was taken to Calgary and sold. The brand had been cut and marred, but the double rowlock of Eckford showed plainly through the cuts. McLaughlin and two others were implicated in stealing two other horses, but, though the evidence was strong against the others, McLaughlin, when convicted, pleaded guilty of these new charges and completely exonerated the other men.

A rancher named A.A. Barre, ten miles west of Claresholm, reported his brother, A.E. Barre, for stealing a "44" steer, and the brother received a term sentence of five years in the penitentiary.

Jesse Hinman, a western American product of the ranges, took his large six-gun from his belt and shot very seriously close to a gentleman who enjoyed the distinguished appellation of Rattlesnake Pete. These two worthies had met at Pincher Creek and entered into a game of cards, a well-known Western pastime. As often happens, especially in romances, a gentle altercation arose as to the legality or honesty of one or other of the players, Hinman dragging out his gun and shooting intently, but with poor aim, afterwards endeavoring to fit his hardware over the cranium of Rattlesnake Pete. Arrest followed, and cells, and trial. At the trial, where Jesse was tried for shooting with intent, he stoutly and legally swore that he did not hit Pete over the head with the revolver. This outrageous declaration, proven wrong by abundant witnesses, and also by abundant marks on the Rattlesnake's head, resulted in a charge of perjury, conviction, and a five-year sentence. The times of the Western gunman and card sharp were about gone.

The principal offenders dealt with at the March sitting of the Supreme Court at Macleod were John Elgin and his wife, who were convicted of killing a Waldron steer. An elderly lady named Broulette, mother of a number of sons who were well known among the Southern stockmen, had been deprived of support from this offspring by the interception of the law, which took several of them away from her and sent them to the penitentiary for irregularities with regard to other people's stock. Perhaps Madame Broulette suggested some of these irregularities, but that was not known. Upon her sons leaving her roof-tree for the more restricted surroundings of a penitentiary, she had no one to support her, so, when John Elgin, an elderly settler of over fifty years, came into her life, she fell madly in love with him and they married. Winter was coming on, and the fires of love, even at the half-century, needed something more than kisses to keep soul in body, so the bride and the bridegroom hied them forth on a short moonlit honeymoon trip and garnered the Waldron steer for the cold weather. This being absolutely *infra dig*, especially if found out, the couple were arrested, and each received five years in jail.

For some years the district around Halfbreed Creek, in the extreme south, had held an enviable reputation as the most prolific breeding ground in the West, It was here that Morgan, starting with eighteen cows, increased his herd to nearly a hundred head in fourteen months, making no purchases in the meantime. These eighteen cows had held the record to the present, but this year of 1904 saw the Morgan record smashed to flinders. James Knowles, a small rancher west of Stavely, owned thirty-two cows, his calf crop that spring totalling sixty-eight head. This remarkable increase was not known until Knowles' hired man, Daniel Dixon, confessed to the Police in order to save his own hide. Knowles was convicted of stealing a calf from W.R. Hull and given ten years, Dixon being let down with two.

Daniel S. Duncan, a stockman who lived near the boundary, directed suspicion toward himself by his habit of loaning saddle horses to Bert Edsall, a well-known Montana horse thief. Edsall was the partner of the Gaibraith who had done similar "borrowing" with the unfortunate Eugene Hasson the year before, and it would seem that this lending and borrowing was his special trait. Anyway, Duncan used to loan Edsall horses, which so aroused the suspicion of the Police that they watched Duncan and finally secured evidence sufficient to send him up for three years for stealing a cow from J.T. Taylor.

A bad gang of horse thieves who had been operating out of Cardston for some time, stealing from the country north of that point and getting as far up as north of Macleod, was broken up by the Police this year, the leader Scott getting four years, while the informer Gustin, who turned against Scott to save himself, was allowed to get away. Two other men were implicated and left the country, and the worst gang of thieves in the southwest part of the Province was broken up.

A few Indian incidents would seem to hark back fifteen years to the glorious days of The Dog and his partner Big Rib. In fact, one name is familiar, that of Eagle Rib, the Blood who was given four months in the latter eighties for interfering with a Mounted Policeman who was trying to arrest The Dog. Though the Eagle Rib of the present instance is designated as a Blackfoot, yet it might well be the same bullying buck of the Blood reserve.

On May 8, Eagle Rib and six other Blackfeet met C.B. Ewing driving a bunch of cattle across the Blackfoot reserve and demanded a cow in payment for such use of the tribe's land. Ewing, though nervous, hedged a bit, saying he would make it all right by settling with the agent. But the bucks, who saw they could go a little farther, and who doubtless enjoyed exercising a chance to frighten a white man, demanded a cow at once, and Ewing wilted, telling them to go ahead. He did not know but his scalp might decorate some lodgepole before night if he did not give in, and anyway he did not own the animals. So the hilarious bucks cut a cow from the bunch, slaughtered it on the spot, and divided the meat, allowing Ewing then to proceed with the remainder of the stock. The Mounted Police, upon hearing of this outrage, went to Johnson, the owner, and wanted him to prosecute, but he demurred, stating that he intended living near the Indians and wanted to remain on good terms with them, as he expected to hire some of them every year to help in haymaking and on the range. The Mounted Police officer in the district then laid the charge himself, and the Indians were tried before a local magistrate, who simply made them pay for the cow. As the Mounted Police knew this sentence would have no effect upon them, instructions were given from headquarters to bring the charge again, which was done, and the gang was tried before Mr. Justice Scott, who sentenced each to two years' suspended sentence.

Another Blackfoot who came into troublous times was Sun Calf, who picked up a pet cayuse near Okotoks belonging to a man named McLeod. Sun Calf, who had already served a stretch in the penitentiary, was sentenced to three years, but after serving eight months of the term, he made a successful escape and was not recaptured.

Stony Indians came before the Police also, owing to some slight differences of opinion among some of their own numbers. Big Joe went hunting with a fellow tribesman named Stony Joe, and during the heat of the hunt they came to blows over a deer, Stony Joe being extensively messed up by the big fellow. The vanquished hit the trail for Pincher Creek, where he complained of the occurrence to the Mounted Police, but Big Joe was not to be found until the following summer (1904), when he appeared at the races at Cowley. One lone Mounted Policeman clamped himself on the husky native and succeeded in arresting him after a most interesting melee, during which another Stony named Dixon, who appeared eagerly desirous of assisting Joe, plucked the revolver from the policeman's holster. Dixon was then arrested with the help of bystanders, and both Indians came to trial. As Big Joe stolidly denied ever striking his compatriot Stony Joe, and as there was nothing but unsupported declarations on either side, Big Joe was dismissed. Dixon, on the other hand, who had butted into a broil where he had absolutely no business, was sentenced to one year. Shortly after his release, he died miserably in a rancher's shack in the hills, from overtaxing his system with a sudden flood of whisky.

By the beginning of the year 1905, Lethbridge district was entirely fenced within a radius of twenty-five miles of the town, and the stockmen were moving northward, while some were selling their herds in recognition of the undisputed fact that the day of the open range had gone forever. Great numbers of fat stock, steers, heifers and cows were sold, and the price of beef was generally low. The Calgary district was feeling this same preponderance of farmers, and many ranchers were reducing their herds in that district also. Of all the wide range land of Alberta south of the main line of the Canadian Pacific Railway, the part thus far left most untouched by the farmers was the wide open land to the south of Medicine Hat. The markets throughout the year were fair, and sometimes local prices were as high as forty-two to forty-six dollars per head, but generally forty dollars was a good figure to get for a prime steer. Late in the fall an overstocked British market caused some of the exporting ranchers to actually lose money on their shipments; 49,991 head were exported, and over 78,000 were shipped altogether, out of the recognized stock districts of Medicine Hat, Calgary, Lethbridge and Macleod, including those exported and those for Eastern markets.

There was a great falling-off in Mexican imports, which cut the stocker industry down considerably. Dr. Rutherford was not personally displeased that this was so, because he favored excluding the Mexican stock altogether until the medical health officers of that country insisted upon more rigid inspections of the exported herds.

The sheep industry decreased by 50,000 head in Lethbridge district alone, but the sheepmen in turning their herds onto the markets were not meeting the low prices that the cattlemen were, mutton being well up.

The policy of the annual Provincial bull sale was proving its value. At first, when the Dominion Government established these sales in 1901, there was some small local dealers' jealousies to contend with, and this hampered sales a little, but, with time and demonstration, the value of the sales won their own place. Excellent results were being obtained throughout the districts among farmers and ranchers, there was a noticeable improvement in size and conformation of Western cattle, and the bull sales were credited with much of this improvement. Three hundred and forty head of good bulls were sold that year in the annual Calgary sale.

It is quite cheering to note, in the face of this wide decline of the grand old ranch industry that was fast giving place to mixed farming and small ranchers, that all of the old-timers bad not given up hope. E.H. Maunsell, the cattle king of Macleod and now one of the greatest individual owners of range herds in the Province, extended his interests tremendously. In association with John Cowdry of Macleod, he purchased the cattle of the Cochrane Ranch, 10,000 head of mixed stuff, with about 2,000 calves, paying $240,000 for the herd. Further, they secured the privilege of running the cattle one year on their old range by paying the interest on the ranch. Cowdry then secured 60,000 acres of closed leases, and 140,000 acres with the two-year clause, between the Bow and the Belly in the Grassy Lake district, and they put the herd there, Maunsell buying Cowdry out during the next few years. Again in the Medicine Hat district, another hopeful rancher drifted in behind 6,000 head of wild cattle, searching for some place where he could settle. This was Lem Pruitt, an old Texas cowman who had drifted north and north before settlement, until he took his last stand sixty-five miles south of Medicine Hat.

Walter Huckvale was elected president of the Alberta Stock Growers' Association this year, with George Lane vice-president and A.J. (Archie) McLean, second vice-president. The Association continued to endeavor to secure closed leases, grazing reservations, and water rights in districts, but failed as usual, and were forced to plod along in the old way, with the farmers ever edging in. As happened every year, the CPR trains killed stock on the tracks, and the Association endeavored to arrive at a satisfactory arrangement concerning these losses. The animals were often killed and done away with before their owners knew, and the loss was often never discovered, owing to the ignorant sectionmen burying the carcasses and not knowing enough to remember or keep account of the brands. The Stock Association endeavored to have the railway company promise to rectify this, but the company would make no definite promises.

There was a noticeable improvement in mange, the results of the compulsory dipping being most evident and satisfactory. This year Calgary district dipped 175,000 head, Medicine Hat, 171,260 head, Macleod and the South, 139,571 head, a total of close to half a million horned cattle.

Among the Indians there were marks of improvement, the Bloods even showing an increase in numbers, which was quite out of the ordinary, as the deathrate among the tribes usually exceeded the yearly birth record since the buffalo had gone and the bucks followed the war and hunting trails no longer. The tribe had 6,000 cattle, 3,000 horses, and some of the men were working for wages in the beet sugar fields at Raymond and Cardston.

The Government adopted an advanced policy this year, which was that of giving compensation for horses destroyed for glanders. This terrible equine disease had been in the country for some years, and though a number of cases were found every year by the Mounted Police, and the animals promptly shot, yet they could not discover all cases, because horse owners did not want, and often could not afford, to lose the animals outright. Government compensation did away with this reluctance to a great extent, and 942 head were destroyed in Saskatchewan and Alberta at a total cost of $59,308 to the Government. Of this number of horses, only a small proportion was killed from the southern districts of Alberta, fifty-seven, in Medicine Hat district, sixty-four in Calgary district, and twenty-nine in the Macleod and Lethbridge districts. From November 1, 1904, to October 31, 1905, there were 292 head of horses killed in maladie du coit quarantine at Macleod, and at the end of that period there were 207 still in quarantine.

Bad prairie fires between the Rosebud Creek and the Red Deer River destroyed many cattle in the fall of the year, with thousands of tons of hay, while many ranchers narrowly escaped with their lives. The fire, coming as it did after haying was completed, and wiping out all of the made hay, forced the ranchers to look for range at the last moment before winter, placing them in a very dangerous position.

Thefts of stock grew but little less, though the numbers of stolen animals seem to be much reduced. Perhaps this was due to the fact that the big herds were reducing, and perhaps it was due to the Police gathering the culprits before they had time to take more. In the Macleod district, cattle-stealing cases were reduced from twelve the year before to three this year, though cattle killings increased from three to seven.

Charles Wright, of Morley, owned no cattle at all, but managed to sell one to the field party of a surveyor named Miles. This one beef animal cost him three good years of his life, for that was the number that Judge Sifton sentenced him to.

Teddy Keg, a Blackfoot youth with a school education, returned to the camps and the ways of his fathers, stealing a horse from another Indian, which he sold at Langdon, stealing another and selling it at High River, and continuing his accumulative way to Lethbridge, where he was finally caught practically in the act of taking a Lethbridge equine.

Seven Blood Indians had organized a lucrative little business of stealing Mormon horses and running the stock over into the Kootenay, where they disposed of them to a rancher named Chaput, who shipped them to British Columbia and made big money. Fifty head of this stolen stock was recovered by the Police, and the Indians all received sentences ranging from five to six years each, Chaput getting off on the plea of innocence, declaring that he never suspected he was receiving stolen steeds, even though he was getting big, strong horses for ten and fifteen dollars each.

Jean Baptiste Faucheneuve, who had just come out into the world after doing five long years for stealing horses, picked up a few head and pleaded guilty when the Police dragged him before a judge. Owing to his apparent insatiable appetite for other people's horses, he was sent down for ten years.

Jack Payne, alias Coyote Jack, alias Coyote Bill, was acquitted of a charge of horse stealing, the Police report ingenuously adding that one of the crown witnesses had left the country before the trial came up.

One of the most pleasing cases of the year from a Police standpoint was that which occurred in early September in the Medicine Hat district. Corporal McLean, of the force, wanting supper and a bed and being a long way from home, dropped into the ranch of a settler who had been considerable of a thorn in Police skin for some time. This settler was a man who boasted that he was too smart for any police, that he had been arrested on eight different occasions and had come through without a blemish every time. He was not at home when McLean stopped, and the latter went into the corral, where he noticed a mare and a young colt, the mare being without doubt no relation at all to the little fellow, as she would have no dealings with the hungry youngster, and her hind feet were hobbled so that she could not display too vigorous opposition to the colt's attentions. It was a colt that needed a mother badly, and the corporal was deeply interested. Shortly afterwards, the rancher came in with a load of hay, saw that the policeman had noticed the colt, and remarked that it was a maverick and the owner was unknown. McLean climbed up to help the man unload his hay— and uncovered the carcass of a newly butchered mutton, the prompt explanation of its presence being that it had been bought from a herder while en route home with the hay.

Next day McLean discovered the owner of the colt, and in company with this man and the anxious mother, he repaired to the ranch. There was absolutely no doubt in the minds of the beholders that the joyously nickering mare and the starving colt were close relations, for the fence was scarcely strong enough to keep them apart once they caught a glimpse of each other. The ranch owner was arrested, inquiries were made of all sheep herders for miles around if any had sold a sheep to him, none had, and the boastful prisoner was tried at Medicine Hat and sentenced to two years' hard labor for his attempt on the colt, and probably his success with the sheep.

Few people know that the Hon. C.R. Mitchell, Minister of Public Works in Premier Sifton's Cabinet, has been a large ranch owner for some time, his herd of horses being among one of the best in the Province. Mr. Mitchell came to Alberta in 1897, and seven years later put a bunch of horses on the range twenty miles out from Medicine Hat, the ranch being under the management of Charles E. Sherman. This ranch is still in existence, and a large herd of horses are on the range, the animals being of very high grade.

About 100,000 head of cattle and 12,000 horses were shipped out of the southern or ranch portion of the Province in 1906, thus further decimating the great herds. The necessity of cutting down was being impressed more forcibly upon the stockmen, and they were reluctantly selling on only a fair market. The section in which ranching was carried on undisturbed the longest was Chin Coulee and the Little Bow, but this year bitter complaints of homesteaders fencing those ranges were frequently heard.

All stock experienced a good winter and came through in good shape, though glanders was still among the horses, and maladie du coit was raging, 120 head being killed on account of the latter disease during the period between November 1, 1905, and March 31, 1906, the stations still holding 112 suspected cases. During the fall W. O'Tott and other ranchers along the Little Bow bunched their marketable stock and shipped it to the British market, but owing to glaring errors in management, lack of space on shipboard, and other misfortunes, the venture was a sad failure.

Loco-weed, that dread danger to horses, was just commencing to make ravages in certain districts in the Porcupine Hills. This weed is supposed to utterly ruin an animal, and it often does, though experienced ranchers and veterinaries declare that in some districts it proves harmless, due, they believe, to the fact that some poisonous ingredient of soil is absent in those particular immune parts.

An interesting event was the visit of Sir George A. French, who visited the West again for the first time since he had commanded that first force of hardy Mounted Police men who came into the raw country to establish order. Great cities, great ranches, large farm settlements, and wonderful development greeted the eyes of the visitor, where there had been nothing but the primitive wildness of the prairie when he saw it last. Sir George French was particularly interested in the old Police points, and in the general Police work, expressing satisfaction at the splendid results obtained by the force.

Various small fires netted considerable damages, the originators of the blazes often giving themselves up voluntarily and paying for the cost of their carelessness.

Maladie du coit produced one good result in scaring all of the ranchers in the south country and causing them not only to keep up all of their stallions, but also to hold their mares in closer touch with their home ranches. This in itself assisted the work of stamping out the disease, and also tended to an improvement in breeding, the carelessness of the past being thus done away with.

The Provincial bull sale resulted in less sales than the preceding year, but the animals were all splendid specimens, bringing some very good prices that averaged better than any previous sales, the price being $108 per head. A total of 170 head were sold to ranchers and farmers, the Indian Department purchasing quite a number.

The feeling that mange was stamped out was general, and the compulsory dipping order was rescinded. Some reports of mange in certain districts came in, and there was a belief that the eradication of the disease would not be assured until the stock had passed through another winter. Some ranchers continued dipping their stock even after the compulsory order was withdrawn, the big stockmen as a rule knowing well where their best interests lay, the worst carelessness and neglect coming from among small owners. Dr. McKay, Dominion veterinary inspector for the Calgary district, took up his duties during the year and assisted greatly in the fight against the various diseases, mange, glanders, etc. George Lane, of the Bar U, as he was always known, even after he had severed connections with that ranch in 1889, imported a spraying machine for treating the vast herds that he was interested in, which machine, it was said, furnished a much more efficacious treatment than any other system.

Along the boundary, the Customs Department was kept busy by smugglers and American herds, though the latter cause of worry had been reduced a very great deal. Doubtless considerable stock was successfully smuggled, for the boundary is wide, and few men guarded it, while nearly every foot was open for smugglers to cross. Forty-seven smuggled horses were seized at Stettler in the possession of William Bain, of Oldham, Montana. Bain, who had a ranch on the Canadian side, went north with 175 head of horses, selling them in the various towns as he journeyed and disposing of all but forty-seven, when the police nabbed him. His detection was out of the ordinary, for he had passed successfully through the hands of a dozen Mounted Police men before he was caught through the superintendent at Lethbridge. Constable Tucker, RNWMP, happened to be at Lethbridge, and was asked by the superintendent there if there were many horses going north for sale. Tucker told of the bunch at Stettler, giving the brands, and the superintendent checked up his reports and found such animals were not shown. Bain, who had stilled all Police suspicion heretofore by displaying a pass signed by Staff-Sergeant Greenwood, was at once held by the Police at Stettler on a telegram from Lethbridge, and a squad was sent from the latter place to arrest Greenwood. This policeman was reduced to the rank of constable, given a month's imprisonment, and then dismissed from the force, it having turned out that he never even saw the horses, but had simply given Bain the pass to save him from further bother from other veterinary inspectors. Bain was assessed duties on the forty-seven head amounting to $1,992, the cost being brought up to $2,100 by other expenses.

Cattle rustling continued to be annoying, small farmers and herd owners suffering about equally, though the farmers were more open to attention, owing to the fact that many did not brand. A Mounted Police report of the year deals with this subject generally and at length, reading as follows: "I cannot say whether cattle-rustling has increased during the past year, but I am inclined to think it has, judging from the rapid increase of some of the herds in this district and from the numerous reports received; this is not the same class of work as is done by the horse thieves, who take chances and drive their ill-gotten gains north for sale. The cattle-rustler rides the ranges with a running-iron strapped to his saddle, and generally in stormy weather, and picks up calves which have arrived at the age to be easily weaned from their mothers. It is only a work of a few minutes for these experts to rope the calf and drive it to some place where it is held till it would not be claimed by the mother or recognized by the owner. Fortunately

Interior of P. Burns and Co. butcher shop in Calgary, *circa* 1900–1903. Patrick Burns was the cattle King of Canada by 1891.

for these rustlers, and unfortunately for the settlers, a number have settled in the district with small bunches of unbranded cattle. In this connection I have issued orders to patrols to warn all such settlers to have their cattle branded. The using of running-irons should be prohibited and a penalty be provided for anyone in possession. These irons are a convenience in the hands of honest men—where in a round-up a rancher's cattle are found a long way from their range, the increase is branded by the round-up using a straight iron or running-brand; but in the hands of the dishonest, and it is well known that this is the class who carry them, it is most dangerous."

The temporary successes of the seven young Bloods who did the rushing horse business with the rancher on the Kootenay had, instead of putting the fear of results in the breasts of the other young men on the reserve, induced them to try and improve on the undertaking. Four boys, Philip Hoof, Yellow Creek, Charlie Davis and Willie Crow Shoes, worked out from Slide-Out and did well for a short time, stealing horses and driving them about thirty miles north, where they easily disposed of them at ridiculously cheap figures to some farmers who had recently arrived from the States. "There is a certain class of these people in this country," remarks Superintendent Primrose, referring to these American settlers, "who encourage this class of offence, as a man who has been accustomed to handling stock for a number of years must know that when he is offered a $120 horse for fifty dollars, there must be something crooked about the transaction, and I am seriously contemplating proceeding against some of these people for receiving stolen property."

The four energetic Indian lads were eventually swept into the Police net and haled before Judge Sifton, who sentenced them to terms of two, three and four years.

W. Roper Hull, who had been engaged in ranching and the meat business in Alberta since 1884, sold his string of meat shops this year to the big P. Burns Company. P. Burns was now the acknowledged beef king of the West, controlling the butcher and packing industries throughout the Western Provinces, running huge herds of beef stock, and shipping great numbers west and east. Starting twenty or twenty-five years before "on a shoestring," he had grown to be the richest man in the western part of Canada, controlling an immense business that stretched out over thousands of miles of prairie and mountain land.

CHAPTER XIX

1907-1909

IF one were to ask an old cattleman which was the best winter he had experienced in Alberta, he would scratch his head and ponder unsuccessfully, because he would think of so many good ones; but if he were asked which was the worst, he would remember keenly and say 1886–7 and 1906–7. The winter of eighty-six was a terrific one, snow piling all over the country in deep, crusted drifts, blizzards screaming across the open, cattle humping up and perishing from the bitter, bitter cold. It was without doubt the very worst winter season ever seen in the Alberta ranch business, for though the loss was heavier in 1906–7, there were circumstances leading up to it that accentuated the inclemency of weather. Everything seemed to conspire to place range stock at a disadvantage, and the results were tremendous losses, estimated at ten million dollars.

Fall started poorly, with late rains, frozen, uncured grasses, and early cold. The range was fenced, the grain fields spread over the land, barbwire held straying stock between the narrow sixty-six-foot confines of the country roads. Prairie fires and crowded ranges took the grass off and left little for winter rustling—in fact, in some districts the range was so overstocked that cattle went into winter in very poor condition, even hay being insufficient to strengthen them against the cold. Mange, which was still among some herds, assisted in making some stock more susceptible, for with impoverished bodies, the animals could not resist the inroads of the parasites, and numbers of cattle were absolutely hairless when New Year came. Lack of food, poor condition and exceptional storms, snow and cold demanded a fearful toll from the range stock and depleted the herds of the Province by about half, forcing many stockmen out of business, and throwing a great number of cattle on the summer and fall markets, when the herds—with the exception of those of such big stockmen as George Lane, E.H. Maunsell, J.H. Wallace and P. Burns were "stripped," and the ranch business suffered the worst setback it ever felt, especially as there remained no more open range to encourage those interested in the industry to attempt to recoup their losses.

Fence corners, railway tracks, coulees, river bottoms, all contributed their tale of dead animals when spring opened the iron-bound land, heaps of rotting bodies showing everywhere, cattle even dropping dead in towns where hunger had driven them from the open. One day in January the citizens of Macleod saw what appeared to be a low, black cloud above the snow to the north, which drew slowly, draggingly nearer until it was seen that a herd of thousands of suffering range cattle were coming from the north, staggering blindly along the road allowances in search of open places in which to feed. A steady, piteous moaning filled the air as the suffering creatures drew close, feeble, starving, skinned from the knees down by sharp snow-crusts and by stumbling and struggling to arise, hair frozen off in patches—naked, mangy steers, tottering yearlings, and dying cows. Straight into the town this horde of perishing brutes slowly crawled, travelling six and eight abreast, bellowing and lowing weak, awful appeals which no one was able to satisfy. They were Bar U and other Northern cattle, and their numbers were so great that it took over half an hour for them to pass a given point. Right through the town they dragged themselves—exhausted animals dropping out every minute to fall and die as they lay, the route through the town being marked with a string of carcasses—past the hotels, the stores, the staring people on the sidewalks, out into the blackness of the prairie beyond, where they were swallowed up and never heard of again, every head being doubtless dead before the week had passed. The morning after they passed, the town authorities dragged forty-eight dead carcasses out of the streets.

Pitiful stories came from everywhere; starving stock brought in from the range stood and took no interest in the hay before their very noses, standing till they fell over and died; on the range they died singly, in scores, and in hundreds, and the road and fence corners were lined and heaped with them. Not a mile of CPR right-of-way fencing but showed in spring the bodies lying where the fence had barred the drifting and caused the dazed and suffering brutes to stand helplessly until the end came. In the railway tracks, where the trains and section hands had cut deep grooves in the snow, the cattle crowded for the shelter of the banks, many dying under the wheels of the trains, though one of these trains was the direct cause of saving the life of one wild steer. A band of Maunsell's cattle had piled and huddled into the right-of-way near Grassy Lake, when the train came along. Failing to move the cattle with threatening blasts from the whistle, and being reluctant to crash amongst them, from pity and a fear of results to the train, the crew decided to chase the animals away. So a brakeman with a lighted lantern was sent to do the chasing, and a wild steer proceeded to liven things up considerably, going right after the trainman with so much vigor that one sweep of his long horns caught the wire handle of the lantern, tore it from the hand of the fast-departing brakie, and fired the animal's heart with a frenzy of fear. No shaking, bucking or plunging would take that terrible glaring object from where it hung, and the steer broke away on a long gallop, flitting like a gigantic firefly across the expanses of snow. Cowboys saw the strange spectacle, pursued, headed it, drove it into a corral and took the lantern off, the steer thereupon staying around the stables and picking up a living through the winter. The lantern is still an object of interest at the Grassy Lake Ranch.

In the Calgary district the loss was estimated at sixty percent, and in some districts the owners lost practically every head that was at large on the range. Farmers suffered to a lesser degree. Half the stock in the Lethbridge district died, and Pincher Creek district was fortunate in suffering only a twenty-five percent loss, due to a great extent to the ranchers having more hay than in other sections.

The severe winter losses that came to the Little Bow country on the heels of the disastrous pool to the British markets the fall before entirely wiped out a number of independent ranchers there. One man named Powlett counted 640 head on the fall roundup, and thirty-three head in the spring. In the fall, Robert Patterson had sold to E.H. Maunsell the balance of his herd left over from the shipment to England, selling on a basis of 1,500 head. When spring came, he had only 506 head to turn over to the purchaser.

Walter Huckvale and his partner Hooper, ranching on Many Berries Creek, lost twenty-five percent, this loss being entirely due to their being eaten completely out of summer range, the cattle consequently going poor into the winter. Though they had plenty of hay, the old cows simply could not strengthen up to face the hard weather, and the deaths were chiefly among these.

Walter Wake, who had moved to High River in 1900 and was now running cattle in partnership with Tom Henry, suffered the usual loss. He and Henry had 800 head with George Lane's cattle on the Bow River, and all they collected in the spring was 300 head. At the fall roundup the Bar U had counted 24,000 head, and their loss was proportionate to that of Wake and Henry, some estimating 15,000 dead Bar U stock, and the owners practically admitting a loss of 12,000. These cattle had drifted south from the Bow when the northern storms came, leaving their open range and crowding into the fences and road allowances, drifting ever before the storm, searching for food. They came finally to the open country across the Belly, just about the time a providential chinook opened that district and saved the remaining stock.

P. Burns was more fortunate than other ranchers, possibly because most of his herds were mature animals. In the fall he turned 2,100 head of steers out to range near Strathmore, and when the storms came, five men, Walter Wake, Justin Freeman, Leonard Gates, Ellis, and Chadwick went out and collected 1,600 head, taking them to Burns' great feeding stations east of Crossfield. The remainder of this herd was rounded up in the spring, and, though poor, had survived the winter in remarkable style.

The Glengarry Ranch, better known as the "44," went into that winter with 4,000 head and came out with 2,500; E.H. Maunsell, of Macleod, lost fifteen percent of his herds on the Piegan reserve, suffering a much heavier loss in Grassy Lake and the Pot Hole districts. The big American ranchers who had come into the country south of Medicine Hat sustained heavy loss, and the Creswell Cattle Company went out of business that year. This company, composed of Tony Day and Creswell, had come into Medicine Hat in 1902 with nearly 25,000 head. Lessening range, wire fences, and this bad winter about convinced them it was time to sell out, and then, during the year, the death of Creswell occurred. Day sold out every hoof and quit the business, though he remained interested in horses, 1,200 head of which he had running on the lease in 1912.

In the Cochrane district, where the winters of the early eighties killed so many of the unfortunate Cochrane Ranch cattle, the loss was practically nothing, cattle coming through in splendid shape. There was not so much snow as in other districts, and there was all the hay that was required. On the range of the Bow River Horse Ranch, almost the exact spot where the first "home" ranch house of the Cochrane Ranch was built, the loss was nothing, though in order to bring this to pass, G.E. Goddard had to feed much more hay than usual to his young stock and cows. The matured steers did well enough rustling in the big pasture; in fact, when spring came, some fat animals were sold out of there that had never been fed a spear of hay all winter.

Knowing full well the great loss suffered by the ranchers, the Canadian Pacific Railway Company did what it could do to assist the hard-hit men. For some years the company had been leasing huge tracts of range land at a rental of four cents per acre, but they removed this rental because of the terrific losses suffered by their tenants. This action of the big corporation was the more prominent because it was glaringly different from the action of the Provincial Government under Premier Rutherford. While the herds were dying in thousands, and the cattle losses were running up into the millions of dollars, the Provincial Government launched their school tax of one and one-quarter cents per acre on all lands not organized in school districts. This was a direct application to the leaseholders, and was partly responsible for putting some particularly unfortunate ranchers entirely out of business.

Government Ranch Inspector Albert Helmer reported to the Dominion Government that fall that the established ranches—those that had not been wiped out by winter losses—were continuing.

"Lessees," reported Mr. Helmer, "have wholly or in part released their holdings. The cause is partly the dread of a Provincial tax, partly on account of the winter losses of 1906–7, but chiefly because the stockmen no longer fear the encroachments of other ranchers, as the ranching business does not appear to present the same attractions to the new investor as formerly. The established rancher is not abandoning the business, but appears to be taking advantage of these conditions in order to curtail expenses."

The horse loss during the winter was not nearly so severe as that among the cattle, this being due to the superior rustling abilities of range horses. Cattle nose the snow away when searching for grass in winter, and naturally, when it is deep or badly crusted, they cannot secure food; horses more intelligently paw the obstruction away, and consequently are usually able to feed where a steer would starve to death.

By the time the fall of 1907 arrived, the surviving range stock had entirely recovered from the terrible strain of the preceding winter, entering the next winter season in splendid condition.

Compulsory dipping was again established by the Government, being a continuation of the regulations passed in June of 1904. Up until the end of June 1907, the Mounted Police carried out all quarantine regulations, but upon that date the Veterinary-General's Department took it over. Maladie du coit was greatly reduced, there being only fifteen cases found.

Nearly two hundred head of purebred bulls were sold at the Calgary sale, an indication that farmers as well as some ranchers were preparing for the future, most of this stock being purchased by farmers, who were raising small herds of pastured and winter-fed cattle of very high grade. Shipments of stock, though not so heavy as the year before, were proportionately heavier when one considers the losses of winter; 1,831 carloads of cattle were shipped from Calgary and district; 961 carloads from Medicine Hat; 966 from Grassy Lake and other Crow's Nest points; 60 carloads from Lethbridge; 268 from Macleod. As a car will carry an average of eighteen head, these figures total high. One hundred and ninety additional carloads were shipped over the Great Northern to Chicago and other American markets.

The next year, ranchers throughout the land who had enough cattle left, and who had not sold out, made huge quantities of hay, which some simply kept as an insurance of safety, and others fed. Some thousands of steers were fed throughout the cold weather, while all weak cows and all calves were fed from the first snowfall to the first appearance of green grass in spring. The Government experimental farm at Brandon made thorough tests of the relative merits of outside winter feeding as compared with the winter feeding of sheltered stock, in an effort to learn if it was advisable for ranchers to shelter as well as feed the hardy, matured cattle of the range. The results of these experiments proved conclusively that outside feeding was the best for the animals from every standpoint.

One departure from the old style of ranching became quite marked this winter, and indicated that the big ranches were coming to what they now admitted they should be, mixed farms conducted on a huge scale. More tame hay, more fodder and more grain was grown, and particular pains were taken to fit their steers for market. Some took their grass-fat steers off the range and kept them up, "trimming" them to prime condition by feeding hay and grain. This style of handling the stock showed positively that an animal not only held its grass flesh when hay-fed in winter, but actually gained eighty to one hundred and twenty-five pounds, while on a grain ration in addition, it showed in some cases a gain of four hundred pounds. In central Alberta alone, from High River north to the Red Deer, six thousand head of range cattle were winter-fed by small individual ranchers, while two thousand were fed by large concerns. In Medicine Hat, Lethbridge and Macleod districts, the majority of range stock, with the exception of "weaners" and weak cows, rustled throughout, though sufficient hay had been put up to pull all through if needed.

All of the south country from the Red Deer to the boundary was practically under quarantine for mange yet, an Order-in-Council ruling that all cattle must be inspected for mange, coming and going. Inspectors from the Department of Agriculture made these inspections and gave certificates of freedom from the disease on all they admitted in or out. Cattle in transit through the affected areas were kept in separate corrals so as not to pick up the infection, and the stock was watched closely throughout the south. J.C. Patterson, the "Jim" Patterson who came into the country about the time George Lane did, the foreman of the Waldron and Oxley ranches of the old days, was appointed stock inspector at Winnipeg.

The Dominion Government was again granting grazing leases, but these were so unsatisfactory and uncertain that they did not induce ranchers to extend. They were granted for a period of twenty-one years, but they also carried a clause that allowed their cancellation with two years' notice if the Government judged they were ripe for homesteading or other cause. In fact, they were the same unsatisfactory leases that had been offered for some years.

An old-time cattleman passed away when Sam Spencer, of the Spencer Brothers' ranch, south of Medicine Hat, died this year. Their Canadian herds had at one time, since they settled in the Province, reached the respectable figure of fifteen thousand head, but this had been cut by selling and weather to nine thousand. William Taylor, a son-in-law, then took charge of the ranch and continued the business, but on the reduced scale.

Prices were fairly good, horses averaging one hundred and twenty dollars, rough cattle thirty-eight, "tops," or export stock, fifty, and sheep six dollars. Sixty-one carloads of cattle were shipped to the Chicago market over the Great Northern Railway, and the Canadian Pacific Railway Company shipped from the entire Province 6,623 carloads of sheep, cattle and horses. Of this number, 4,932 carloads were cattle from Calgary and southern districts. Medicine Hat shipped 1,169 head, Calgary 1,867, the district between Medicine Hat and Lethbridge along the Crow's Nest branch shipped 1,121 cars, Lethbridge city and near district 129, and Macleod 650. The ranchers were still selling their herds as fast as they could. A remarkable development in small shipments from among farmers with small herds who sold ten or twenty head a year was noticed, these farmers shipping in such numbers that they held up the averages even from districts that had been formerly purely ranching and grazing lands. In ranching days, it was figured that it required thirty acres of native grass to bring a "cow-beast" through twelve months; the more intensive methods of mixed farming, grain, tame grasses and no waste energy in ranging reduced this very much, and the predictions that the export of cattle from Alberta would increase instead of diminish when the big ranches went out of business was becoming true.

One or two of the criminal cases tried during the year were of interest, particularly to old-timers. Tom Purcel, the whisky smuggler and the murderer of old Dave Akers, had served his time for that crime and drifted back into Alberta again, in the country near where he had formerly lived. In October, a man named Johnson living near Nanton lost a black filly and sent his son, J.A. Johnson, in search of it. The latter, accompanied by Fred Skinner, rode across the country and finally encountered Rev. Mr. McLean, who, upon being questioned, told them he had seen the colt at Tom Purcel's ranch on the Little Bow. McLean said that on the previous Sunday, a lad who had been out riding on a pony returned home with the black colt following, the animal having been very badly cut in the barbwire fences. McLean placed the colt in the barn belonging to the lad's father, treated its wounds and left it there. Next day a half-breed who worked for Purcel came and claimed it, taking it away with him to Purcel's place.

Johnson and Skinner rode to Purcel's and met the old man, who, with prompt Western hospitality, at once invited them to alight and eat, it being then dinner time. The visitors accepted, putting their horses in the stable and telling their host they were buying horses. As they talked, they saw the black colt they were looking for, tied in a stall, Purcel taking it out and putting it in another place, as it was too near the strange horses. When it was being led away, Johnson remarked that it was a nice beast, and offered to buy it, but the old man refused to sell. After they left the place, they hurried to lay an information against Purcel, who in the meantime must have suspected something, for when the police descended upon him a short time later, the black filly had vanished and never was heard of again, doubtless having been killed and buried. Purcel and Davies, the half-breed, were tried, Purcel being convicted and sentenced to a term of two years, Davies getting one year.

Charles R. Johnson, the "Hippo," who so brazenly escaped from the Macleod guardroom while "waiting" for his attorney to arrange to find some important witnesses a few years before, was apprehended in Montana and came cheerfully to trial, waiving extradition. Judge Harvey sentenced him to four years in jail, but the ranchers of the district thought this too severe, as many were convinced, that though "Hippo" was surely guilty, he had been "jobbed" by the stock detective who secured the evidence against him. Most Western people like open work, and underhand methods did not appeal. A largely signed petition, headed by E.H. Maunsell's name, was presented, and Johnson's sentence was cut down to two years only.

During the warm July weather, two young men created quite a furore in the Macleod district, disturbing events so thrillingly that some of the old stock-men who sat in the cool shade of the porches were reminded of older times, when thrills were everyday occurrences. Carlos Montay and Jack McDonald, two youths with well-developed traits of wild Westernisms, two lads who longed for guns, gore and glory, started to paint a luridly brilliant trail of wickedness across the prairies. They were "bad" men and they were "gunmen," and they would rather "shoot up" a peace officer than play with a running-iron. These two needed an outfit to start their career, so they stole a horse and bridle from a man named Fairweather of Macleod, stole chaps from the Great West Saddlery store, and broke into the Macleod Hardware Company store and outfitted with revolvers, knives and ammunition. Two days later they broke into the farmhouse of Frank Speth near Yarrow, and then stole a horse from Linhoff of Macleod. McDonald, who was riding without the dignity of a saddle, did not appreciate his horse's trusty back, and so they stopped at Shipley's ranch and stole a saddle. Arriving at Twin Buttes, they raided the residence of Percy Gregory, who returned in time to see that his home had been entered and that two men were just mounting to speed away toward the south. Gregory, a man of practical mind and quick action, looked for his revolver and found it gone, seized his rifle and discovered it was unloaded, grabbed his shotgun, shoved serviceable shells into it, swung up on a fast horse, and went right after the lawless pair. Montay and McDonald, "gunmen and bad," did not hasten particularly, having heard of the efficacy of the threatening muzzle of a revolver in such times of stress. When Gregory rode up and demanded that they stop, they simply turned and swung their heavy revolvers in alignment with his head—but as they did, they saw the gaping muzzles of the double-barrelled shotgun and the intent and practical face of Gregory peering over the hammers. The bore of that shotgun was the biggest thing that McDonald or Montay had ever looked into, and the weight of the heavy revolvers at the ends of their extended arms dragged so forcibly that they lowered their hands to rest—and then raised them empty in response to a more or less polite request from the shotgun man.

Gregory herded his pair of prizes to the place of a man named Thornton, where McDonald made a desperate break and got away, Gregory deciding it best to keep the bird in hand. Leaving Montay in Thornton's charge, Gregory rode for police, and with them rode fifty miles in search of McDonald, failing to locate that worthy. Returning, they dropped into Thornton's place, only to learn that Montay had experienced a revulsion of feeling when Gregory and the shotgun had vanished, and, seizing a club, had threatened to kill everyone who interfered between him and his horse. No one at Thornton's desired to be killed, apparently everyone believed that Montay would and could if he wanted to, so he was again at liberty. But pursuit was too close, the "bad" men were out-of-date, obsolete, not even so well mounted as Police, and Montay was caught at Mountain View, and McDonald at Pedersen's ranch. Montay was given a sentence of six and a half years, McDonald received four, much of the stolen property was returned to rightful owners, and not a drop of blood was spilled. Either Mr. Gregory's methods were more Western and convincing than those of the bizarre pair, or else the West had experienced a change. Anyway, a shotgun tears an awful hole, and messes a fellow up so that no one could look like a hero, even in a well-lined coffin.

About this time, too, the cattlemen and ranchers were chuckling to themselves over the discomfiture of some homesteaders who had experienced hospitality from an old-time character who might be designated as "Bill." Cattlemen never especially loved the homesteader-farmer, but when the latter came looking for land, the ranchers always treated them with the well-known hospitality of the West, fed their horses, gave them meals, gave advice as to good homesteads in another district, and in fact did all they could to assist the strangers. Bill was not too rich, and he was situated beside a well-travelled trail about a good half day's journey from town, so his place became a sort of stopping-house, and land seekers visited him in herds, advised through word of mouth one to the other, or simply because of the natural convenience of the thing. Bill fed them as they came, giving food until his finances really began to get strained; none offered to pay or even help with the preparation of meals. They would just drop in, ask if they could have dinner, hang around until it was ready, despatch it in haste, and speed on to find that particular quarter-section they desired. Bill grew disgusted. He did not want pay, but he did want appreciation—and a land seeker was not really a pleasing object to him, anyway. One day five or six dropped in, some of whom had eaten there before.

"Hullo," they said. "Can you feed us?"

"Sure," declared Bill, beaming hospitality, "dinner'll be ready in jest uh minnit." The self-bidden guests filed into the shack and made themselves comfortable against the wall, while Bill cheerily proceeded with his preparations. Taking a great iron pot, he filled it half full of water, built a roaring fire in the stove, and sat down to wait. After a time he went to his corral, came back with an armful of hay, crowded this into the bubbling cauldron, and waited some more. The heat was stifling. Impatient land seekers arose every minute and wandered out to cool; hunger drove them back; they looked at Bill and longed for him to commence cooking, but Bill was engrossed with his kettle, poking the contents with a stick, sniffing the steam, adding a pinch of hay, a dash of salt, filling the firebox with new wood, and stirring up the mess in the kettle again. Finally, with a satisfied sigh, he sank back and wiped his flushed face, reached for his pipe, and—

"Say," broke in an impatient and hungry "guest," "when'll dinner be ready, ennyhow?"

"She's ready now," replied Bill, his features wreathed with the smile that comes only to the cheerful giver. "Pitch right in, fellers," and he placed the cauldron on the floor, where the visitors could help themselves.

In 1909 Dr. J.G. Rutherford, Veterinary Director-General and Live Stock Commissioner of the Dominion Government, prepared a splendid report on the cattle trade of Western Canada, dealing particularly with the ranches, a report that applies today as sharply as it did then.

"Ever since July, 1906, I have been quietly investigating the conditions surrounding the commercial livestock trade of Canada. To this subject comparatively little attention had previously been given, my predecessor having devoted more time and effort to the interests of the breeders of pure bred stock than to those of the ordinary farmer and feeder.

"This was doubtless both proper and necessary, the pure bred herd or flock being the fountain-head of all profitable stock-keeping, and therefore of prime importance to the whole industry.

"It is, nevertheless, a fact that in Canada, as elsewhere, the breeders of pure bred stock are more independent and less needful of Government assistance than any other class in the farming community, excepting perhaps the original settlers on the Western prairie, who, certain of a rich return, and reckless of the future, too often exploit the virgin soil with a fine disregard of all the principles of husbandry.

"The breeder is independent of Government aid for two reasons: firstly, because he is a breeder and, therefore, as a rule, a man of more enterprise and wider knowledge of business methods than the majority of his fellow tillers of the soil; and, secondly, because being united with others equally intelligent in one or more thoroughly organized and active breed associations, he is in a position to reach a definite decision as to what his rights and requirements are, and to apply to those in authority the pressure or persuasion necessary to obtain them.

"On the other hand, the breeder or feeder of ordinary live stock pays but little attention to the commercial aspect of his business, and being as a rule without organization, is at the mercy, to a large extent, of the dealer, to whom he is practically forced to sell and who is generally more than a match for him in experience and acumen, and, besides, often in a position to dictate his own terms as to price and delivery.

"In view of these facts, I deemed it my duty to endeavor to ascertain and present to you a summary of the facts as to the conditions under which our commercial live stock trade is being carried on, so as to enable you to take such steps as might appear to be necessary or advisable in the interests of the producers. The present report is confined almost entirely to the cattle trade of the Western Provinces, as, of all branches of the business, this appears to me to be subject to the most numerous and serious disabilities and disadvantages.

"You will recollect that in 1902, at your special request, I prepared a brief statement regarding this trade, dealing specially with transportation, which was published in your annual report for that year. Since that time conditions have been somewhat bettered, but there is yet much room for improvement, particularly in the matters of transportation and marketing.

"During the seasons of 1907 and 1908 special officers were employed to investigate all phases of the Western cattle industry, beginning with the animal on the ranch and ending with his marketing, either on this continent or in Europe.

"The reports of these officers, which deal very fully with the details of the trade and especially with its transportation features, contain much valuable information, and will, I trust, be of great value in enabling the department to undertake intelligently, either by special legislation or otherwise, the improvement of existing conditions.

"As is well known, the Canadian West is now experiencing the same change in cattle-raising methods as has already taken place in much of the country south of the line, formerly devoted to ranching purposes.

"The incoming of settlers, many of them from the dry belt, has transformed large areas of land, formerly considered only fit for ranching, into fertile farms growing great crops of grain and fodder. While there is yet much territory untouched by the settler and on which the cattle still range as formerly, its area is being yearly curtailed, and, as a natural consequence, the free, easy and somewhat wasteful methods of the rancher are gradually giving place to those of the farmer and feeder. That this change will, instead of lessening the output, eventually result in a large increase in the cattle production of the transformed districts, needs no demonstration. Under ranching conditions, twenty acres is the usual allowance for each head of cattle, while the losses from exposure, from lack of food and from wild animals constitute a heavy drain on the herd.

"The farming settler raises an abundance of feed of all kinds which he cannot use to better advantage than in fattening cattle. With the aid of his fences and with cheap buildings, or even with none, he can keep his cattle under constant observation and control, with the result that loss is reduced to a minimum. At the same time the cattle, being at least partly domesticated, and generally to some extent grain fed, handle and ship infinitely better than do the grass-finished range steers which often, on the long journey from their native prairie to Liverpool or London, shrink the profit from their bones, and go to the butcher in such a condition as to fairly justify the Scottish feeder in his persistent opinion that Canadian cattle can only be fattened in his sheds and courts.

"Again, the winter feeding of steers will abolish the heavy handicap which the rancher, pure and simple, has always had to carry in being compelled to market his cattle off the grass and before the advent of winter. Under the new order of things, demand will, to a much greater extent, regulate supply, and the element of compulsion being removed, prices will be more even, while much of the present difficulty in transportation, due to the seaward rush of cattle and other produce in the fall, will also disappear.

"The close farmers are, as yet, however, in the minority in the less thickly settled portions of Alberta and Saskatchewan. There is still much open grazing land available, and many settlers let their cattle run at large during the summer, thus, for the present as it were, combining ranching with farming. As time goes on and the land becomes more generally taken up, this condition will disappear, as it has already done in many districts in Manitoba, as well as in the newer West, and the farmer will have to depend for his feed on the output of his own acres.

"The ranching industry in Canada is rapidly passing. In Saskatchewan and Alberta the handwriting is already on the wall, and in these Provinces it is only a matter of time until even the districts still regarded as unfit for general agriculture will, through modern methods of dry farming or by means of irrigation, be brought under cultivation. In the Peace River country ranching may persist for a time, but there, as elsewhere on the continent, the settler will soon be its undoing and the cowboy will disappear. This being the case, a brief history of the industry during the thirty years since its inception may be found interesting.

"In the beginning fences were unknown, the cattle being controlled by herders, but about 1885 the proprietary instinct began to assert itself and many of the larger holdings were put under fence, although, needless to say, the smaller owners continued to prefer the open range system.

"The big concerns used almost exclusively pure bred bulls of the beef breeds, and, as the grazing was good and not over-stocked, usually held their steers until at least four years old, the result being that a most excellent type of beef animal, full grown and well finished, began to find its way from Alberta to the Eastern market. Being mature and well furnished with fat in the fall, driven slowly, feeding and hardening on the way through a rich grazing country to the railway, distant often many days' journey from the home ranch, these cattle stood the trials of the export journey fairly well, and landed in Britain, somewhat shrunken it is true, but still yielding a reasonable profit on the comparatively small cost of production. In the early days of the industry, only the best were exported. The lighter and rougher stock went for local consumption and to fill railway construction and Indian contracts.

"As time went on the country became more heavily stocked, many men without adequate capital or experience began to keep cattle, cross-bred bulls became commoner on the range, carelessness in breeding methods lowered the natural increase, the purchase of stockers, first from Manitoba and later from the Eastern

Provinces, introduced many very inferior animals, and a general deterioration both in quality and value became only too evident. '

"The climax of this deterioration was reached about the year 1902 when, tempted by the low prices of Mexican cattle, some of the larger ranchers began to make importations from Chihuahua and Coahuila. These degenerate descendants of the ancient Spanish breed, although hardy and exceeding in length of horn, as in length of wind and in speed, anything ever before seen among our Western cattle, did not recommend themselves to the intelligence of the Canadian rancher, and, after a few years of trial, the trade practically died out in 1905.

"About the same time the effects of the policy of this department in establishing annual provincial auction sales of pure bred bulls began to make themselves apparent.

"These sales, although to some extent hampered by the jealousies of local breeders, as well as by the indifference of many of the less intelligent and less progressive ranchers, have done an excellent work in raising the standard of our Western cattle, as regards size and conformation.

"There is still much room for improvement in this direction, and it is to be hoped that as diversified husbandry takes the place of ranching, the necessity for using a better class of bulls than those hitherto employed will be more generally recognized and appreciated.

"The advent of the Mormons and others familiar with dry farming, and the experience of a few of the more progressive ranchers themselves, especially in the Pincher Creek district, having demonstrated the suitability of much of the country for general farming, a strong tide of immigration set in about the year 1900, and since that time many of the old ranches have been divided, cultivated and built upon, and now form populous rural areas, rapidly beginning to resemble in appearance similar districts in the older settled Provinces.

"At the present date, while many of the larger ranches have closed out, the cattle industry is by no means at an end. It is true that many cattlemen, seeing the inevitable end of ranching, have been rapidly 'beefing' out their herds by selling cows, spaying heifers and disposing of bulls, but this is only a link in the chain connecting the old with the new and better condition of the industry. The determination to beef out has temporarily increased the output of cattle of range quality, but, while this is going on, the incoming settlers are stocking up, not to return to the old system of selling their cattle off the grass in the fall, but to follow the more profitable method of finishing beef throughout the year for the good markets, as is done in other progressive countries, where beef raising is recognized as a legitimate and useful adjunct to mixed farming.

"The condition of the range industry was described in striking terms by a representative Western cattleman at the National Live Stock Convention in February, 1908, who said: 'No one at all familiar with the ranching industry will hesitate to state that it is in a condition of rapid decline, dying as decently and as quickly as it is financially able to do.' It is not yet dead, however. There were still in force in the four Western Provinces, on April 1, 1908, 939 grazing leases, involving 3,259,271 acres divided as follows: Manitoba, 12,642 acres; Saskatchewan, 632,493 acres; Alberta. 2,132,718 acres; British Columbia, 281,418 acres. The average area under lease is 3,481 acres. It would therefore appear that there are still a good many cattle kept under the old conditions, even when the sheep and horse leases are taken into consideration.

"From its very inception the ranching industry was subject to winter losses, more or less severe according to the nature of the weather, as well as that of the rancher himself. Even in the worst winters those herds whose owners had made reasonable provision for bad weather conditions escaped, as a rule, with comparatively little loss, although they also occasionally suffered heavily through sudden storms, which, coming early in the season, drifted the cattle so far away from the stores of fodder prepared for them that it was impossible to get them back before the advent of spring, or until a timely chinook enabled the cowmen to collect from far and near the remnants of the herd.

"The winter of 1886–7 was almost fatal to the industry, being unequalled in severity by any season, either before or after, until the memorable year of 1906–7, when approximately fifty percent of the cattle on the range were lost.

"In the year first mentioned, however, there was much more grass and many fewer cattle, while on the ranges then occupied there was considerable natural shelter, so that, although badly hit and sorely discouraged, the ranchers did not abandon the field, but, investing new capital and energy, soon regained the ground they had lost.

"As stated above, the rancher who makes adequate provision for a bad winter, may, through unforeseen circumstances, lose heavily in spite of his foresight; on the other hand the careless and improvident owner, who trusts to luck and stores no hay for winter use, is certain to be seriously hit, should the season prove exceptionally rigorous.

"Apart from the mere question of money, the practice followed by too many owners of facing the possibilities of the winter without laying in at least enough fodder to sustain life, is cruel and reprehensible to a degree, and should, I think, be made the subject of drastic legislation.

"It would be possible to go much more deeply into the question and, in fact, to practically show by a consideration of its various demerits, that while in its own time and place it served a useful purpose, the ranching industry has properly had its day, and that its early disappearance from southern Alberta and Saskatchewan need, except perhaps from the standpoint of sentiment, cause no deep or lasting regret.

"The export trade in Western range cattle, as hitherto carried on, has been sinfully wasteful, unbusinesslike and unprofitable to the producer. Cattle wild, excitable and soft off grass, are driven to the railway, held sometimes for days on poor pasture waiting for cars, and finally, after more or less unavoidably rough handling, are forced on board. Once in the cars, they are, not unfrequently, run through to Winnipeg without being unloaded for feed or water. It is 840 miles from Calgary to Winnipeg, and as many shipments originate beyond the first-named point, it may be readily seen what this means, even when the run is a good one. Some shippers unload at Moose Jaw, 440 miles west of Winnipeg, but others claim that it is alike more humane and more profitable to run through, as the cattle, being still wild, excited and unaccustomed to handling, not only refuse both feed and water, but suffer much more in the unloading and reloading than they do when left in the cars. On arrival at Winnipeg they are always unloaded, fed and watered, being by this time hungry, thirsty and fairly quiet from exhaustion. After being rested they are inspected, culled and reloaded, the next stop being, as a rule, at White River, 678 miles further east. There they are again fed and watered, and after another stage of 755 miles arrive at Montreal. Here for most of them the land journey ends, although when navigation is closed at that point it extends to Portland, Boston or St. John, New Brunswick, as the case may be, very rarely to Halifax. At Montreal, however, all are unloaded, fed, watered, rested, and carefully inspected by the veterinary officers of this department, whether they are to be shipped by water from there or from some other port. If the latter, they are on arrival rested and again inspected before going on board the steamer.

"While the facilities for loading cattle on the ship at St. John are excellent, those at Montreal are not of the best, and this necessitates more and somewhat rougher handling than would otherwise be the case. Even on the ships there is much room for improvement in conditions. The regulations as regards space, fittings and similar matters are, oddly enough, drawn up and enforced by the Department of Marine and Fisheries, and although these might, in my opinion, be revised with advantage, this is scarcely the proper place to discuss them.

"One matter, however, I must mention, namely, the class of men employed to look after and care for cattle on our Canadian ships. These are, as a rule, picked up indiscriminately, through agents at the port of shipment. These men, known in the trade as 'stiffs,' are often returning emigrants who have failed, through drink or other causes, in making things go in Canada, or sometimes simply men looking for a cheap passage, decent enough, perhaps, but with no knowledge of cattle, and in many cases quite unaccustomed to the sea. Such men are frequently incapacitated for duty through seasickness, and, in other cases, simply refuse to work, with the result that any who may be capable and industrious are overwrought, and the cattle suffer accordingly. In rough weather, especially, the feeding and watering are apt to be irregular and insufficient.

"Is it a matter for wonder that after a journey of five thousand miles, made under such conditions, our grass-fed range steers arrive in British lairages gaunt and shrunken, looking more like stockers than beeves, that our Scottish friends think we have no feed, or that I should declare a business so conducted as sinfully wasteful?

"And still it is profitable; profitable to the middleman who, coolly reckoning on the shrinkage, fixes accordingly his price to the producer; profitable to the commission man, who pockets in commission what the middleman takes in profits; profitable to the railway companies; profitable to the steamship lines and profitable to the British butcher, who pays only for what he gets and not even that much if, by combination or sharp practice of other kinds, he can manage to keep prices down. To the producer, however, and therefore to the country, it is the very reverse, and the odd feature of it all is that if conditions were so amended as to make it profitable for them, the others mentioned above would gain, rather than lose, by the change.

"No wild, grass-finished cattle should be shipped for export. In a country like Western Canada which, one year with another, is full of all kinds of material for winter-feeding, there is no excuse for the sending forward, for immediate export, animals which, owing to their lack of domestication and the nature of their food, cannot, under ordinary circumstances, reach their destination on the British market without a woeful depreciation in both quantity and quality of flesh.

"Our friends in the United States long ago realized the folly of shipping to Europe, alive, steers direct from the range. Their range cattle are brought to the Middle West, dehorned, if this has not been earlier done, fed for at least sixty days on a ration comprising a liberal allowance of grain, then sent to market, generally in Chicago, and carefully inspected and culled. Those deemed fit for export are then taken to the seaboard by fast trains and in cars specially fitted for feeding and watering en route. They are loaded on these cars under careful supervision, no over crowding or rough handling being permitted. The men in charge are almost invariably regular salaried employees of the shipping firms, and the same is true of the foremen on the ships and of those working under them.

"As a result of these superior methods, United States cattle, even when originally from the Western ranges, arrive in Britain in much better condition than Canadian range cattle and, of course, command correspondingly higher prices.

"Domesticated Canadians, properly finished, land, as a rule, in excellent condition, and compete closely in price with the best States cattle of the same class. There is no reason why our Canadian range cattle, if treated on similar lines, should not compete as closely with steers from the Western States.

"As a matter of fact, considerable improvement is already taking place in the finishing of Western cattle, as year by year more winter feeding is undertaken. Many thousands of good steers are, in the autumn, put on a hay or grain ration for the winter. When the feeding is liberal and judicious and good water available, the grass flesh is not only held, but gains on hay alone of from 80 to 125 pounds, and from hay and grain up to 400 pounds, are not uncommon. The cattle thus wintered are ready for the spring market, on which they usually sell well, prices always being better at that season, the demand good, and, as but few cattle are being handled, shipping facilities much better. Winter feeding is now systematically carried on by some of the largest operators in the West.

"Arrangements are yearly made by one firm with individual farmers throughout the country to feed during the winter at a fixed price per head per month. The cattle are handed over to these men on the approach of hard weather and taken from them when wanted. While many are slaughtered for home and coast consumption, a large number may now be seen during May and June at the Winnipeg yards on their way to the British market, where, needless to say, they get a much more favorable reception than do those which come direct from the range.

"A carefully prepared estimate of the number of cattle on feed in central Alberta during the winter of 1908–9, gives six thousand head being fed in small lots by individual farmers and two thousand head by large concerns. It is believed that seventy-five percent of these were receiving a grain ration and twenty-five percent hay alone. In the southern part of the Province additional large numbers, of which reliable statistics are not available, were also fattened.

"The growth of the practice of finishing cattle on dry feed (hay or hay and grain) in the three Western Provinces is indicated by statistics of shipments received at Winnipeg from January to June (fed on dry feed), as compared with the shipments from July to December (grass fed) for the years 1906, 1907 and 1908, as follows:

Number of cattle shipped east from Winnipeg,

January to June, 1906: 9,435

Number of cattle shipped east from Winnipeg

July to December, 1906:...81,609

Number of cattle received for local use,

January to June, 1906: 9,135

Number of cattle received for local use,

July to December, 1906:...<u>31,462</u>

131.641

Number of cattle shipped east from Winnipeg,

January to June, 1907: 1,487

Number of cattle shipped east from Winnipeg,

July to December, 1907:...50,062

Number of cattle received for local use,

January to June, 1907: 16,397

Number of cattle received for local use,

July to December, 1907:...<u>32,254</u>

100,200

Number of cattle shipped east from Winnipeg,

January to June, 1908: 19,531

Number of cattle shipped east from Winnipeg,

July to December, 1908:...86,593

Number of cattle received for local use,

January to June, 1908: 22,342

Number of cattle received for local use,

July to December, 1908:...<u>41,622</u>

170,088

"The above tables show the percentage of dry fed cattle arriving at Winnipeg for the past three years to have been as follows:

1906.. 16.37 percent

1907.. 21.62 percent

1908.. 48.67 percent

"The shipments via Winnipeg in no sense include all the cattle produced in the three Prairie Provinces. To these must be added the large quantity of beef consumed in the local markets, in addition to that shipped to British Columbia and the Yukon. It appears safe to infer that the percentage of winter fed cattle that have gone to Winnipeg, as shown by the above tables, indicates fairly accurately the relative proportion of these to grass-fattened stock produced in the three Provinces. These tables further indicate that within a few years comparatively few lean, or rather half-fed, cattle will be shipped from Western Canada for immediate killing.

"This is an excellent showing, as far as it goes, but I am satisfied that, one year with another, a profitable business can be done by farmers in the grain-growing districts of the three Western Provinces, in finishing for the market the big growthy grass-fed steers from the range country. In seasons when rough grains are scarce or dear it would not, of course, be so profitable as when these were cheap and plentiful. There is almost always roughage in abundance. In many districts good prairie hay is procurable at small cost, while straw is always available and can, as Mr. Bedford and many others have shown, be fed with profit, when intelligence and some other things are in the combination. Once in a while, too, there is a little frozen wheat in the country, and in years when this is the case the best market for it is usually to be found among the live stock, if one is fortunate enough to have them.

"With the object of encouraging the proper finishing of range cattle in the West this branch has for two seasons offered to a number of selected farmers in Manitoba and Saskatchewan, who have suitable locations and would undertake the finishing of range steers in winter on their farms, a bonus of two cents per pound of gain on such cattle fed by them. It is not desired that the cattle be housed, but fed either in open sheds or naturally sheltered locations.

"Sufficient evidence is at hand to demonstrate that profitable finishing can be done without the use of expensive buildings and upon such feed as is now being wasted on many wheat farms. The bonus offered was not in any case accepted, farmers intending to feed preferring to utilize the semi-domesticated cattle available in most districts rather than undertake the feeding of range steers to which they were unaccustomed.

"Outdoor feeding was, however, undertaken at the Experimental Farm at Brandon, where in the tests made in 1907–8 it was found that the cattle fed outside made more profitable gains than similar cattle fed under the usual stabling conditions. The experiment is being continued on a larger scale this winter. Following is Superintendent Murray's report of the 1907–8 experiment:

> The feeding of cattle out of doors for the production of beef has been receiving considerable attention of late at the hands of Manitoba cattlemen. The strongest advocates of this method of producing beef are men

who have been successfully practising it for a number of years and those who have seen it in operation. The conditions of outdoor feeding are so radically different from those that have been generally considered essential that the majority of cattlemen are sceptical about it, while many others look upon the practice as ludicrous, and aver that it must involve a wanton waste of feed.

Last fall some work was started to secure definite information as to the comparative economy of making beef in a comfortable stable and in the open with comparatively little shelter. The first lot of steers, thirteen head, has just been marketed and the results are available.

Thirteen were purchased in late November and divided into two groups as nearly alike as possible in size and quality, eight being dehorned and put outside and five (as many as we had accommodation for) tied in the stable. The steers were domestic, purchased in the neighborhood of Oak River, and cost 3⅛ cents shrunk. The inside group were started on September 3 on a standard ration that has given good results here for a number of years for beef production and consisted of silage, straw, hay, a few roots and grain. The grain ration at the start consisted of two pounds of a mixture of oats, barley and feed wheat, and two pounds of bran per steer. This was increased from time to time until by the first of April each steer was receiving daily ten pounds of grain and two pounds of bran. The steers were not out of the stable after being tied up until they were sold.

The eight steers outside had a range of about one hundred acres of rough rolling land, some of which was well sheltered with scrub. Water was available in one of the coulees, the ice being cut every day. No shelter by way of sheds was provided. Grain was fed in a trough about three feet wide and high enough off the ground to prevent the steers getting in it. Straw was always kept before them in an enclosure of stakes that would hold about a load, arranged so that the straw could not be wasted by tramping over it. On December 3 they were started on a ration consisting of two pounds of mixed grain and two pounds of bran, this being increased from time to time, so that by April 1 each steer was getting nine pounds of grain and two pounds of bran. For about six weeks rough hay was fed instead of straw. This is charged for at the rate of $2 per ton, which is its full value.

Both lots of steers were sold on April 22 for $4.25 per hundred, with four percent shrinkage. Following is a statement of the transaction:

	OUTSIDE	INSIDE
Number of steers in lot	8	5
First weight, gross	8,854 lbs.	5,695 lbs.
First weight, average	1,106 lbs.	1,139 lbs.
Finished weight, gross	10,630 lbs.	6,950 lbs.
Finished weight, average	1,328 lbs.	1,390 lbs.
Total gain in 138 days	1,776 lbs.	1,255 lbs.
Average gain per steer	235 lbs.	251 lbs.
Daily gain per steer	1.6 lbs.	1.81 lbs.
Daily gain per lot	12.8 lbs.	9.05 lbs.
Gross cost of feed	$100.76	$77.95
Cost of 100 lbs. gain	$5.67	$6.20
Cost of steers fed out of doors, 8,848 lbs. at 3⅛ cents	$276.50	—
Cost of steers fed indoors 5,695 lbs. at 3⅛ cents	—	$177.97
Total cost to produce beef	$377.26	$255.92
Out-of-door steers sold, 14,135 lbs. at 4¼ cents, less 4 percent	$433.71	—
Indoor steers sold, 6,950 lbs. at 4¼ cents, less 4 percent	—	$283.56
Profit on lot	$56.45	$27.64
Net profit per steer	$7.05	$5.52
Average buying price, per steer	$34.56	$35.59
Average selling price, per steer	$54.21	$56.71
Average increase in value	$19.65	$21.12
Average cost of feed per steer	$12.59	$15.59
Amount of meal eaten by lot of steers	8,892 lbs.	5,390 lbs.
Amount of straw	8 tons	5,680 lbs.
Amount of hay	6 tons	2,840 lbs.
Amount of millet	1 ton	—
Amount of ensilage and roots	—	25,850 lbs.
Amount of corn fodder	1 ton	—

The net profit as shown here, $5.52 on those fed inside and $7.05 on those fed outside, makes no allowance for interest on investment or labor involved in tending the cattle. For the outside lot the only investment was the price of the steers and the value of lumber for troughing, a total of $286. The labor incident to attending this lot, including the drawing of straw, feeding grain and cutting ice, would at the outside not amount to more than the time of one man for one hour per day. The extra expense in attending fifty head would have been not more than the time required to draw the additional straw—a small item.

In feeding inside the investment is necessarily very much greater, no matter how economically the building be done. Provided a building suitable for stabling thirty steers could be erected for $1,000, an additional gross profit of $2 per head would be required to meet interest on the investment. The labor required to attend to the cattle fed inside was fully four times as much as that required when the feeding was done outside.

The point has been raised in discussions on this subject that a large part of the food consumed by the cattle fed outside must be utilized to keep up the animal heat, and since those fed in a comfortable stable do not have the same waste of heat to provide for in the food consumed, they should on that account lay on fat more economically. It must be borne in mind, however, that cattle that are not stabled grow a coat of hair more resembling in its density that of a beaver than that of a steer, and that this provision aids greatly in conserving the animal heat. During the coldest weather that we had this winter, when for a week the temperature averaged twenty-nine below zero, the steers did not seem to suffer in the least, and were not standing around the straw pile with humped backs, as one might imagine.

The cattle were always ready for their feed and none of them went off during the winter. The abundance of fresh air has no doubt a salutary effect in keeping the digestive system in tone.

The work carried on this winter is intended as introductory to more extensive trials. Experiments of the sort above outlined must be continued for a number of years, when different kinds of seasons are encountered, before the results can be considered of any great value. The past winter's results may be taken as representing what may be expected in an unusually mild winter free from severe storms or prolonged cold spells. How these results will compare with what may be obtained in a more severe winter remains still to be seen.

"The results achieved at the Brandon Experimental Farm in the one season tried have been verified over and over again, year after year, on Manitoba farms. The following description of a number of years' feeding near Newdale prepared by Mr. Grayson, Mount Pleasant Stock Farm, of that place, and published in the *Nor'-West Farmer,* shows the method to be a profitable one even in severe seasons:

Some fifteen years ago Mr. John B. Cook, of Newdale, in connection with the late Dr. Harrison, built a large barn and started somewhat extensively into the business of winter feeding of beef cattle. After about three years' experience, during which time the balance was always on the wrong side of the ledger, another bunch of cattle was bought and fed hay in the shelter of the scrub which extends along the north side of the farm, the intention being to bring the cattle to the barn as the weather got colder. The cattle had access to open water in the ravines and appeared to be doing so well that they were left out all winter. A small allowance of grain was added to the hay about March 1. These cattle were sold early in the summer and were the first cattle to net their feeders a profit. Since that time Mr. Cook has continued to feed from sixty to one hundred head of steers each winter, and the writer, as well as others, has done something along the same line with satisfactory results.

In this article I propose to give some idea of the work carried on here. In doing so I know I shall say things that are at variance with what most of us believe to be essential to the production of beef, but I would ask readers to remember that what I am writing is actual experience and not theory. Years ago Mr. Cook's plan was to buy in the fall a bunch of cattle, big, lean steers and thin cows and heifers, almost anything with a large frame that might be made to carry meat. But today nothing is selected but steers of good beef conformation and weighing from 1,100 to 1,300 pounds in the fall, steers that carry a considerable amount of flesh. Experience has proved that the fleshy steer is the most profitable to winter and makes better gains than the leaner one in the bunch, and we rarely find a steer so fat from the grass that he will not stand a finishing spell with grain. These steers have usually been bought from some regular cattle buyer, a premium being paid for the privilege of selecting suitable feeders.

The steers are usually bought during October and allowed to run on the farms until winter sets in in earnest. As early as convenient after the steers are bought they are dehorned. Clippers are used for this purpose and a handful of lime is pressed on each stub to assist in checking the bleeding. With the approach of winter the steers seek the shelter and straw is drawn to them.

I have noticed from questions that have been asked me and from criticisms that the generally held idea regarding shelter is that the cattle retire into the bottom of some thickly-wooded ravine or into some heavy bush where they would be almost as much shut in as they might be in some sod building without windows. Instead of this the cattle prefer the high open spaces, with just enough scrub to prevent the snow from drifting over the straw. The cattle enjoy the life, and especially enjoy the sunshine, so long as the winds are broken from them.

Here I may speak of another point, and that is the manure. One of my critics of a previous article seemed to think that it would be out of the question to gather the manure from among the scrub. Now, if straw is fed in a comparatively limited open space, until it reaches a depth of two or three feet of straw and manure, I fail to see the difficulty of getting it gathered up. And I contend that I know of no better way to convert large quantities of straw into useful manure than by feeding it liberally out of doors to grain-fed cattle. In feeding straw it is necessary to use much more than the cattle will eat up clean, as by this means the cattle can always have a comfortable bed, and we aim to have them comfortable.

About the first of December, or earlier, if the weather is severe, the cattle are given about four pounds of grain each day. The grain is all fed in the evenings in troughs about three feet wide, eight inches deep, and raised about two and a half feet from the ground. The grain ration consists of a mixture of oats and barley chopped (barley principally) and bran, about one-third bran by weight. Finely ground chop gives best results and is most appreciated by the cattle. It is our plan to feed about sixteen hundred pounds of grain per steer during the feeding period, and the ration is increased in January to about eight pounds per steer per day and about April to ten pounds. This is continued until about June 20, when the steers are sold. If the grass becomes good in June less grain is needed at the finish.

In feeding cattle on such a dry ration, watering is of considerable importance. Those who are so situated that cattle can have access to open water at all times are especially favored for this work; the cattle need to drink frequently and in small quantities. Where water is not so easily available it must be kept in the trough as much as the severity of the weather will permit, as a large drink of cold water following long abstinence would chill any animal and cause a temporary check to digestive processes. In regard to salt, we usually place a barrel in some convenient place and knock the head in.

In carrying cattle until June, it is a great help, if one has hay enough, to feed for about a month after the snow goes, and by confining the cattle so that they will not ramble too far they can be made to at least hold their own during this trying period. The object in carrying cattle until June has been to wait for a profitable market. If the market on, say the first of April, was anything like equal to the market of June, I am sure that good results would follow the feeding of the same total quantity of grain in the shorter period.

Now as to our business methods. The steers are bought when cattle are at about the lowest, a premium over the market being paid for the privilege of selecting steers of approved type. In working out our balance sheet we have been in the habit of charging the grain fed to the cattle at the rate of eighty cents per one hundred pounds. This we consider a fair price in an ordinary year. We charge interest, wages, and all necessary expenses, and have been able with a margin of one and one-half cents per pound between buying and selling price to have a balance on the right side of about an average of seven dollars per head.

"The question of outdoor winter fattening was discussed at considerable length at the National Live Stock Convention. The view held by many Western grain growers, that winter fattening cannot be profitably done in the Prairie Provinces, was freely expressed, but it was just as readily refuted by those who spoke from experience. A delegate stated that he knew of a carload of cattle fed in the open air during the winter of 1906–7 on prairie hay and water, the gain averaging 100 pounds per head. Another speaker explained that 90 head, averaging 1,250 pounds in the autumn, were made to weigh 1,400 pounds by spring, fed in a ravine in Manitoba. The feed consisted of straw and chaff, that would otherwise have been burned, with grain chop. Charging for the grain and the labor, the steers made a clear profit of sixteen dollars per head. After summing up the various arguments presented, the chairman of the convention pointed out that it was simply the old story—some men could make it pay, while others, too careless or too lazy to do the thing properly, would fail in the fattening of cattle as they would in any other undertaking.

"There are thousands of wheat growers who spend their winters in idleness after marketing the season's harvest. Continuous good crops, desirable as they are, have very great disadvantages for the farming community. Already are to be found in these new Provinces districts yielding little more than half the returns per acre they did some years ago, and while the yield, following continuous cropping, is going down, the land is becoming foul with weeds, whereas a system of mixed farming, including the feeding of the straw and other rough feed to cattle, together with a suitable system of rotation, involving spreading the manure on the land, builds up the soil, keeps it clear of weeds, and hastens the ripening of the grain, thus reducing the danger from early frost.

"There are in certain sections of the West farmers who finish their cattle during the summer and ship them to the British market. An example of this may be seen on a farm near Moosomin, where Mr. R.J. Phin is devoting his attention to this work. He handles about nine hundred head each year, sometimes shipping direct to the Old Country. These cattle are gathered largely around Moosomin and in the Moose Mountain country, where there is abundance of water and grass. The chief points of interest regarding his operations are (*a*) the finishing on rape of cattle not otherwise fit to ship; (*b*) winter feeding.

"(*a*) Finishing cattle on rape.—The land intended for this purpose is treated as a summer fallow during the early summer, and about the first of July is sown to the forage crop mentioned, two pounds of seed per acre being used, sown in drills. After the sowing is done, manure is applied with spreaders; surface cultivation is followed about once a week, thus keeping the weeds under. The cattle are turned on about September 15, and kept there until the frost sets in; in addition some chopped grain is fed. The cattle come off the rape in prime condition and ship well. The grains fed consist of oats, barley or frozen wheat, depending upon the price at which these may be obtained. Not only are the steers thus turned off in good condition, but the land is cleaned and made to bear a profitable crop of wheat, the straw being strong and the heads well filled. The packing of the soil seems to have the effect of preventing a rank growth of straw and also hastens the maturity of the crop. In 1908 sixty-five acres were under rape, but some years double this quantity has been sown. This course of husbandry has teen followed now for five years with satisfactory results.

"(*b*) Winter feeding outside.—During the winter months, from one to two hundred steers are fed on cut straw and chopped grain. The equipment is not expensive, consisting of cheap wooden troughs, up about two feet from the ground on the leeward side of the buildings. Adjacent to the buildings is a yard with cheap sheds, but the steers fed there do not seem to make any greater gains than those altogether in the open. As remarked by Mr. Phin, 'A big well-fed steer seems to take little heed of the cold.' The cattle fed are practically all Shorthorn grades, which are preferred, as, in addition to being good feeders, they have size and weight.

"The following statement by Mr. W.F. Puffer, MLA, of Lacombe, Alta., who is, in every sense of the word, a practical man, will be found both interesting and instructive:

In the district around Lacombe and Red Deer and, in fact, in that part of the Province generally spoken of as Central Alberta, the winter feeding of cattle is becoming more general.

There is still plenty of grass throughout this district, but the farmer is already occupying considerable areas. The country is somewhat rolling, with abundant water, and dotted with frequent groves of poplar and some spruce, affording excellent opportunity for winter feeding in the open without the expense of stabling.

The method of feeding which is now being generally followed and which, after an experience of twenty years of cattle feeding, the most of the time in Alberta, I have myself found to give the most satisfactory results, I will describe briefly.

First, let me say that I strongly favor feeding in the open, and that I am convinced that many of those who attempt feeding cattle do not feed grain with sufficient liberality to obtain the best results. This, I believe, is one reason why Canadian cattle are generally quoted on the Liverpool market one cent per pound lower than United States cattle. In the United States feeding districts, cattle are put on a full feed of corn almost from the start, which is kept before them constantly for six or eight months. One hundred bushels of corn is reckoned as the requirement of an ordinary steer during the feeding period. This method gives rapid gains, producing better cattle, which make better prices, than where limited grain rations are fed. The disposition of a thoroughly fattened steer is changed; he becomes docile and contented, ships better, and thus brings a better price at the end of his life's journey. We have just as good cattle here as in the United States. Chopped barley, wheat and oats are fully equal to corn as a fattening ration, but we must give the cattle all they will eat of it, and when we learn to do this, I contend that our cattle will not sell at a lower price on the British market than United States cattle.

I have been pleased to note that some good work is being done by the superintendent of the Experimental Farm at Brandon in outdoor cattle feeding, and I have read with interest reports of other Manitoba farmers who are experimenting along similar lines. I cannot help but think, however, that all these experiments would be better if they would adopt the method I here attempt to describe. At the time the Experimental Farm cattle were sold at Brandon last spring for 4½ cents, which I fancy was about their value, a good many cattle were being sold here for 4¼ cents, but our best feeders were getting 4½ to 5 cents for cattle for export, and they had to contend with the long rail journey, extra freight and shrinkage and other expenses which would make cattle cost to the dealer in Montreal from 6 cents to 6½ cents per pound.

Where there is no natural shelter, a corral with a tight board fence about seven feet high, with a rough, straw covered shed for stormy weather, is necessary, and even where there is good natural shelter, cattle will do better with a roughly improvised shed in which to lie down during stormy weather. The rest of the equipment consists of racks for holding hay or rough feed, which should always be kept filled, and the cattle allowed access to them at all times. The grain feeding bunks should be placed in the centre of the corral or in the open, where the cattle can get all around them. They should be about 2½ feet high, 3 feet wide, with 8-inch sides to keep in the chop, and if made 16 feet long will be found convenient. With cattle not dehorned, and until they are on full feed, about one of these bunks to every eight head is necessary; after they are on full feed a bunk would accommodate more cattle. Self-feeders may also be used and are very satisfactory.

It is perhaps needless to say that attention to the smallest details is absolutely essential to obtain the best results in the feeding of cattle, and this applies just as emphatically with cattle that are being fed in the open, as under the most artificial conditions. They must be provided with plenty of bedding, good clean straw a foot deep; all frozen lumps of manure should be regularly removed so that cattle may have 'solid comfort.' Remember that when cattle are lying down, quietly and contentedly chewing their cud, they are making the most money for the feeder.

As above stated, the feed racks should always be kept filled, and I always like to supply the best hay at the first of the season before the cattle have got on to the full grain feed.

I find, like Mr. Grayson, of Newdale, that finely chopped grain is best, being more easily digested. Barley and oats ground together are what is usually fed; sometimes oats and wheat, but I have had better results from feeding barley alone. I like to put in three-year-old steers weighing about 1,200 pounds. I begin feeding, about December 1st, five pounds of chop once a day, gradually increasing this till the 15th of the month to four pounds twice a day, which is still further increased until, by the end of the month, six pounds twice a day is being fed. This is gradually increased for the next ten days or so, when a little chop will be left over in the bunks; they should then be filled up and never allowed to get empty. I find more grain is eaten the third month than the second. Steers, such as referred to above, will sometimes average two pounds per head per day when on full feed, depending on the size of the steer and the quality of rough feed, and also, to some extent, on the weather. Steers of good breeding will gain in weight in five months, from December 1st to May 1st, from 350 to 500 pounds. Such steers will continue growing after the date mentioned until sold, and I am sure no one ever yet experienced any difficulty in getting a good price for such cattle in the spring.

I suppose objection would be raised to the amount of grain fed, but I contend that half-way methods don't pay, and in my experience the results obtained justify the extra quantity of grain. On limited rations, steers do not become contented; they remain on their feet too much of the time, playing and fighting, thus wasting a certain amount of the feed consumed, whereas when put on full feed even the wildest cattle soon become lazy and lie down a great deal of the time, when, as I have already said, they are making flesh economically.

Another important item is the water supply, and it is most essential that water should at all times be available. If water is supplied from a well, a tank heater is a necessity to keep the water in the troughs from freezing, and it will pay for itself in a short time if twenty head or over are being fed. If the water is supplied from a lake or a stream, then ample water holes should be provided, and attention should be given that these are made convenient for drinking from, so that the animals can stand comfortably. This can be done by making a long opening in the ice, say not over twelve inches wide, and as long as necessary. A little ledge should be left all around the edges of the water hole to keep their feet from slipping in, and the ice should be chopped away at the back so that their hind feet are down almost on a level with their front feet. The ledge round the water hole will also prevent the water from becoming contaminated on warm days. Barrel salt I find best, and it should not be allowed to get lumpy or hard.

"Many other Alberta feeders are adopting intensive fattening methods, A representative of this Branch, travelling in Alberta, reports the operations of one firm that had, in the winter of 1908–9, 1,400 head on feed at three points— Carbon, Midnapore and High River. At High River, where 485 head were feeding, the cattle had only a bush shelter on the banks of the river. In January when visited they were getting all the hay they could eat, and a meal ration of sixteen pounds daily. The meal consisted of a mixture of two-thirds oats and one-third barley ground fine. This meal was fed in self-feeders, of which there were thirty, these being filled every second day. The cattle when seen in extremely cold weather appeared comfortable and contented. They were eating comparatively little hay— about four tons per day, or sixteen pounds per head, which is about equal to the weight of meal consumed. They had free access to salt and to High River water. While they had not been weighed, they appeared to be putting on weight rapidly.

Three men were able to look after this herd of 485 head, including the work of grinding the grain by means of an engine and chopper.

"There is perhaps no greater loss in the entire ranching industry than that arising from the usual methods of wintering calves. Not only are many promising calves lost from exposure and shortage of feed during severe periods, but practically all that have come through the winter have lost weight and become stunted for future growth. Those who have taken the trouble to weigh their calves in fall and again in spring have been surprised to learn that fully 200 pounds of flesh per head have been sacrificed by allowing the youngsters to take their chance on the range along with the herd. Calves that weighed 600 pounds at the beginning of winter had actually shrunk to 400 pounds by spring, losing just one-third their weight, and this all flesh, as neither bone, hide nor horn had been reduced. Any stockman can readily imagine the time it requires for such animals to regain the lost ground. It is fair to estimate that fully a year is lost in the animal's growth and a year delayed in the time the ranchman must wait for the price of his crop of steers.

"Is there a better way practicable? That is the question which concerns the cattleman. Housing is not an easy problem and help is expensive, but something must be done to prevent or reduce the enormous loss from fatalities, shrinkage and stunting, that goes on from year to year. A year's saving of time and feed would do a great deal towards a provision for caring properly for the calves, especially since it can be done without expensive housing, or even the feeding of grain.

"Many up-to-date ranch owners are recognizing the importance of proper shelter and feeding and make special provision for the calves during the first winter. Rough sheds are constructed in which they are run loose and fed on hay and oat sheaves or other suitable feed. Others bring their calves through successfully without the sheds. Mr. W.E. Tees, of Tees, Alberta, the owner of a herd of good cows, winters his calf crop satisfactorily without buildings. Describing his experience and system, Mr. Tees writes as follows:

> I will try to give you my plan and experience in wintering calves. I have never weighed before or after wintering, but I am sure I can bring them through the worst winter in very satisfactory form. During the hard winter of 1906 I had to change my usual plan, as the snow was too deep for grazing, so I held them in a yard or corral. That winter I had some forty head, and fed them on wild hay and green cut oat feed, and I certainly had a fine bunch of calves in the spring, with no loss. I will try and explain my usual plan, describing what I am doing this winter:
>
> First, I have a good amount of pasture land under fence; in the fall I cut and bunch all the available hay on wild land and leave it in the

bunch for calves to run to. Then there is usually some fall wheat or rye stubble land, as I do not fall plough, and I have plenty of straw stack for them to run to. About the last of October, I take the calves from cows to wean. I place them in a large pasture of twenty or more acres, under high pole fence, securely built, so that it is impossible for them to get to the cows, but still have a good range. There I give them the best hay, with either a straw stack or some green feed. In forty-eight hours after separation I let them to the cows again, but this is really to benefit the cows, as by letting them drain the cows at this time there is no danger to the cows' udders. This plain I have always followed. I do not try to drive the cows to another enclosure, as some do, but they are separated from the calves only by this pole fence, therefore they are near each other all the time. It is surprising how little they worry and fail. The weaning is accomplished without perceptible shrink or falling off in flesh. In about two weeks' time I can turn them into the stubble field, where they have access to wild hay and where they will remain till grass comes in the spring. However, should the feed mentioned not hold out, I am careful to take them plenty of wild hay. I do not feed any grain and have no build-ings for them, only the bush and straw stacks.

"The best of beef may be raised and finished in our Western Provinces, but unless it can be marketed in good condition, and at reasonable cost, its produc-tion is not likely to be continued. The home demand will of course grow as population increases and towns and cities multiply, but farming is certain to remain the chief industry and beef production will undoubtedly always exceed local requirements. Outside markets will therefore be necessary and the means of reaching them must be duly considered.

"The transportation facilities furnished to Western cattle shippers have, for long, been declared altogether inadequate. It is charged that the supply of stock cars is irregular, uncertain and inefficient, their construction faulty, their equipment defective, that engines are overloaded and the speed of trains thereby greatly lessened, and that as a consequence of these conditions cattle in transit undergo much need-less suffering and their owners serious financial loss. While there is doubtless good ground for these complaints much of the trouble unquestionably arises from the fact that until within the last year or two export shipments have been confined to a period, little, if any, exceeding three months, during which one railway company has had, in addition to meeting the demands of ordinary live stock traffic, to do its best to move from 50,000 to 80,000 head of cattle over an average distance of 2,000 miles. The cattle shipping season in each year also overlaps the great eastward grain movement, during which every effort must be made to get the crop to the terminal elevators before the close of navigation. In spite of these extenuating circumstances, however, there is both need and room for improvement, and although the adoption of winter feeding will change and extend the shipping period, and the near advent of railway competition will doubtless greatly better existing conditions, the reasonable demands of the present day trade must be given reasonable consideration.

"At the National Live Stock Convention held here last year the Western cattlemen present declared that without prompt and radical reforms in transportation methods their export trade could not, under the altered conditions of beef production, be any longer profitably carried on. As a result of the statements made by these gentlemen, the convention passed unanimously a resolution that the matter should be referred to the Railway Commission for action, and it is gratifying to know that, on the request of the Western stock growers, that Board is, with characteristic promptitude, now actively engaged in remedying as far as possible the faulty conditions which have caused so much dissatisfaction and given rise to so many complaints.

"In shipping cattle practical experience is of immense value, and if the shipper himself is lacking in this qualification, he should endeavor to secure the services of a reliable and trustworthy man, especially if he intends doing business on an extensive scale. By following this course he will save himself much time, worry and money. This is particularly true in the case of shipments to distant and above all to foreign markets. Unless one knows the ropes, he is certain to find himself often at a loss and so driven into the hands of commission men and others who, whatever they may do for their regular customers, seldom show much compassion or consideration for the chance wayfarer, who is trying to do business on his own account. Loading must be carefully watched—overcrowding in a single car of a train load may mean a heavy loss. Cars should be clean and well bedded or sanded to prevent slipping; they should be in good, sound condition, and each should be closely examined inside to ensure that there are no projections such as splinters, bolts or nails, likely to injure the stock.

"Where hay is fed in transit, its distribution should be carefully supervised, and at any time when car doors have been opened they should be properly closed before the train moves.

"At feeding points the shipper must insist on ample time and space being allowed for rest, and must see that the feed and water supplied are of good quality and that each animal has an opportunity to get its reasonable share of both.

"Undue delays in furnishing cars or in the movement of trains, as well as all cases of injury to stock through rough handling, violent shunting, or otherwise, should be promptly reported to the proper railway officials, who are generally more interested than their subordinates in seeing that satisfactory treatment is afforded to shippers. By looking sharply after their own interests in matters of this kind shippers will avoid much annoyance as well as financial loss.

"It is almost superfluous to say that cattle ship much better when dehorned. This should, however, be done some time beforehand, preferably when close feeding begins, or better still when they are calves. The dehorning of range cattle which are to be winter fed is especially advisable, as it tends to make them quieter and much more peaceable than when the horns are left untouched.

"Fully aware of the disadvantages attending the present methods of marketing, the more advanced thinkers among our Western stock growers have, for a long time, been earnest advocates of the establishment of a dead meat trade. There is no doubt that if the enterprise were properly financed, started on a sound basis and conducted in an honest and business-like manner in the general interest of the producer, there would be far less actual wastage than at present. It is altogether likely that, had it been possible to secure the required capital, the trade would have been inaugurated years ago. For such an undertaking on a scale sufficiently extensive to furnish effective relief, however, a great deal of money is necessary and as our Western ranchers are, like the Eastern farmers, not much in favor of the co-operative principle, while several large interests have been rather opposed to any change in existing conditions, nothing definite has yet been done. A number of packing establishments in which both beef and pork are prepared for local and Pacific coast trade are now in operation in Alberta and Manitoba, but no serious attempt has ever been made to develop and build up an export industry in meats or meat food products. It is true that in recent years some members of the great American Meat Trust have established outposts in the Canadian West with results, so far at least, beneficial to the stockman, and it is possible that this action on their part may be only preparatory to larger operations, provided the field is found to be sufficiently promising. It is questionable, however, bearing in mind the methods usually followed by these gentlemen once their grip is assured, whether the establishment of a Canadian dead meat trade under their auspices is a consummation devoutly to be wished.

"Such an enterprise, to be productive of the greatest benefit to all concerned, should be under effective public control, and it is to be hoped that in the not too far distant future some practicable scheme will be evolved which while affording a better and more reliable and regular market for our Western live stock will still leave the producer free from the trammels of any trust, whether foreign or domestic.

"The advantages to be gained from the establishment of an export trade in dressed meat are, in the opinion of those who have most fully and carefully considered the subject, quite beyond question.

"In the first place, as has already been shown there is a very serious loss from the unavoidable shrinkage which occurs in the carriage of live cattle by land and sea over the enormous distance which separates the original seller from the ultimate buyer. While this shrinkage will, no doubt, become proportionately smaller with the general adoption of improved methods of handling, finishing and transporting the stock, it can never be entirely eliminated, and even when reduced to a minimum, it will, I think, be found to constitute the determining factor in establishing the superiority of the dead meat trade from the profit point of view, at least as far as concerns all cattle except those of the very best quality and finish.

"As will be shown later, there is good ground for the belief that animals of the class last mentioned will continue to be profitably disposed of on the hoof.

"Secondly, the competition which would be afforded by a sanely established, honestly conducted, and properly controlled dead meat trade would have a marked steadying effect on the prices paid to producers. With such a trade in constant operation, we would not see so often the fluctuations in values which now occur, and which are often undoubtedly due to friendly arrange-ments between buyers, many of whom unfortunately appear unable to resist the temptation to feather their own nests unfairly by unduly cutting prices when stock is plentiful and easy to obtain. Dealers in Canada as well as in the United States and other countries never seem to learn that tactics of this sort cannot be counteracted by the payment of high prices when stock is scarce and when, as a rule, but little remains in the hands of the producer. Scarcity of this kind is almost always attributable to the discouragement and disgust of the farmer or feeder, who, feeling that he has not received fair remuneration for his feed and labor, disposes of all his stock and ceases to be a producer. If buyers of live stock, which, to a greater degree than any other farm product, suffers from petty price manipulations, could only be made to grasp the fact that the time for small profit margins is when prices all round are low, they would soon begin to reap the benefits of self-denial in the form of a steady supply, and a regular if perhaps not excessively profitable trade. So long as they continue as at present to shake the confidence of the producer by scheming for unjust profits when stock is plentiful, so long will they continue to suffer, as many of them are now doing, from a shortage of raw material, not only disastrous to themselves, but involving great national loss.

"Another and by no means unimportant reason for the establishment of a dead meat trade is one which has been plainly set before us on two different occasions within recent years.

"In 1902 and again during the winter just passed foot and mouth disease made its appearance in the United States, with the result that large areas were in each instance at once debarred from participation in the export live stock trade. While this was serious enough for those portions of the United States concerned, it was, for geographical reasons, of trifling importance, when compared with the results which would inevitably have followed a similar outbreak in Canada.

"The United States has a long Atlantic coast line, and many different seaports, situated far apart, and served by numerous widely separated lines of railway. They have also, in constant operation, a complete system of fully equipped modern abattoirs, refrigerator cars and ships, which enable them, on the shortest notice, to convert their export live stock into dressed meat, which can be sent forward without let or hindrance.

"We, in Canada, are in an entirely different position; our Atlantic seaports are few in number, and the railways leading to them pass in convergence through a narrow neck of land, measuring only a few miles from north to south.

"We were on both occasions fortunately successful by efforts much more strenuous and exacting than is perhaps realized by the majority of Canadians, even those most interested, in preventing the introduction to the Dominion of this notoriously infectious and easily transmitted disease. Had these efforts failed our export live stock trade would have been stopped at once. The British authorities would undoubtedly, and from their point of view very properly, have prohibited the importation from Canada of live cattle, as well as sheep and swine. As a matter of fact, it was only with the greatest difficulty that they were induced to refrain from scheduling Toronto and a large portion of Western Ontario during the last outbreak in which the States of New York and Michigan were involved. This attitude on their part was due to the fact that in the advices from Philadelphia, the origin of the outbreak in Pennsylvania, which was the first to be recognized, was wrongly attributed to a shipment of cattle from Toronto. I was fortunately, at the time, in close personal communication with the British Board of Agriculture, and it was only by the strongest representations that the action above indicated was averted. The Board, however, insisted on a farm to farm inspection of the whole of the area to which any suspicion could possibly be attached, and it was therefore, at its direct insistance, that this particular line of work was undertaken and carried out.

"Canada is practically without abattoirs equipped for the slaughter of cattle, except to a very limited extent, for the home market; she has no system of refrigerator meat cars, and has, entering her ports, very few ships fitted for the carrying of chilled meats. In view of these facts, it is scarcely necessary to dwell on the risk which she is constantly carrying. At any time, in spite of the best efforts of her veterinary sanitary service, the appearance within her borders of one or other of the diseases scheduled by the British Board of Agriculture is within the range of possibility. As matters now stand, were such a thing to occur, especially during the short period in which our Western cattle are shipped, or at the time when our winter fed steers are being marketed, the consequences to the producer would be disastrous, while the whole trade would receive a blow from which it would require many years to recover. For this reason, if for no other, the establishment of a chilled meat trade on sound business lines and under proper control may fairly be termed a matter of national importance.

"It must not be forgotten, however, that there is a constant paying demand in Britain for home-killed dressed beef. This demand is certain to continue and as it can never, under existing conditions, be fully met by the British feeder, it is likely to remain profitable to those countries which, owing to their freedom from disease, are permitted to land live cattle in Great Britain, and are at the same time so situated geographically as to be able to transport such cattle at a reasonable cost and with not too great a risk of loss.

"In these two respects Canada occupies, and will probably continue to occupy, a most favorable position. Many countries which, under other circumstances, would be our keenest competitors, have been compelled, for one reason or another, to abandon their export trade in live stock for that in chilled or frozen meat. As they are year by year improving their facilities for the carrying on of this trade, the supply of dead meat in the British markets is likely, in the near future, to exceed the demand. In the United States, the only country at present in a position to compete with Canada in the live cattle trade, the home consumption of meat is increasing so rapidly that the surplus for export is likely soon to be a negligible quantity.

"It would thus appear that while the establishment of a chilled meat trade is necessary and advisable, it would be a short-sighted policy to contemplate the complete abandonment of our present export business in live stock. It should, therefore, in my opinion, be not only continued, but fostered and encouraged, by making the conditions surrounding it as nearly perfect as possible. This can best be done by the maintenance of strict Government supervision, involving full control of the methods adopted in transportation and the establishment of some comprehensive system of inspection, which, in addition to the present examination for health, would include the rejection of any animal of inferior quality or condition.

"It is, to my mind, somewhat doubtful whether it would ever be possible, in the face of the keen competition of an honestly conducted dead meat trade, to profitably ship grass-fed cattle on the hoof from Western Canada to the British market. There is, however, no question that, given better transportation facilities than at present exist, a profitable business could be done in grain-fed Western steers, as well as in the stall-finished cattle from Ontario and other Eastern Provinces.

"In any event it is well to have two strings to one's bow, and as each line of trade would steady and balance the other, it is to be hoped that, in the near future, we shall see both firmly established on a solid and paying basis."

CHAPTER XX

1910–1913

THE last few years of the ranching business saw no increase in ranches, and little increase in the few big herds that remained. Ten thousand farmers were breeding and feeding more cattle than the thousand ranchers of the early nineties, in many cases supplying better beef because they were feeding grain to trim their marketable stock. Hundreds of ranchers had quit the business and retired to live on the proceeds, some returned to England, some bought homes on the Pacific coast, some returned to Eastern Canada, and others to their old home districts in the States.

In 1909 the herds of the remaining ranches were inspected for mange, nearly two hundred thousand head being examined, out of which there were found to be fifteen thousand affected, some very slightly. Very few cases of maladie du coit were found, and thanks to the energetic way that Dr. Rutherford and the ranchers went after it from the first, the danger from this disease was now practically nil. Five thousand four hundred and twenty-two carloads of stock, of which about five thousand were cars of cattle, were shipped from the southern districts over the Canadian Pacific lines, eight single carloads going to Chicago. By the end of the year there were only a couple of ranches of any size in the country west of Cardston, where at one time thousands of head ranged at will. H.M. Hatfield, an old-time rancher in that district, and one of the last to quit, closed his business and retired, selling his cattle to Gordon, Ironsides & Fares. Mr. Hatfield had been ranching in the south Porcupine Hills for many years, but the Mormons, their farm settlement and their filling of the great ranges with settlers, had cut his range until it was no longer sufficient.

The price of horses soared, and there was not nearly a sufficient supply to meet the demand, though ten thousand head were sold from the southern districts. Only one prairie fire of any particularly destructive quality was suffered, this being in the Beaver Hills, the entire country between the Canadian Northern and the Grand Trunk Pacific Railway lines being burned over, and scores of homesteaders who had just gone in during late fall being burned out. Hay, feed, shelter went before the flames; cattle and horses perished, though the ranchers and settlers fought bitterly. One young girl perished in the flames while going out across the blazing plains to take refreshments to her father and the other men fighting the fires.

Another stockman went out of business the following year, Lem Pruitt, of Medicine Hat, who had trailed in five years before in search of range and located south of that point. Settlement hedged him in, his range grew less and less, he reduced his herds as necessity forced him, and at last had to sell every head and quit.

Spring fires ranged in the Porcupine Hills from the Old Man's River to Fish Creek, threatening the entire summer range. At one time it seemed that the whole West was in flame. Hundreds of settlers and ranchers fought night and day to stem the advance of the ever-advancing element, working in gasping relays, their women cooking night and day to feed the fighting men. Horses, ploughs, and men worked through the murky darkness of the smoke-clouded days and the thick blackness of the nights, ploughing fireguards, whipping out crawling snakes of flame, saving property and life. What can be done under stress was shown at the Bar U, where Herb Miller and his men were fighting desperately to save a big stretch of pasturage. They ploughed and dug and finally had it entirely protected excepting up the side of one big hill, which no cowboy had ever ridden, and which was considered insurmountable for horses. This hill was the open gate for the fires to enter into the otherwise protected tract of feed, and the red tongues were licking ever closer. Miller took a team of mighty Percherons that were harnessed to a plough and faced them up the hill. The ploughshares cut the sod, the speed of the galloping horses threw clods for yards, and the great brutes strained on the drag as they faced that sheer grade and set their big feet for the surging, heaving struggle of the climb. No rider had been up it, but this pair, urged by the yelling, driving Miller, cut a clean furrow from top to bottom and saved the grass.

This year 154,340 head of beef cattle were shipped out of Alberta over Canadian Pacific lines, in 7,978 cars, 4,038 carloads going out from the country north of Calgary. Mixed farmers supplied goodly numbers of these shipments, the big ranches shipping not over one thousand carloads.

The Indians, with some few exceptions, were very orderly, attending to their own business on their reserves, and going out in fall and early winter on hunting trips. The Stonies slaughtered much game in the foothills and mountains, and proved to be constant annoyances to game wardens; one man who went into a part of their country to hunt deer north of the Red Deer and west of the track, started into the brush country on the opening day of the season, and had not journeyed a mile beyond the Raven River before he met Stonies. They told him there were fifty in their party and they were hunting in a long drive, one man every two hundred yards. They had just started out, it being the first day of the season; they had not gone a mile north of the river, and yet they had already killed eleven deer, and two bears. Passing on, they disappeared, and the popping of guns sounded all day. Their drive would take in all the country between the Red Deer, the Saskatchewan, the Raven and the James rivers. Other parties of the same tribe were working northward along the foot of the mountains, killing as they went.

In 1909 the Piegan Indians were much aroused by efforts of white people to throw their reserve open for settlement. Hon. Frank Oliver, Minister of the Interior, was approached and let it be understood that the land could be opened by approval of the sub-chiefs. The head chief and his chief councillors were strongly opposed to this, and succeeded in blocking the proposition even though a more or less satisfactory approval had been given by a majority of the secondary chiefs. This trouble still smouldered when Sir Wilfrid Laurier made his tour of the West in 1910. Arriving at Macleod, the Premier was asked by the Board of Trade to open the reserve for settlement, but Sir Wilfrid replied that this could not be done without the consent of the natives. A few days later the Piegan chiefs rode up in great state to the door of E.H. Maunsell's ranch and told him they had just elected him to a chieftainship in the tribe. They had heard he was going East as far as Ottawa, and they wanted him to represent them there. Sir Wilfrid Laurier had said the Indians could not be forced to give up their land, and the chieftains wanted him to know they appreciated his statement, and that they did not want to leave the reserve.

"And tell Frank Oliver," said Chief Butcher, drawing himself up in all his native dignity, "that I shook his hand the last time I saw him. Tell him I will never do it again. Tell him he is a white man, but he has made me very, very glad that I'm an Indian."

One of the most dastardly crimes in the annals of stock history took place in September of 1911, in the country south of Medicine Hat, when A.R. Marchessault and John Flynn lost a number of horses by having them shot down in a spirit of revenge. Marchessault had foreclosed a mortgage that he held on property owned by Frank Herrington, and the latter craved revenge. Assisted by a Finlander named Hjlemar Bodie, he went to Marchessault's pasture late one night and ran all the horses out of there into the hills, three of the animals belonging to a settler named John Flynn, who happened to be camping at that place overnight. The two miscreants took the animals, and when they felt they had reached a safe place, Herrington rode in front with his repeating rifle and sat on a hillside, where he shot eight of the animals dead as they were driven slowly past him by Bodie. A rancher named Leslie Nantes found the bodies the next day and spread the alarm, and the Police were soon on the trail. In addition to killing the eight head, the men had kept two of the best, a pair of valuable Percheron mares, which were never found. Herrington, to whom suspicion pointed from the first, was located in Montana, lost again, and finally arrested in the lumber camps at Morrissey, B.C., by Sergeant Ashe and Corporal Corby, of the RNWMP, who were watching the place disguised as lumberjacks, having been given a hint that Herrington might drift that way. Herrington confessed and received five years in jail, and Bodie, who was then arrested, was given a similar term.

The Mounted Police Superintendent for the Macleod district reported this year that the cattle industry was practically a thing of the past, and that in the Pincher Creek district there was not enough cattle to supply the local market. Deep snows and cold weather in the Cardston district were the cause of the remaining ranchers losing twenty percent of the stock that was out, and summer droughts caused great suffering among the homesteaders along the boundary. These unfortunates had settled in a district that was quite frequently visited by summer frosts, and just as frequently liable to drought. The drought caused utter crop failure this year, and the Government was forced to supply relief, taking liens on the settlers' properties. Supply depots were established at Medicine Lodge, Coutts, Writing-On-Stone, and Pendant d'Oreille and Grassy Lake, Mounted Police constables riding the bitter, windswept winter ranges in search of suffering families. One constable named White was so badly frozen while on this relief detachment that when he pulled off his gloves, the skin came too. And on top of this hard experience of the fall and winter of 1910–11, there came extensive and destructive hailstorms in the Lethbridge country, by which millions of bushels of grain were ruined.

The full result of the reduction of the range herds was not felt until 1912, when prices reached unheard-of heights. Beef steers that used to sell for twenty-five to forty dollars were being eagerly bought at prices ranging between seventy-five and eighty dollars. The Waldron ranch sold twenty-two hundred head, and a number of small ranchers, who still existed west of Nanton, furnished small numbers to the markets. Thousands of stockers were brought in by J.H. Wallace, Gordon, Ironsides & Fares, and other big ranchers who remained. That year there were shipped out of Cayley, the greatest shipping point of ranch cattle between Macleod and Calgary, the shipping station of the Bar U and other North Porcupine Hills ranches, 7,677 head of beef cattle. Gordon, Ironsides & Fares brought in 3,000 Mexican stockers, which they put on the Blood reserve. Huckvale and

A party of travellers in an early Packard touring car in 1912 include driver Malcolm McKenzie, Chief Butcher sitting next to him, and Edward Maunsell in the back seat. Also along for the ride are Big Plume and three unknown men.

Hooper, on Many Berries Creek, sold the balance of their cattle and all of their horse herd to Ryan & Fares, Huckvale going into extensive mixed farming on a well-irrigated ranch of three thousand acres in the Medicine Hat district. The old Alberta Stock Growers' Association remained in existence and retained much of its vigor, the remaining ranchers knowing the value of co-operation and association. P. Burns was president during 1912, and George Lane was elected to the presidency the following year, with W. Roper Hull as the vice-president. D. W. Gillis, of Calgary, was appointed to the position of secretary-treasurer, which had originally been held by R.S. Mathews, of Macleod, who relinquished the position when the practical headquarters of the Association were established at Calgary.

Lafayette French, the old trader, friend of Crowfoot, and one of the first white men in the West in its wildest days, came to an unfortunate end about Christmas of 1912. He had been living for some time up High River in a small cabin owned by George Emerson. He was an old, feeble man, and in some unknown way his cabin took fire, burning completely, and so badly injuring French that, though he escaped to the prairie and was there picked up, he passed away in the High River hospital a short time later.

The day of the open range was a thing of the past when 1912 first showed his smiling face, the little cowtowns had become large towns and large cities, the hitching-posts in front of the hotels had given place to automobiles, street cars clanged through the paved streets where galloping cow-ponies had kicked up the dust, cowboys were subject to street curiosity, and the range was a thing of the past with the exception of a few large fenced leases in the south and east.

The old-timers were passing away, the industry was secondary to farming and other industries, and the cattlemen were spoken of as of the past. Here, there and everywhere grain farms grew green and then yellow, the range stock had passed, and the fat, sleek barnyard animals browsed placidly in pasturages of cultivated grass. George Lane, A.E. Cross, P. Burns, Hon. A.J. McLean, old ranchmen all, decided that their industry, the story of the cattle, the cowboy and the rancher, should not slip out unforgotten, so they "put up" the financial sinews to provide a wonderful Western pageant and contest, an exhibition compared to which the ordinary Wild West Show would be very small peanuts indeed. These four men, rich in cattle and lands, made wealthy by the ranch industry, loving the occupation, were assisted in the practical staging of the great show by H.C. McMullen, livestock agent of the Canadian Pacific Railway, and a cowman of thirty years' experience on the ranges of the West, and by Guy Weadick, a young cowboy, who had the knack of planning big things, and a world-wide acquaintance with horsemen, ropers and cow-people. The aim of the "Stampede," as the exhibition was named, was to bring together the Indians of the Province in their original costumes, to hold a reunion of old-timers, to show what the country had come to and what from, and to bring the finest ropers and rough-riders of the North American continent in competition. Exceedingly large purses were hung up to attract the best men in the West, and the response was generous indeed. Ropers, bronco-twisters, bull-doggers, all came, and to supply the wants of these competitors, the herds of the Western States and Canada were combed for their wickedest animals, while the agile, fleet-footed steers of Mexico were shipped in to test the ropers' utmost qualifications.

The "Stampede" was held at Calgary, and thither in September of 1912 the range men flocked, while from all over Eastern Canada and the States the spectators came in thousands. His Excellency the Duke of Connaught, Governor-General of Canada, accompanied by the Duchess of Connaught and Princess Patricia, attended the exhibition, and were interested spectators at two days of the sports. The champion ropers, riders, and bull-doggers of Old Mexico entered the contests, the best riders and ropers of New Mexico, Oregon, Utah, California, Wyoming, Montana, Colorado, Oklahoma, Arizona, Alberta and British Columbia came in scores, ready to ride, rope, or wrestle with the wild range animals. Roping and riding are known to every one, but bull-dogging is not and deserves a word of explanation. It is somewhat brutal sport when the steers have long horns, for very frequently the animals suffer the agony of having one snapped off. A wild steer is turned out of a corral chute and is given fifteen feet start of a waiting vaquero who pursues, overtakes, reaches across and leaps from his saddle, swinging on the frantic steer's horns and throwing him helplessly to the ground. With horse and steer going at their fullest speed this is a dangerous, thrilling and exceedingly stirring spectacle.

Wild horses were driven into every central corral in Alberta, where they were tried, and if found vicious enough, were purchased and taken to Calgary, there to be fed oats and strengthening foods in order to make them have the strength and endurance to fight their fiercest. Splendid riders were employed to test and try out these outlaws, and one of these men was Joe Lamar, son of the John Lamar who had the gun discussion with Mackay at the Waldron ranch many years before. Young Lamar was a master rider, considered the best in Alberta, and he was expected to come pretty near winning the thousand dollars that went to the best rider. But an unfortunate end came to the young man just a day or two before the contests commenced. He was in the big ring of the horse-show building at Calgary, trying an outlaw called "Red Wing," when an especially vicious surge of the fighting horse shifted him and brought him across his cantle, his back being broken by the wrench. Grinning cheerfully to the anxious cowboys who rushed to his aid and tenderly lifted him, he said, "It's all over now, boys," and passed away.

The contestants, big-hearted range folk, decided to do what they could to assist the dead man's wife, and gave an exhibition one evening and turned the entire gate proceeds over to Lamar's widow, some two thousand dollars in all. At this exhibition an Oklahoma cowgirl, Goldie Sinclair, rode and conquered the savage "Red Wing."

Throughout the days of the exhibition, it was a wonderful picture of Western sports and the old times. Indian tepees were pitched in the showgrounds, reproductions of old Fort Whoop-Up and of a Hudson Bay post were erected, Fred Kanouse himself being in charge of the Whoop-Up fort. Traders, trappers, cowmen and policemen of forty years before were present and talked over the stirring days of the early times. Lafayette French, the old trader, refused to attend because there was to be a pioneer present who had promised some thirty or forty years before to shoot him on sight, and French believed he was still unregenerate and determined.

The pageant—two thousand Indians in warpaint and feathers, guiding their mounts by leather thongs tied round their under-jaws, travois, papooses, buxom squaws, hundreds of cowboys, Hudson's Bay carts, half-breeds, stagecoaches, a squad of Mounted Police of 1874 and a trim body of the Mounted Police of the present—made a thrillingly interesting picture of progress. Scarred warriors, painted and half-naked, rode in line with Indian cowboys in chaps, spurs, and Western saddles, or beside a wagon that carried some sturdy young Indian farmer who had erected a display of the crops off his little farm. The past and the present were plainly shown, and the watching thousands received in that great scene more education as to old times and development than they would get from a hundred books.

The contests of bucking horses and wild steer-roping were exceedingly good, the bad horses from Wyoming, British Columbia, Alberta, Saskatchewan, Montana, and Colorado fighting wildly against the clever riders. Ed Echols of Dragoon, Arizona, who had ridden in to the train from a ranch where he was working and decided to journey the thousand miles to Calgary to attempt to annex the prize in roping, won it in twenty-three and four-fifths seconds, a remarkable record for catching, throwing, and tying a steer. For years Mike Herman and John Ware had been the master ropers of Alberta, doing their work in the neighborhood of fifty seconds. In 1889 the Arizona cowboy tournament produced a champion roper named Juan Levis who roped and tied his steer in one minute, seventeen and a half seconds. Echols' time was marvellous, considering the fact that he was roping the surest-footed, strongest and quickest breed of steers known on the Western ranges, the lithe greyhounds of the Mexican hills. If old John Ware had only lived, he would assuredly have appreciated the work of Echols and of Echols' horse, for the horse had much to do with it. "Ribbons," the little rope-horse that the winner rode, was a learned animal who possessed the wisdom of the serpent and the speed of light. He followed his steer as though both were in a groove, he tossed the mighty mass of beef high in the air when the rope tightened, and he then stretched the animal helplessly on his back, leaning sturdily against the taut lariat until his owner had dismounted, run to the steer, tied it, and thrown both hands in the air as signal to the judges that the work was done. Then "Ribbons" knew his work was over, so he slacked the rope and stood negligently and sleepily while judges rode close to assure themselves the work was well done. Echols won, through the steadiness of his own rope-arm and the reliability of his mount, a total of fifteen hundred dollars, five hundred for the best time on any one steer, and a thousand dollars for the best average time on three steers.

In the roping competitions, the men of the southwestern States proved to be markedly superior to those of Canada, due, to a great extent, to the difference in the animals they worked amongst. The range stock of the South-West, of Texas, New Mexico, and Old Mexico, is quick, light, and as speedy as most horses, while the range animals of Alberta and Montana are grade Shorthorns and Herefords, huge, clumsy, well-fed brutes, whose best gait is a lumbering gallop, and whose agility compared with that of the Mexican steer is as a tortoise to a hare. Canadian stock was half as big again as the South-Western range beasts, twice as slow, and not a quarter as agile. A Canadian rope-horse was picked for strength and weight to oppose the weight and strength of the heavy steers; the Arizona and Texas and other American steeds were picked for speed, sure-footedness, and dodging ability. Many a "longhorn" actually outran the Canadian rope-horses, and many out-manœuvred and out-dodged Canadian horses and riders, the Americans easily winning the palm in roping. But the riding was a different tale, as will be seen.

Moving-picture men moved through the arena endeavoring to take pictures of the fast-moving Westernisms, and in one instance the picture men moved as fast as any of the cowboys. A yellow steer, angered by the jerk that snapped the rope that a cowboy dropped on his horns, headed straight for a moving-picture camera around which were grouped three interested photographers. It was an opportunity of a lifetime—a mad steer charging straight into the lens—and the energetic cameramen lost it. The proprietor fled madly to a suitcase that stood nearby and was filled with photographic plates, flung himself behind it and palpitatingly watched the approach of the avalanche of angry range beast; his two assistants were equally prompt; one who had been cranking the machine sped with enviable alacrity to a telegraph pole, and the other, ostrich-like, held the camera cloth between himself and the animal and thus concealed himself. It only took an instant, for by that time the pursuing cowboy had chucklingly inserted his horse between threatener and threatened and saved the day. The photographers came close together again, and the proprietor could be seen angrily shaking his hands before the face of the crank wielder, who had deserted his post at a time when there was a positive assurance of a film that would have paid all expenses. The proprietor had thought of these things after the yellow steer had gone.

Riders from all the West, the top-notch men of the range, the winners at Cheyenne, Tucson, Pendleton and other great Western exhibitions, were congregated to ride for the gold and the glory of themselves and the "Stampede." Americans, Canadians, king-riders, all were present, and among them was one Indian, a Blood, named Tom Three Persons, a tall, handsome young man who wore his hair short and dressed as the cowboys did. The riders had ridden the bad horses, the good being gradually weeded from the excellent until at last only a half-dozen remained, this Indian, Henry Webb, the winner at Cheyenne, and a few others. Among the bad horses was one outlaw, a black brute that had already "piled" several cowboys during the exhibition, and held the proud distinction of

Blood cowboy Tom Three Persons won the championship in the bucking horse competition at the first Calgary Exhibition and Stampede in 1912.

having thrown one hundred "top" riders and never been ridden. The distinctive appellation of "Cyclone" was the name this bad horse bore, and he was worth much money to his owners, a worth that was accentuated by his wickedness. Henry Webb was given a nasty little sorrel, Tom Three Persons was given the black demon—the horse who had the day before thrown Clem Gardiner, best of Calgary's cowboys, using Gardiner to make his hundredth victory. Glen Campbell, inspector of Indian agencies for the Dominion of Canada, the man whom the Stampede management must thank for the presence of the thousands of Indians, personally assisted Three Persons to saddle Cyclone, and then anxiously watched the battle between the red-shapped red man and the bawling, roaring black horse. Straight in the air the animal reared until the Indian's body stood almost horizontal, then, like a dog coming from water, Cyclone shook himself with a mighty shake that would have dislodged a panther; but the saddle cinches withstood that awful strain, and Three Persons stayed firmly in the seat, waving his hat as he "fanned" his mount to force it into the buck-jumps. Johnny Franklin, judge of the buckers, rode close and watched with eagle eye as the big brute sunfished, swapped ends, zigzagged, and pitched, but the rider rode, by balance and skill, without spurs or roll or reins. A hundred times a minute, the thousand pounds of hardened horse muscle humped five feet in the air, and hit the ground with every joltable twist and jar that he was capable of, but the Indian was waving his victorious hat to the thundering grandstand thousands. Finally the animal grew discouraged, bucked a few final plunges, and gave up, while his owner, Bertha Blanchett of Arizona, wept her disappointment, and Glen Campbell threw his arms around his ward in his joy. For the native of Alberta, lineal descendant of the original horsemen of the North-Western plains, had won the championship of the rough-riding world, and won it from the winners of a hundred state competitions. Tom Three Persons was a protégé of E.H. Maunsell's, and he would never have entered the contests had it not been for the urgings of Maunsell, who knew a rider when he saw one and who knew no horse was born that could pitch this red riding star.

The names of the various winners in the contests, which are given below, will enable one to judge from how wide a field the Western best were picked:

Bucking horse riding—Tom Three Persons, Blood reserve, Macleod, one thousand dollars, saddle and championship belt; second prize won by Harry Webb, Cheyenne, Wyoming; third prize won by Charles McKinley, of Plattville, California.

Cowgirls' bronco-riding—Fanny Sperry, Mitchell, Montana, won first prize, of a thousand dollars, a saddle and championship belt; second prize went to Goldie Sinclair, of Merrimac, Oklahoma; third prize won by Bertha Blanchett, of Phoenix, Arizona.

Cowboys' steer-roping won by Ed Nichols, of Dragoon, Arizona; second prize by George Weir, of Monument, New Mexico; third prize by Joe Gardner, of Sierra Blanco, Texas.

Special steer-roping prize won by Charles Tipton, of Denver, Colorado; second prize won by Art. Acord, of Portland, Oregon; third prize won by C.P. White, of Brooks, Alberta.

Steer bull-dogging—First prize won by O.K. Lawrence, of Sulphur, Oklahoma; second won by Charles Tipton, of Denver, Colorado; third prize won by Senor Estevan Clemento, of Old Mexico.

Cowboys' bareback bucking-horse riding—First prize won by James Massey, of Texas; second prize won by "Doc" Pardee, of Stillwater, Oklahoma; third prize won by William Rooke, of Dewey, Alberta.

Cowboys' relay race won by John Mitchell, of Medicine Hat; second prize won by Dug Wilson, of Claresholm, Alberta; third prize won by H. Bray, of Medicine Hat.

Cowgirls' relay race won by Miss Bertha Blanchett; second prize won by Mrs. H. McKenzie, of Crossfield, Alberta; third prize won by Miss Fanny Sperry.

Cowboys' fancy roping contest was won by "Tex" McLeod, of San Antonio, Texas; second prize won by Senor Magdelena Bamos, of Mexico City, Mexico; third prize won by L. Walsh, of Kew, Alberta.

Cowgirls' fancy roping won by Florence LaDue (Mrs. Guy Weadick), Calgary, Alberta; second prize won by Lucille Mulhall, of Mulhall, Oklahoma; third prize won by Bertha Blanchett, of Phoenix, Arizona.

Cowboys' trick and fancy riding won by Otto Kline, of Livingstone, Montana; second prize won by Art. Acord, of Portland, Oregon; third prize won by Jason Stanley, of Los Angeles, California.

Cowgirls' trick and fancy riding won by Dolly Mullins, of Ingle, New Mexico; second prize won by Bertha Blanchett, Phoenix, Arizona; third prize won by Hazel Walker, of Los Angeles, California.

Canadian championships, which were held especially for Canadian cowboys and cowgirls, were won as follows:

Cowboys' steer-roping won by F. Dey, of Medicine Hat; second prize by William Pendlan, of Medicine Hat; third prize by Clem Gardiner, of Calgary, Alberta.

Flores (or Florence) La Due of Calgary won first prize for fancy roping at the first Calgary Stampede in1912. Florence was married to cowboy Guy Weadick.

Cowboys' bucking-horse riding won by Thomas Gibson, of Calgary; second prize won by "Red" Parker, of High River; third prize, a tie between Clem Gardiner and Joseph Bradwell.

Twenty thousand dollars was divided among these winners of the last big Western show in Alberta, and a hundred thousand people saw them compete. George Lane, Hon. A J. McLean, P. Burns, A.E. Cross, H.C. McMullen, and G. Weadick, the men who were responsible for the great picture of Western days, had presented to the later comers to the Province a vivid and never-to-be-forgotten illustration of Western life, of the work and the play of the cowboys and ranchmen when the ranges were open and the cattle were wild.

CHAPTER XXI

The Future

ANY country that can produce such animals as have been developed in Alberta is a stock country second to none, and the future of the industry rests with the methods employed for the future. John A. Turner, of Balgreggan Ranch, has sent his Clydesdales and Shorthorns to every great Canadian exhibition for many years, and his prizes number scores. He exhibited fourteen times at Winnipeg, winning fourteen championships; he exhibited at every Western Dominion fair and won championships at each. The horses and cattle from this ranch are a splendid tribute to the country and the breeding. At Chicago, New York, Spokane, Seattle, Vancouver, and other points, George Lane's Percherons swept everything before them in later years, and Lane owns the biggest herd of purebred Percherons on the continent. They range all winter, they are native-born, Western bred, and are admitted to be as fine representatives of the breed as can be found anywhere in the world.

For feeding, one need only to look at the result obtained by J.D. McGregor with the magnificent Polled Angus steer that he exhibited in the Chicago fat-stock show in 1912. Mr. McGregor took this animal from Alberta, fitted it for the show to enter into competition with the finest corn-fed and stall-raised animals of the United States, and won the championship from them. As a feeding and breeding country, Alberta has proven her grade, and the future of the livestock industry rests upon the methods adopted. Alberta is like a great mixed farm, with huge wheatfields, large woodlots, and great pastures. Naturally the soils are not uniform; some are good wheat-producing areas, some are better for grass; some should be given over to stock, and some to the farmer. Large tracts of good pasturage that are unsuitable for wheat could be turned over to the grazer.

No rancher injures the farmer; on the contrary he should prove a benefit to him. Throughout the wheat-belt, there are every year tens of thousands of tons of straw burned up, an utter loss, because there is no use for it. On the leases there are thousands of head of stock that have to rustle hard to keep their hearts throbbing until the green grass of spring. The ranchman would willingly pay a few dollars per head if his stock could be turned into the straw stacks, the price being in the nature of an insurance. There are perhaps a hundred or two hundred farmers within ten miles of every lease in Alberta, and each of those farmers has straw to waste. It would be to the farmers' and the ranchers' advantage to arrive at some definite arrangement; fifty head there, a hundred head in another place, assurance for the stock and an income for the farmer where he formerly suffered a total loss. A great industry could again spring up if such arrangements were assured, for the ranchmen could put double the stock on their leases, using them only for summer range and turning the herds among the adjoining farmers when fall came down. P. Burns has done this to a limited extent for some time, E.H Maunsell has commenced it, and other large ranchers would do it if they could. It is the logical solution of the ranch industry. Another point in favor of the ranchmen and farmers thus co-operating would be the added assurance for the marketing of the farmers' grain. If grain was low or damaged, they could buy ranch stock and feed it, making more per bushel thus than they would by selling to the elevators. Grain rations put three to four hundred pounds on a range steer with three or four months' feeding, and four hundred pounds of beef is worth a number of bushels of oats or barley. It is not often that four, five, six, and seven cents is paid for a pound of barley or oats, but it is not an exceptional price for beef.

If the Government were to reserve grazing lands, giving ranchers the assurance of permanency of leases and business, if the farmers would take up the work of joining with the rancher, Alberta would be the most prosperous community of mixed farmers that the world knows. Each farmer, with a small band of his own stock, wintering a larger bunch of range stock, money from both, help to the ranchers, would mean the development of the industry and an assured future for the Province. If the good farming land could be separated from the bad, and if the ranchmen could be given the bad while the grain districts proceeded with the harvests, it would be a wise solution and one that would meet commendation from every hand.

Some such result is being obtained in small measure by the independent endeavors of a few ranchmen. The Canadian Pacific Railway Company is encouraging every kind of mixed farming, the Provincial Government has undertaken to encourage this line of industry, and the Dominion Government has this year (1912) taken up the serious consideration of making certain districts in southern and central and eastern Alberta pure, unalloyed cattle and range districts, an action, which though late, is appreciated, and has added a fillip to the development of the livestock industry. There are in the Province among the farmers, breeders and ranchers numerous specimens of as fine purebred stock as can be found anywhere—native-born purebreds, the equals of the best. The evils of promiscuous breeding have been experienced and the lesson taught, the reward of fine breeding, of good conformation and good distribution of flesh has been learned through a sometimes bitter school, and the breeders are educated to their own best advantages, which is to the advantage of the public and the markets. The markets want the best, the public causes the market's demand, and the stockman supplies the demand of the market. He has learned that there is more money, satisfaction, and general success in producing the best than there is in the slipshod, careless method of selling anything and feeling it was "good enough." Alberta beef has therefore won a high mark in the markets, the prime animals of the Province commanding a special price. George Lane shipped large numbers of "tops" to Chicago in the fall of 1912, prime steers from off the ranges, fat, well-formed, well-fleshed animals, that brought higher prices than any other range beasts ever taken there. Mr. Lane shipped from Alberta, paid freight, duty and expenses en route, and received a price that netted him a larger profit than he could have made on the British, Winnipeg, or home market. The fact that the Chicago buyers, who receive stock from every State in the Union, who know good beef when they see it, and who often saw it, were willing to pay more for the Alberta range stuff than for their own top cattle of the West, established the place of the Alberta ranches, established the grade of the stock and fixed the best Alberta stock as the equal of any on the continent.

Alberta's markets are now scattered wide: Chicago, Winnipeg, British Columbia, England, the Yukon, and at home. Chicago and England want only the prime animals—the fat three- and four year-olds—and there is a shrinkage and considerable loss in every shipment, especially of grass-fattened livestock. P. Burns and Company have, at Calgary, a great packing establishment, and thus, to a considerable extent, they reduce their own loss in shrinkage, but the farmers and ranchers do not share the full benefit of this. With the tens of thousands of head of livestock that are shipped every year, increasing as time passes, there should be an opportunity for other great plants situated in central distributing points such as Calgary, where the chilled meat industry could become the greatest of all the meat branches of the Province. If export and all beef should be sent to Calgary for butchering and dressing before being placed on the markets, it would mean added riches to the stockmen, and added prosperity to the entire country; practically no lost weight in shipment, the ability to hold the product until markets were more favorable, a wider field in which to dispose of the surplus, and more satisfaction generally. There is room for a packing plant of generous proportions, and when it comes, there will be more satisfaction, not only to the breeders, but also to the public and the merchants.

Farmers working with ranchers, poor grains fed to cattle, adequate packing establishments, and the same attention paid to breeding that has been given during the past few years, and the industry of livestock—beef, mutton, pork—will again be one of the most important and most profitable in the Province.

At the annual meeting of the National Live Stock Association, at Ottawa in February of 1912, W.F. Stevens, Live Stock Commissioner of Alberta, delivered an interesting address on the horse, cattle and sheep industries of the Province, including the mixed farms and the range stock. Following him, George Lane of Calgary, and R.J. Phin of Moosomin, Saskatchewan, spoke on the subject of Western beef cattle, Mr. Lane's speech applying particularly to the range stock of Alberta. R.J. Phin spoke at length on Western cattle conditions, and was followed by George Lane, who led the discussion. Lane's speech summed up in brief and sharp words the cause of the decline of the ranches, and the remedy for strengthening it. He had lived through the life of the Alberta ranches from start to finish, he had grown wealthy in the industry, and he still held large interests. Perhaps no man in Alberta knew more of the cause of the fall of the ranching industry, its weak spots and its strength, than Lane, and his words are consequently employed to close this volume.

"During the past twenty years," said Mr. Stevens, "there has been a complete revolution in the horse industry of Alberta. A score of years ago Alberta had horses to burn. Everything was favorable to production, but for the want of a profitable market the business languished. The first event worth noting in the chain of circumstances which revolutionized the business was the rush for the Yukon. This created a demand for a large number of pack horses, which at that time were obtainable for $5 each or $50 a dozen. It stimulated also the trade in saddle horses and animals suitable for freighting and cartage purposes.

"Many of the adventurers who started for the gold fields were unable to endure the hardships of the far North, and turned back. On their journey southward they took time to note the agricultural possibilities of the country through which they were passing, and a considerable number decided to remain and engage in farming, and in so doing, they created a demand for horses suitable for agricultural work. Coincident with these events there began the tide of immigration from Europe, the Eastern Provinces and the United States, which by the close of the last century had more than doubled the value of farm horses. Then followed the Boer War, which stripped the country of its surplus animals suitable for saddle purposes. With each new demand prices rose and the horse ranchers of

Alberta looked into the future with a feeling of confidence they had not known for years. Their ranges were still intact, the ever-increasing tide of immigration to the north assured a permanent market and their chief concern was how to derive the greatest possible benefit from it.

"Their good fortune was, however, destined to be short lived, for almost coincident with the close of the Boer War it became known that winter wheat could be successfully grown in the southern portion of the Province. This event sounded the death knell of horse-raising in Alberta as a ranching proposition on a large scale.

"As early as 1905, females began coming to market in large numbers, showing that many ranchers were either tempted by fancy prices offering or were forced by the rapidly increasing mileage of barbed wire fences to sacrifice their breeding stock. Then immature animals and even foals began to be in evidence. Every such shipment proclaimed the fact that pastures were being converted into grain fields.

"Although the records of shipments showed more animals going to market each year over the preceding one, the fact that so large a percentage were females and immature animals was a sufficient warning that the time must come when there would be an abrupt falling off in Alberta shipments and importations would begin. The records of shipments since 1905 are as follows:

1905	9,310 head
1906	13,302 head
1907	11,924 head
1908	14,419 head
1909	22,752 head
1910	27,887 head
First ten months of 1911	13,113 head

"The total for the year 1911 may safely be reckoned at sixteen thousand head, or nearly twelve thousand head less than was shipped the year before.

"Railway construction and municipal improvement, which in 1905 began to assume large proportions and have steadily increased since then, drew heavily on our stock of draft horses, and by 1908 it became evident that the local supply of this class would soon be exhausted and that prices would mount to higher levels than had yet been known. In self-protection, contractors began importing draft horses from Ontario and mules from the United States.

"The importation of mules was especially large during 1909 and 1910, and while I have not the figures showing the exact number, they may be fairly estimated to be about two thousand five hundred teams. There are many evidences that large importations of draft animals will continue for many years. The farmers of the older settled districts in Central Alberta are beginning to have horses to sell, but the number is far below the demand. Prices are on an importing basis and are consequently high. A team of thirteen hundred pound horses, sound in wind and limb, finds ready sale at $600, and I have known instances in which $650 has been paid, and heavier animals command a premium of from $30 to $50 for every hundred pounds in excess of this weight.

"Importations are no longer confined to draft horses, but eleven and twelve hundred pound animals are now being brought in. To my own knowledge, there are at the present time several Alberta buyers in the States of Montana, Idaho and Washington looking for horses suitable for farm work.

"This condition is placing a heavy charge on our agricultural development and is greatly retarding it. It accounts for the enormous increase in the number of tractors now in use on the prairie and it compels men of limited means in the bush country to use oxen or quit. Naturally, this condition will in time right itself; the question at the present is by what means can this process be hastened.

"A great deal of educational work has been and is being done by the Provincial Department of Agriculture, through its institute meetings and short-course schools, to induce farmers to raise more and better horses. Our farmers are today better judges of horses than they were five years ago, and are more exacting in their demands of what a stallion must be that they will consent to breed to. The results are noticeable in the improved class of foals being exhibited at our country fairs, but with the settler on the frontier it is still largely a matter of 'Hobson's choice.'

"Considered from the standpoint of climate, water, feed and markets, there is every inducement for the farmers of Alberta to embark more largely in the growing of work horses and of mules. For those who understand it, the former may be made as profitable in Alberta as they now are in Ontario and Manitoba, and the latter as remunerative as they ever were in Kentucky or Missouri. These are the facts that the superintendent of fairs and institutes is trying to impress upon our farmers.

"But we have a class who do not need this fact to be brought home to them in order to induce them to embark in the business. They are the men of the past generation who, like Othello, find their occupation gone. They are too young to quit and too old to learn a new trade. They are the ranchers whose ranches have been homesteaded and put under fence. The question is, to what extent would encouragement to this class promote the general welfare? There are in Alberta, in the north as well as the south, small areas of doubtful value as farm lands. My idea is to grant closed leases of these lands in tracts not exceeding ten sections to any one individual, for a term of years sufficiently long to induce him to stock it, and provide such buildings, fences and watering-places as will enable him to live in comfort and conduct his business economically, and at the expiration of his lease, permit him to purchase, not to exceed two sections of his leasehold, at a price stipulated in his lease. I know, personally, of townships which would be much better off if they were occupied by four or five prosperous ranchers than they now are with a dozen or twenty homesteaders struggling against starvation, and the Province as a whole would be benefited if settlers were directed to lands well suited to agriculture, of which there are still hundreds of thousands of acres available, and these rough, sandy and stony areas devoted to grazing purposes.

"In the raising of beef cattle, Alberta is now rapidly approaching that low ebb in production which, so far as I can learn, has characterized every country that has changed from the ranching to the farming system. Even in the farming and dairying districts, those who were growing beef cattle ten years ago did so more after the methods of the rancher than after those of the farmer. They either had large tracts of land of their own not yet brought under cultivation, or there were unoccupied tracts adjoining their homesteads which provided them with pasturage at little cost in money or attention. The straw which would otherwise have been burned, was used as a substitute for the rancher's winter range, and the result was that a matured steer had practically no cash and but little labor charged up against him. But as these lands gradually came under cultivation, each man found that the pastures on which his animals were grazing had a cash rental value for grain growing, and it was then that he realized that he could not afford to raise steers and sell them at the prices which prevailed, while our markets were on an exporting basis. It was then that he went out of the business and turned his attention to other lines of agricultural effort.

"That the process of reducing herds was gradually going on among the farmers of the north as well as the ranchers of the south could be easily seen from the number of females and calves which were each year being sacrificed in all parts of the Province. Although the yearly returns showed no reduction in the number of animals going to market, as in the case of horses, the class of animals marketed told plainly that sooner or later the evil day must come. As in the case of horses that evil day came in 1911, during the first ten months of which only 46,074 head of cattle were reported by the Provincial brand inspectors as against 184,229 for the year previous, and during the last half of May and the first half of June, fat cattle and dressed beef were actually imported from the East.

"The sudden rise in prices incident to this condition, as well as the unsatisfactory returns from grain farming during the past two years, is causing our farmers to turn their attention again to beef production. Tempted by the higher price of cattle and discouraged with grain farming because of the high price of labor and of horses, many are seeding down their farms to the cultivated grasses. This fact is eloquently told by the returns from Alberta, which show that in 1911 there were 165,000 acres of land seeded down to hay and clover as against 65,000 for the year previous. I may add that during the past year I have received more requests from farmers for information regarding the cultivated grasses best suited for both hay and pasture, than I have received during any year since undertaking to perform the duties of Live Stock Commissioner for the Province. These areas seeded down to grasses, taken individually, are usually small, but since 'Mony little maks a muckle,' they are sure to be reflected in the live stock returns of the future.

"Besides this, the higher prices prevailing during the past two years on beef cattle and dairy products have made it possible to utilize districts for stock-raising in which the Indian, the moose and the deer were, a few years ago, unmolested. Small ranches are being established wherever hay and pasture are obtainable, as far as seventy-five miles north of the Saskatchewan River, from the eastern boundary of the Province to the foothills. The most serious drawback to the development of these small ranches, and I may say to the general improvement of the live stock industry in the north half of the Province, is the fact that we haven't the kind of cattle especially suited to the country and to the needs of the people who are going there. When determining the class or breed of cattle a settler should buy, due consideration should be given to the use to which they are to be put, to the environment and general conditions under which they are to be kept, and to the skill of the person in whose hands they are to be placed. Unfortunately, most people, when giving advice on this question, ignore the last two factors in the problem. They make the broad statement that if one is going in for beef he should have cattle of one of the special beef strains, and if he is going in for dairying he should have cattle of one of the special dairy breeds. But here we have a class of settlers who have not the means to wait for returns from beef cattle, yet they can raise a few steers at little greater cost than the hay necessary to feed them through the winter, and they want steers that are worth feeding; they can't go in especially for milk, yet they have to milk enough cows to supply their daily wants, and they ask for cows that are worth milking, and that kind of animal is practically non-existent in Alberta today. I knew them in Ontario when a boy, and in the Western States in later years, and they played an important part in the pioneering days of both countries, but the movement toward specializing in the older Provinces has prevented their appearance in our newer ones.

"An important fact that is too frequently overlooked is that the special dairy breeds give good returns only when in the hands of a special dairy man and in a special dairy environment; if these two be lacking, my observation has been that the returns fall below those of animals of less highly specialized breeding.

"With increased attention on the part of our farmers to stock-raising and the utilizing, for grazing purposes, of areas hitherto unused, there is certain to result, within a few years, a marked increase in our output of beef cattle. How soon it will overtake consumption in Alberta, British Columbia and the Yukon, I shall not undertake to predict; but this I do not hesitate to say, that it will not far exceed consumption in these three markets unless some method of exporting better than we now have has been provided. It is impossible to raise two steers on the farms of Alberta for the price of one on the Smithfield market. A charge of $30 per head for transportation, feed and attendance, a shrinkage amounting to $10 and a loss in quality amounting to another $10, in addition to a reasonable profit to the men who engage in the export business, are burdens that the industry will not stand, and the farmers of Alberta will again quit raising beef cattle the moment they are subjected to them.

"The sheep industry of Alberta, as a ranching proposition, is, for the want of pasture, rapidly on the decline. (Author's note: Since Mr. Stevens delivered this report a large sheep ranch commenced operations in Alberta with ten thousand head of good grade Merino ewes and some hundreds of purebred rams. This is the Western Horse Ranches of Cluny and Bassano, an organization composed of Messrs. Hoople, Honens, Peterson and Renton, of Calgary. The ranch is fenced, the range is owned by the organization, immense sheds have been erected to winter the stock in, and the ranch is in charge of an experienced Australian sheepman. Splendid results have already been realized.)

"The areas formerly set apart for sheep leases, though among the poorest of our grazing lands, are being invaded by the homesteader, and the flock masters are looking about them for other ranges. They have little to hope for on the prairies and their eyes are now turned towards the foothills. They say if they were permitted to take their flocks into the unoccupied mountain valleys, for the summer months, they could establish winter headquarters in the foothills and perpetuate their business indefinitely. In company with a representative of the Department of Agriculture at Ottawa and a committee appointed by the Southern Alberta Wool Growers' Association, I examined, during the summer of 1910, the mountain valleys, beginning at the International Boundary and continuing northward to township fifteen. We avoided all districts in which there were evidences of cattle and horses being kept, and still we found areas which were estimated by the committee of wool growers to be sufficient to pasture, from June to October, fifty thousand head of sheep. Speaking personally, I have no hesitancy in recommending that the requests of the wool growers be given favorable consideration by the Department of the Interior, and that a limited number of sheep be permitted to graze in these unoccupied valleys, preference being given to small holders who combine to make up a flock sufficiently large to justify the employing of a herdsman, and where winter quarters are nearest the desired summer pasture.

"As a farming proposition, interest in mutton production is on the increase. Even in the districts devoted to wheat growing, the farmers are beginning to realize that sheep can profitably be made to enter into their system of agriculture. They see that it is cheaper to let a flock of sheep attend to the business of packing the soil and killing weeds than to do it with teams, especially while the wages of a four-horse outfit are from eight to ten dollars per day.

"The principal detriment to a rapid expansion in sheep-raising on the farms of Alberta is the cost of building suitable fences. If some means could be devised whereby farmers could secure coyote-proof fencing at a moderate price sheep-raising in Alberta would at once enter more largely into the general husbandry of the Province. Providing foundation stock and purebred rams at low cost does not greatly interest the man who is conscious of his inability to protect and care for a flock after he has got them."

"In considering the question of Western beef cattle," said Mr. Phin, "one's mind naturally reverts to the range steer. Unfortunately, the range steer, owing to a number of causes, will not be available for export in any large numbers in the future.

"In the first place Western beef consumption has very largely increased. For instance, it requires about 40,000 cattle for the mountain and British Columbia trade and, naturally, consumption has very largely increased west of the lakes. Export and local transportation facilities have been poor, prices up to the last year or two unsatisfactory, and finally the settler with the plough and fencing in of the drinking places has put on the finishing touches in forcing a large number of ranchers out of business. The Western beef steer of the future must, therefore, come from the farms.

"It is a fact that there are not so many cattle on the farms per farm in the West today as there were a few years ago. Farmers in the mixed farming districts of Manitoba and Saskatchewan have not found cattle profitable and thousands of head of breeding stock have been disposed of in the past five years. I will illustrate to you why beef-raising has not been profitable to the Western farmer. A few years ago I brought my first winter-fed lot of cattle to Toronto—three cars—in March. Export cattle were worth 6 cents in Toronto, and I received 5½ cents for mine of equally good quality when they left the West. It cost me 1¼ cents from Moosomin, Saskatchewan, to Toronto, for all freight and expenses including shrinkage, with a fairly good run, as runs go. That is, I was at a disadvantage with the Ontario feeder of 1¾ cents. On some other shipments I was from 2 cents to 2¼ at a disadvantage, and 2 cents would be a fair average difference. That is, with cattle 6 cents in Toronto they were worth 4 cents in the West, and with 5-cent cattle in Toronto, 3 cents at Moosomin. These prices are not very alluring, and this is one of the main reasons why our farmers will not grow cattle. It cost me from Moosomin 1 cent for freight, feed, selling and other expenses. The other cent is lost from depreciation in quality, shrinkage in actual beef, and the necessary shrinkage of cattle in transportation. The necessary shrinkage of cattle in transportation that distance, weighed full at both places, should not exceed forty to fifty pounds, being less than one-quarter of a cent per pound on the value of the animal. The depreciation in value should not exceed one-quarter cent per pound with good transportation. That is, we are losing one-half of a cent per pound for actual loss of beef and consequent loss through depreciation in value and weight owing to slow transportation. Since that time I have on three different occasions taken my cattle to Liverpool and London and the same depreciation in quality is seen there. Our cattle, which, in Winnipeg, are fine and fat, present a sorry spectacle compared with American cattle. Once having seen them one is not at a loss to know why Americans are quoted so and so and Canadians are quoted so and so and ranchers so and so, arid that the Liverpool *Meat Trader Journal* comments yearly at the end of the season, 'Canada still continues to send a large number of unfinished cattle.' The truth is, a large portion of them are unfinished en route, and I believe I am very conservative in my estimate when I say the unnecessary loss to the cattle industry of the West from slow transportation amounts to fully half a cent a pound on every hoof going out of that country, or seven dollars a head.

"The railway company claim that eighteen miles per hour is the best they can do with stock trains. Local stock is handled on ordinary freight trains at from five to fifteen miles per hour. Why, when we took our stock and effects up to that country thirty years ago over the American lines these appeared to have no trouble to make twenty-five miles per hour, and the change was very noticeable as soon as we crossed the border, as thousands of farmers will tell you. The whole trouble is, as far as stock trains are concerned, although we may put up a train of thirty cars, the railway company immediately adds another thirteen to fifteen cars of heavy freight, making a train that is too long and unwieldy to make fast time even if the engine is powerful enough to handle it. A train of forty-five cars has too much slack in the coupling (about 30 feet) and will jar stock very badly if run at a high rate of speed. It loses too much time at all stops owing to its length, and must take the siding when within fifteen to twenty minutes of a superior train, causing half an hour's delay, thereby spending far too much time on sidings for meets and passes. Local stock is picked up by way freights and heavy freight trains, and receives practically no consideration whatever over dead freight, and averages, probably, eight miles an hour,

"The railway company claim to reduce tonnage on all trains having a certain number of cars of stock by 20 percent. This is practically no reduction at all of the length of the train since stock cars are 20,000 pounds capacity and average freight cars 60,000 pounds.

"The whole situation appears to be that the railway companies have had more traffic than they could handle at most seasons in the past, and have been met with increased demands by their men and have resorted to the slow, heavy freight as the most economical and practical means of handling the traffic offered. The live stock interests have suffered in consequence, and the service is less satisfactory today than it was years ago.

"They have greatly improved their passenger service and have enormously reduced the cost of heavy freight transportation to themselves by better roadbeds, heavier steel, heavier engines, cars of twice to three times their former capacity, the use of the airbrake, etc., but I am bound to say they have done little or nothing to improve the stock car or transportation of live stock.

"They claim to wish to stem the tide of all-wheat production and encourage the production of stock to keep up the fertility of the farms, and to lessen the enormous grain traffic thrown on them in the fall and winter months, and offer to furnish demonstration trains to our governments to encourage mixed farming. Let me say to them and to you that the first essential to get farmers to raise cattle and go in for mixed farming is good transportation for our stock. What we must aim at, I believe, and I have given the subject a good deal of consideration—have been over the road a good many times to Toronto, Montreal and London, and know the conditions fairly well—is for our stock to be handled on passenger time. If they will reduce the length of stock trains to from twenty-eight to thirty cars I believe they can easily make passenger time, in fact they have done so at various times in my experience when stock has been delayed and the mischief was done and I could reach the proper official.

"What I would suggest is that the live stock interests combine in an effort to secure stock trains to be handled on passenger time, and on all main lines a daily scheduled train to run at passenger time to pick up local stock on all divisions on which stock were offered.

"Then we should have a chilled meat system established as early as possible. Although the British people have been very prejudiced against frozen and chilled meat, yet they are year by year using it in larger quantities at higher prices. The demand is less and less for home killed, and it takes a very limited number of live cattle to break that market below a profitable basis today.

"Next as regards production we should show farmers the advantages of better summer and winter-feeding their cattle, and thereby reduce the flood of half-fat stock on our markets and bring stock in to winter or for sale in good shape.

"A few years ago I was forced to consider this matter of better summer-feeding to act as a sort of safety valve in my business. One fall, after a poor cattle season, I found myself in this position. I exported three hundred head at a loss of ten dollars each and had four hundred head left. Of these I wintered one hundred and fifty, and winter-fattened two hundred and fifty head. I succeeded in fattening them, but owing to shortage of water and the large number I was forced to market them in February and March. Owing to low prices and very bad transportation of one shipment they were not profitable. I wished to provide against this condition of affairs.

"In that country almost all farmers bare-fallow their land every third year. I fenced my fallows yearly after that until the farm was fenced, taking in any adjoining broken land, and grew rape on them and have never had a bare fallow since and sometimes had as much as two hundred acres of rape.

"I plough my land, if possible, in the fall, and if not, as early as possible in the spring, and give it surface cultivation and packing till the last week in June or first week in July. May and June are our best germinating months for seeds, as through July and August it is often too dry and later too cold for germination. Therefore, I get rid of most of the weeds by the first of July, get the soil in a thorough state of cultivation and sow a rape in drills on the flat thirty inches apart with an ordinary grain-drill and two pounds of seed per acre, using scorched seeds or any other material of similar weight to the rape seed in equal quantities so that it is not necessary to set the drill so close as to cut the seed to sow as low as two pounds per acre. We run a one-horse cultivator once through to kill weeds, promote growth and conserve moisture. We sow in drills and cultivate once, because if you were to sow broadcast and allow all weeds to grow your land would be sucked dry. Unless the land was very rank the succeeding wheat crop would be unsatisfactory.

"As the rape grows rapidly, we turn on the cattle about the middle of August. We turn them on at any time except early in the morning when the frost is on. After that we let them go at will, but always try to have some grass pasture as well as the rape, as they require something to hold the rape, and they will be found a large part of the time off the rape. I would advise sowing about one-half an acre per head, and have had no trouble whatever from bloating.

"There are many advantages in having this succulent feed for cattle. Most of our farmers, who have not got a free run, have their cattle in small fields or herded, and the cattle do little good after the middle of August either in growth or fattening. The frosts come on, the grass becomes dry and pasture bare, cattle fail and cows fail in their milk.

"With the rape, the cows keep in full flow and there are no bad results to the milk if the cows are taken off an hour before milking and are not allowed on the rape at night. The young cattle grow and improve in condition, and the stock to be turned off are in prime condition to either winter-feed or sell and the owner is in the best possible position. If he decides to winter-feed, his cattle are in the best possible shape provided he starts early and does not allow them to go back. Every experienced feeder will tell you that it takes the best part of a month to get a beast well started that has once commenced to go back.

"The rape-fed steer, like the steer that has had a free run, will take on flesh at once. He has a big liver in him and his organs and digestive system are in shape to make the most of his feed from the start. All he requires is sufficient to keep him improving and then finish off at the end, which is vastly different from attempting to put on flesh and beef on a run-down steer, which is far too common a Western practice.

"There is also another advantage in the land being packed and the excess moisture taken out of it. The wheat crop from many of our summer fallows is unsatisfactory, owing to too rank growth of straw causing late ripening, rust, lodging and consequent poor filling. Seeding some forage crop and pasturing results in a shorter stiff straw, which ripens early and fills well and has been with me invariably clean.

"It may be thought I have dwelt more fully than necessary on this subject of growing feed on our summer fallows. Our Manitoba and Saskatchewan farmers are wedded to wheat. We have the wide open area and it is through wheat production they are in the prosperous condition in which thousands of them are today. Therefore, it is impossible to induce them to raise cattle unless it can be shown that by having their farms fenced and by growing some feed on their fallows they will save work in the destruction of weeds, pack their land and ensure an earlier wheat crop which can be handled at less expense than the ordinary rank fallow crop and at the same time provide succulent feed for their stock. In this way there is some chance of gradually inducing them to raise more cattle.

"Then as regards the wintering of our cattle; too many of our farmers feed straw, only with the result that the cattle make no growth and are much reduced to what they were in the fall. They look upon cattle as a sort of necessary evil and always appear to me to be trying to get something out of nothing. You may do that in Western real estate, but it won't work out in the cattle business. I always make a practice of feeding my young stock a little green feed and grain with the straw, as we have no roots and little hay, and in that way keep them growing throughout the winter. The practice of failing to supply sufficient summer feed for cattle and wintering young stock on straw is one cause of our markets being flooded with half-grown, half-fed cattle. Our farmers in Manitoba, Saskatchewan and Southern Alberta are the principal offenders in this respect, and only a very small percentage of export stock comes from the farms. Ranchers seldom market their cattle until they have sufficient age, weight and condition to market to the best advantage. They are also more careful about the breeding of their cattle, and usually ranch cattle show good breeding. On the other hand, farmers have been reducing their stocks, looking upon them as being unprofitable anyway, and, consequently, they have become careless about the breeding of their cattle. This is a great mistake— an equally great mistake, too, is the failure to feed sufficiently to make proper growth. The two must go together to obtain the best results. No matter how well you feed you will not succeed if you have not got the breed and proper conformation, neither will you succeed with the best of bred stock without feeding sufficient to make continuous growth. I do not, by any means, mean that cattle should be necessarily pure-bred. In fact, I think it would be a great mistake for the ordinary farmer to start with pure-bred stock unless he was specially adapted to the business and was prepared to give them the necessary care and attention to obtain the best results. What I do mean is that he should use pure-bred bulls of good size and quality and improve his stock in that way. As to winter-fattening, I have done so in almost every conceivable way and find you can feed the big steer in almost any way provided he has water, shelter, plenty of feed and attention. I have fed them tied up and loose in stables, in outside sheds and in the barn-yard.

"One winter I fed one hundred steers in the barn-yard without the sign of a shed and they did about as well as any inside cattle, had better digestion, required little, if any, more grain, but ate a great deal more straw and other roughage. One of the greatest drawbacks to winter feeding in that country is the water supply. Our country is not so well watered as Eastern Canada and we have more difficulty

Perhaps no man in Alberta knew more of the reasons for the decline of the ranching industry, its weak spots and its strengths, than long-time rancher George Lane, seen here in Pekisko, *circa* 1910.

owing to the severe frost. The best system I have found, where water has to be pumped, is to use a small gasoline engine and a large circular drinking tank, say eight feet across and two feet deep, two-thirds of which is in a tight, well-built pump house with a tank heater on the inside and lid to be opened and closed at will on the outside. This is where you have not good barns and cannot have your water inside on the Ontario plan.

"It would be suicidal for any farmer to attempt winter feeding without an ample water supply, because it is usually necessary to feed into the May and June months to get the remunerative price. There are, therefore, to sum up, four main essentials to the increase of the beef cattle industry throughout the farms of the West.

"1st. *Better transportation*, and this includes a chilled meat system.

"2nd. Show the farmers the advantages of growing succulent feed on fallows, both as regards the grain crop and bringing stock in to winter in the best condition.

"3rd. Show farmers the advantage of better wintering the young stock and of having stock to be winter-fed in a thrifty condition to begin feeding.

"4th. Show farmers the necessity of greater care in the breeding of their cattle. It is only the well-bred, well-finished animal that will show satisfactory results in the long run."

"I have been asked by your Executive Committee," began Mr. Lane, "to lead this discussion on 'Western Beef Cattle,' and I have also been asked, as President of the Horse Breeders' Association of Alberta, to protect the horse interests. I am also Vice-President of the Western Stock Growers' Association, and that body has asked me to look after its interests at this convention. In taking up this subject of 'Western Beef Cattle,' I propose at the same time to express my views on mixed farming which is, after all, perhaps the most important phase of the questions under consideration.

"Now, gentlemen, I have come down here from Alberta to try to protect the interests of the poor old cow, the animal which has driven the mortgage away from the farm, the animal which has made so many homes happy, and which, in spite of this, has been and is now being destroyed in the West in such a shameful manner.

"Last year, 30,000 calves were killed in our Western country. All these calves were still sucking their mothers and were all of the very best beef strains that could be produced. In addition to this I want to tell you that, at a very conservative estimate, at least 65 percent of all the cattle that were slaughtered during 1911 in Alberta and British Columbia were she stock, and when I tell you that our Western country is capable of carrying 300,000 to 500,000 head of beef cattle per year, you will, I am sure, agree with me as to the seriousness of the situation.

"We use in Alberta and British Columbia about 130,000 head of cattle per year, and in Manitoba and Saskatchewan, so near as I have been able to gather from the figures, about 150,000 are used.

"The wholesale destruction has been going on for a number of years until we have dropped down to scarcely cattle enough for our own home use.

"I have here the cattle export figures for the six years commencing with 1906. These show a steady drop each year down to 1911, and you will see that in 1911 these exports are less than 16 percent of the number exported in 1906.

CATTLE

In 1906 we exported in all from

Alberta...74,733

In 1907 we exported 42,960

In 1908 61,810

In 1909 67,257

In 1910 51,627

And in 1911, only...11,869

"You will see that this decrease has been going on for a number of years, and that the exports for 1911 are less than 16 percent of the exports for 1906, and in addition to this the export figures for 1911 are less than 25 percent of those for 1910.

"Now these figures show an enormous drop, instead of the gain we should have. When people started in to destroy their herds in Alberta, and by this I mean the spaying and beefing out of their breeding stock, I used to say to them, 'Now, this principle is wrong altogether'; and some of our best cattle men said to me, 'No, we can buy stockers cheaper than we can raise them.' I never believed this, although they used to try to prove it to me with figures.

"Now the day has come when I do not know of any country where a man can buy steers to go on feed or on his range.

"A few years ago I bought a bunch of cattle in Old Mexico, and a short time ago I thought I would try to see if I could buy any more there. The price I was quoted this year for yearlings and two-year-olds was just three times what I paid seven years ago, and they seemed to be hard to get at that.

"I see by the statistics of the United States that in that country there are 7,000,000 cattle less than there were ten years ago in spite of the fact that they have 21,000,000 more people.

"Although we have been talking particularly of cattle, I wish, before going further, to draw your attention to the money we have paid out in the West for horses during the last two years.

"We brought last year from Eastern Canada 21,832 horses, which at an average price of $275 means over $6,000,000 of our money left in the East for horses during 1911. In addition to this 4,240 horses were imported to the West from the United States (without including horses entered duty free, as settlers' effects), and valuing these at the same average price this means that $1,166,000 was paid by the West to the United States for horses during 1911.

"In 1910, 33,571 horses were brought to Manitoba, Saskatchewan and Alberta from Eastern Canada and the United States. At the same average price of $275, a little computation will show you that in 1910 we paid out $9,232,025 for horses from Eastern Canada and the United States, in spite of the fact that there is not either in Canada or the United States (nor in fact in any other country), a place where the horse can be raised as cheaply as he can in Alberta and Saskatchewan, or where he will grow to be a better animal.

"Now, as you know, I have been a breeder of horses for a great number of years; I have always been in close touch with the situation in the West, and I feel quite safe in saying to you that there is not, in my opinion, the slightest danger of overstocking horses in Alberta and Saskatchewan during the next ten years at least, on account of the land that will come under cultivation during that time.

"Now, to take up mixed farming, I want to say, first of all, that I think certain parts of Saskatchewan and Alberta are the greatest mixed farming countries in Canada today, and I greatly doubt if their equal is to be found in the United States. The only State that I know of that equals it in any degree is Colorado, and you could put the State of Colorado in one corner of the Province of Alberta. I was told, however, two years ago that Colorado fed 600,000 lambs and about 120,000 beef cattle per year.

"It was only last month that I went down into the Twin Falls country in the State of Idaho. In the Twin Falls irrigation tract, which is 40 miles long and 12 miles wide, I found 400,000 tons of alfalfa hay, and in the neighborhood of one million sheep on feed, together with a great many cattle. I went down there intending to buy some wethers, and while there met a man originally from Kingston, Ontario, named R.F. Bicknell, who may, perhaps, be known to some in this audience. He offered me 20,000 head of two-year-old wethers, and was in a position at that time to sell me that number or twice that number.

"I may say right here when I got the railway rate and the duty and then figured this against the New Zealand mutton it put the deal right out of business.

"Now, if I am any judge, we have in Western Canada a better country than that, yet we bring 66 percent of the mutton used in Alberta and British Columbia from New Zealand and the United States.

"In addition I may add that we do not raise 25 percent of the hogs used in Alberta and British Colombia.

"My authority for these last statements is my accurate knowledge of the country, and I also have been told this by Mr. Pat Burns himself, and also by a member of the Swift's Canadian Company, and the authorities of the Canadian Pacific Railway.

"Now, if I were to go into details and give you the figures showing how much money we have paid for the hog products that are being brought in, and which should be raised in the country, you would wonder where we were getting the money from, for no man ever saw a business more thoroughly destroyed.

"You may say to me, 'What helped to bring this about?' and I would have to say that, in my opinion, very low markets was the first reason for the wholesale destruction of the beef herds.

"Then when the Government put the two years' clause in the leases it had a great tendency to make cattlemen very restless. These men said they could not afford to take chances on keeping a large herd of cattle and being compelled to go out of business after a couple of years.

"When the farmers commenced to come into Alberta and Saskatchewan, it was said, and generally thought, that there would be more cattle in the country when the farmers came in than when the big ranchers were running it. Now, in my opinion, this has not proved true, and it has never proved true anywhere in the United States. Texas today has 6,000,000 beef cattle, and the only state that has half that number is Kansas.

"Now, on top of all this, there began to arrive in the West the men who made it a business to sell wheat lands. Now, I am sure I am not exaggerating when I say there are over eight thousand real estate men selling land in these Western Provinces, and it is safe to say that at least 95 percent of them talk wheat and nothing else in their efforts to sell land to the people.

"Now a great many people may know something about mixed farming, but they are continually being persuaded that 'Wheat is king.'

"After the land agent has done his part, along comes the steam plough agent, and the gasoline agent and half a dozen other machinery men (all these being, as a rule, clever salesmen), and in turn each of these men advises 'raise wheat.' Why?—the first man wants to sell his land; the second man wants to sell his machinery. Nobody advises the landowner to carry on mixed farming.

"Now, they get their wheat and lots of it, and along with the wheat also they occasionally get frost, which means tons of frozen wheat—but no cattle, hogs nor sheep to feed it to. In this I am speaking only of certain parts of Alberta and Saskatchewan.

Workers harvest a wheat crop using modern machinery in 1913.

"Gentlemen, it is a crying shame that there are going to be hundreds of thousands of tons of feed burned up in our country during the coming spring. This will be clear loss, while the money we are sending out of the country is also lost, and I leave it to your common sense to decide whether or not it is going to need even a greater country than our Western Canada to stand such extravagance.

"You may ask, 'What suggestions have you to offer for the remedying of these conditions?'

"First, I would suggest that the experimental stations should enter largely into the raising and feeding of cattle, sheep and swine, with the view of demonstrating the results possible from mixed farming. This will be bound to do a great amount of good. People have been talked to so much about raising wheat that they need to be shown plainly and with actual results what can be accomplished with carefully managed mixed farming.

"The press can also help this work along by showing clearly, as they can show, where the advantages of mixed farming come in.

"I am also of opinion that it would be to the interests of all the banks which have branches in the West to advocate to their customers the value of mixed farming. Something might also be done to help mixed farming by the banks as well as the loan companies by advancing money on good terms to their customers who carry a certain amount of stock each year.

"We all appreciate very much what the railways have done in reducing the rates on pure-bred stock, and I feel sure they could be induced to assist in the matter of demonstration farms.

"These demonstration farms are, in my opinion, doing a great work for the West. Now and then you will hear a man say, 'Oh, I would not listen to those professors with the stiff collars,' but you will find a great many others writing to these farms for information on different points. When a man gets this information he uses it, and his neighbor will soon be using it as well.

"Now, right here I want to make another suggestion to help to encourage the live stock industry in the West.

"I would suggest that, in order to encourage the cattle, hog and sheep industry in Alberta, the packing companies which are doing business in the West should each contribute the sum of $1,000 each year for five years to the prize lists of the different fairs in the Provinces, this prize money to be given for the best fat cattle, sheep and swine. This should do a great amount of good.

"I would also suggest that, in regard to homesteaders going in for mixed farming, it might be possible to arrive at some arrangement with the Department of the Interior whereby it could be stipulated that men taking up land must have a certain amount of stock on the *leasehold* at the end of the second and third years. This might be accepted by the Government in lieu of some of the other homestead duties now required.

"Now, gentlemen, I have lived in Alberta for twenty-nine years, during which time I have been engaged in the general ranching business. I have seen a great many ranchers start in that country, and I want to say that I have never in that time known of three men who lost their money, where they carried on mixed farming, raising cattle and horses with sufficient fodder to take care of them, at the same time attending to their business, and spending their time looking after its interests, not loafing around hotels, barrooms or billiard halls.

"On the other hand I have known of at least one hundred to one hundred and twenty-five men who came West and started at from $10 to $40 per month, who today are worth anywhere from twenty-five thousand to two hundred and fifty thousand dollars; some of them, indeed, have gone up to the million mark, about ten, of whom I have knowledge, having retired from business and gone to Victoria, while a number of the others have scattered to other Provinces. I think I could tell you the names of every one of these men.

"Just here I think it only fair to say, for the benefit of the mothers and fathers of the boys who have come to us from Eastern Canada, that these are the most successful men we have. While 95 percent of them started right at the foot of the ladder, receiving, as I have before stated, only from $10 to $40 per month, they are men of whom the East might well be very proud.

"I might also add, before concluding my remarks, that I think it is only fair to the different breeds of cattle to name those breeds with which I have had the best success in Alberta. I think most of our Western breeders will agree with me that the breeds giving most satisfaction in Alberta and Saskatchewan are the Shorthorns and the Herefords. I am of the firm belief that certain parts of the country are adapted for the raising of certain breeds of beef cattle just as much as certain parts of the country are better adapted than others for the raising of apples or peaches.

"Now there has been a great deal of talk in our country as to what governs the price of cattle. It has been generally said that the abattoirs and packing-houses govern the prices. Now, I do not think this is true. It may be true in some instances where men get into towns or cities with small shipments, but our market has always been governed by the export trade, as, until last year when we only had a little over eleven thousand cattle to export it was impossible for the packers and abattoirs to tell just how many cattle were in our country.

"Now, I know from my certain knowledge that one packer, who, two years ago, got more cattle than he could use, exported them himself, and his actual loss was more than fifty thousand dollars.

"A great many people here know the live stock export man, and do not begrudge any man the money he makes out of export cattle. I have exported a good deal, and I hope I will never have to export again under the same conditions as before. Now, gentlemen, we never get a steer from our country to the old country for less than $28, and from that to $32. Now is this not an awful tax on any country to have to pay to get to a market? The only remedy I know of is in the hands of the Government, the railways and the steamship people, who, by an earnest effort, might reasonably reduce the expenses of transportation.

"On the other hand, I do not think any man need be afraid to go into the export cattle business, as everything points to a continued shortage of beef in all countries, and especially with our population growing as rapidly as it is.

"Now I do not want Ontario to think that the West is the only wasteful country. At their winter fair at Toronto a short time ago I saw beautiful Shorthorn heifers weighing from 800 to 1,000 pounds being sold to the butcher."

PHOTO CREDITS

＊—＋ ⊫◊⊐ ＋—＊

All photos from Glenbow Archives: Page 5 (detail from NA-789-19), 9 (NA-2006-1), 27 (NA-3929-6), 35 (NA-1071-2), 74 (NA-4035-159), 76 (NA-354-30), 83 (NA-101-27), 101 (NA-1734-1), 113 (NA-266-5), 137 (NA-4365-5), 142 (NA-1508-5), 151 (NA-1087-2), 182 (NA-2520-30), 207 (NA-3684-4), 267 (NA-1508-1), 284 (NA-1149-4), 328 (NA-495-2), 333 (NA-584-1), 336 (NA-628-4), 352 (NA-2520-68), 359 (NA-1497-59).